THE ECOPHYSIOLOGY OF PLANT-PHOSPHORUS INTERACTIONS

Plant Ecophysiology

Volume 7

Series Editors:

Aims & Scope:

The Springer Series in *Plant Ecophysiology* comprises a series of volumes that deals with the impact of biotic and abiotic factors on plant functioning and physiological adaptation to the environment. The aim of the *Plant Ecophysiology* series is to review and integrate the present knowledge on the impact of the environment on plant functioning and adaptation at various levels: from the molecular, biochemical and physiological to a whole plant level. This series is of interest to scientists who like to be informed of new developments and insights in plant ecophysiology, and can be used as advanced textbooks for biology students.

The titles published in this series are listed at the end of this volume.

The Ecophysiology of Plant-Phosphorus Interactions

Edited by

Philip J. White
Scottish Crop Research Institute
Dundee, UK

and

John P. Hammond
University of Warwick
Wellesbourne, UK

Springer

Editors
Philip J. White
Scottish Crop Research Institute
Dundee, UK

John P. Hammond
University of Warwick
Wellesbourne, UK

ISBN 978-1-4020-8434-8 e-ISBN 978-1-4020-8435-5

Library of Congress Control Number: 2008924854

Cover images (from top right, clockwise) courtesy of: John Hammond (Flower of *Leucadendron salignum* × *laureolum* a member of the Proteaceae); Andrea Grundy (Wild perennial Lupins in Norway); Rory Hayden (Tractor spreading fertiliser).

Printed on acid-free paper

9 8 7 6 5 4 3 2 1

springer.com

CONTENTS

CONTRIBUTORS

Rosario Azcón
Departamento de Microbiología del Suelo y Sistemas Simbióticos
Estación Experimental del Zaidín
Consejo Superior de Investigaciones Científicas
C/ Profesor Albareda N° 1, E-18008 Granada, Spain
e-mail: rosario.azcon@eez.csic.es

Concepción Azcón-Aguilar
Departamento de Microbiología del Suelo y Sistemas Simbióticos
Estación Experimental del Zaidín
Consejo Superior de Investigaciones Científicas
C/ Profesor Albareda N° 1, E-18008 Granada, Spain
e-mail: conchi.azcon@eez.csic.es

José-Miguel Barea
Departamento de Microbiología del Suelo y Sistemas Simbióticos
Estación Experimental del Zaidín
Consejo Superior de Investigaciones Científicas
C/ Profesor Albareda N° 1, E-18008 Granada, Spain
e-mail: josemiguel.barea@eez.csic.es

Kathleen M. Brown
Department of Horticulture
Pennsylvania State University, 220 Tyson Building, University Park, PA 16802, United States of America
e-mail: KBE@psu.edu

Nuria Ferrol
Departamento de Microbiología del Suelo y Sistemas Simbióticos
Estación Experimental del Zaidín
Consejo Superior de Investigaciones Científicas
C/ Profesor Albareda N° 1, E-18008 Granada, Spain
e-mail: nuria.ferrol@eez.csic.es

Timothy S. George
Scottish Crop Research Institute
Invergowrie, Dundee, DD2 5DA, United Kingdom
e-mail: tim.george@scri.ac.uk

John P. Hammond
Warwick HRI
University of Warwick
Wellesbourne, CV35 9EF, United Kingdom
e-mail: john.hammond@warwick.ac.uk

A. Edward (Johnny) Johnston
Lawes Trust Senior Fellow
Rothamsted Research
Harpenden, AL5 2JQ, United Kingdom
e-mail: johnny.johnston@bbsrc.ac.uk

Ernest A. Kirkby
Institute of Integrative and Comparative Biology
Faculty of Biological Sciences
University of Leeds
Leeds, LS2 9JT, United Kingdom
e-mail: ernest.kirkby@ntlworld.com

Jonathan P. Lynch
Department of Horticulture
Pennsylvania State University, 221 Tyson Building
University Park, PA 16802, United States of America
e-mail: JPL4@psu.edu

Petra Marschner
Soil and Land Systems
School of Earth and Environmental Sciences
The University of Adelaide
Adelaide SA 5005, Australia
e-mail: petra.marschner@adelaide.edu.au

Karl J. Niklas
Department of Plant Biology
Cornell University
Ithaca, NY 14853, United States of America
e-mail: kjn2@cornell.edu

John A. Raven
Division of Plant Sciences
University of Dundee at SCRI
Scottish Crop Research Institute
Invergowrie, Dundee, DD2 5DA, United Kingdom
e-mail: j.a.raven@dundee.ac.uk

Alan E. Richardson
CSIRO Plant Industry
PO Box 1600, Canberra, ACT 2601, Australia
e-mail: alan.richardson@csiro.au

Gabrielle Thiébaut
LIEBE
Université Paul Verlaine - Metz
CNRS UMR 7146, Avenue du Général Delestraint
57070 Metz, France
e-mail: thiebaut@univ-metz.fr

Holm Tiessen
Inter-American Institute for Global Change Research
Avenida dos Astronautas, 1758, São José dos Campos
SP 12227-010, Brazil
e-mail: htiessen@dir.iai.int

Carroll P. Vance
USDA-ARS
Agronomy and Plant Genetics
University of Minnesota
St. Paul, MN 55108, United States of America
e-mail: Carroll.Vance@ars.usda.gov

Philip J. White
Scottish Crop Research Institute
Invergowrie, Dundee, DD2 5DA, United Kingdom
e-mail: philip.white@scri.ac.uk

PREFACE

The Ecophysiology of Plant-Phosphorus Interactions is the seventh volume in the *Plant Ecophysiology* series. It reviews the current state of knowledge, concepts and research of plant-phosphorus interactions in natural and managed ecosystems together with aspects of the phosphorus nutrition of crop plants, addressing in particular the sustainability and possible environmental consequences of agricultural production.

Phosphorus (P) is an essential macronutrient for plant growth. Plants take up P as phosphate (Pi) from the soil solution. Since little Pi is available to plants in most soils, they have evolved mechanisms to acquire and use P efficiently and foster symbiotic relationships to help them acquire P sources beyond their immediate range. Whilst in agricultural systems P limitations are frequently overcome by the application of P-fertilizers, these may cause environmental pollution and the use of inorganic Pi is unsustainable. The genetic and phenotypic variation among plants adapted to ecosystems with low P availability provides an opportunity to improve our understanding of plant responses to P limitation and this knowledge could be utilized to develop crop varieties with better P use for agriculture.

In the first chapter of this volume, Holm Tiessen places P in a global context. He reviews the geochemistry of P, the cycling of P in the environment, the effects of humans on P cycles, and their consequences. Next, Karl Niklas describes the allometric relationships between tissue C, N and P concentrations among and within plant species, and explores the implications of these for various physiological, ecological and evolutionary phenomena. Gabrielle Thiébaut explains how P is acquired by aquatic plants and how P supply and seasonal fluctuations in P loads affect the abundance and distribution of aquatic plant species, while Philip White and John Hammond summarize the requirements and functions of P in terrestrial plants and the impacts of P availability on their ecology. These authors also introduce the biochemical, physiological and morphological traits that enable terrestrial plants to acquire and utilize P most effectively, and how the expression of these traits might be regulated by plant P status.

Jonathan Lynch and Kathleen Brown focus on the root traits that provide an adaptive strategy for P acquisition by terrestrial plants, which include: greater root biomass allocation, changes in root architecture to exploit local P patches, increased root length density, proliferation of root hairs, symbiosis with mycorrhizal fungi and the secretion of organic acids and phosphohydrolases. This theme is continued by Carroll Vance, who addresses the adaptations for the acquisition and use of P in plants lacking effective mycorrhizal symbioses, concentrating on species that develop specialized complex roots (cluster and dauciform) and on Arabidopsis. These chapters are complemented by those of Jose Barea and colleagues, who describe the nature of plant-mycorrhizal symbioses and their impact on plant productivity, plant community structure and P cycling in the environment, and of Petra Marschner, who provides an overview of the influence of rhizosphere microorganisms on the growth and P nutrition of plants. These chapters describe the major influence of plant species on rhizosphere community composition, and discuss the possible reasons for this. They also discuss the use of microbial inoculants to improve plant productivity.

The role of P-fertilizers in agriculture is reviewed by Ernest Kirkby and Johnny Johnston, who emphasize the necessity of P-fertilizers for crop production and reflect on their environmental and ecological footprint. Against the backdrop of depleting Pi reserves, and the necessity for global food security, they establish strategies for more efficient use of soil and fertilizer P based on knowledge of the behavior of P in soils, the introduction of best management practices and the potential for developing "P-efficient" cultivars of crop plants. These strategies are further explored in chapters by John Hammond and Philip White, who describe how the application of P-fertilizers to crop plants can be optimized by monitoring and modeling the P status of soils and plants, and by Tim George and Alan Richardson, who describe how appropriate breeding and transgenic approaches can be used to improve crop P acquisition. The volume concludes with a thought-provoking perspective by John Raven on the past and future P-nutrition of plants, which includes a checklist of priorities for immediate action to enable the world to feed its burgeoning human population.

It is hoped that this book will be of interest to students and researchers studying all aspects plant-phosphorus interactions: omicists, physiologists, ecologists and all readers interested in sustainable crop production.

John P. Hammond
Philip J. White

Chapter 1
PHOSPHORUS IN THE GLOBAL ENVIRONMENT

Holm Tiessen

INTRODUCTION

Phosphorus is not one of the "global" elements, it does not enter the atmosphere like nitrogen, it does not spread like sulfur by acid rain and its solubility in water is so low that there is only a slow, steady movement of P down-stream as landscapes erode and weather, or P-containing pollutants are discharged. Yet, there are some global trends in the distribution of P. To understand these and their drivers it is useful to review some of the basic properties of P in the environment.

The earth's crust contains about 1,200 mg P kg^{-1}, making it the 11th most abundant element. Common concentrations for total P in soils are between 200 and 800 mg kg^{-1}, with older soils containing lower amounts of P and younger soils containing higher amounts of P. In primary rocks and young soils, P is largely bound to calcium or magnesium, giving P a typical water solubility near 0.5 mg P L^{-1}. The weathering of minerals changes the solubility of P, as Ca is preferentially leached out, the relative abundance of Fe and Al increases and the solubility of P becomes controlled by Fe- or Al-phosphates, which have much lower solubilities than Ca-phosphates. As a result, the sequestration of P in low-solubility Fe and Al-phosphate compounds and the effect of leaching and erosion, many older and tropical soils are P deficient, i.e. the availability of P to plants and other organisms restricts ecosystem processes such as N fixation or C sequestration.

The availability of P to plants is controlled by physical and chemical reactions, including sorption/desorption and precipitation/dissolution and biological processes such as immobilization (uptake by plants and microorganisms) and by mineralization (decomposition of residues). The sorption of P, followed by slower transformations, such as solid state diffusion into the matrix of the sorbent, reduce the solubility of P, sometimes to such a degree that P is said to become "fixed". Strictly speaking, P fixation is a misnomer, since all chemical reactions are to some degree reversible, but the amount and rate of release of "fixed" P may be so low that they are ecologically insignificant.

Over 99% of naturally occurring P is in the form of phosphate, either as inorganic phosphates or as organic phosphate esters. With its four oxygen atoms per P, phosphate has a high negative charge density, so it can readily bond to any positively charged

P.J. White, J.P. Hammond (eds.) *The Ecophysiology of Plant-Phosphorus Interactions*,

cation or surface. This greatly restricts the mobility of P in the environment. When phosphate is bound into relatively large organic molecules, this charge is somewhat shielded. Consequently, organic forms of P are often more mobile in the environment than inorganic P. In most soils organic P accounts for 30–65% of the total P, although some soils contain up to 90% organic P (Harrison 1987). This accumulation of organic P implies low mobility, which is due to the sequestration of P in recalcitrant soil organic matter. The negative relationship between mobility and accumulation of organic P forms is seen in the abundance of inositol hexaphosphate, which may account for half the identifiable organic P in soils. Inositol hexaphosphate is sorbed more strongly than inorganic phosphate in soils due to the high charge density resulting from its six phosphate ester groups.

In water bodies, with few absorbing surfaces and constant mixing, organisms can take up P much more easily than from a soil matrix, where sorption is strong and transport towards uptake surfaces is limited by diffusion. Even low concentrations of P are therefore very effective in increasing the biological productivity of aquatic systems. This makes aquatic systems highly sensitive to P contamination. Phosphate losses from land to water, commonly have significant eutrophying effect on surface waters.

THE BIOLOGICAL IMPORTANCE OF P

Phosphorus is an essential element of biological systems. It is part of the genetic material and the phosphate ester bond is universally used for energy transfer reactions in organisms. Plants take up and concentrate P from near 0.1 mg P L^{-1} in soil solution to 100 mg P L^{-1} in xylem sap, and can accumulate near 4,000 mg P kg^{-1} in seeds. Mammals contain around 25,000 mg P kg^{-1} dry weight. Because of the importance of P in biological processes, changes in P availability can have major impacts on ecosystem function and structure. Often both N and/or P availabilities may be near limiting levels, but the dependence of biological N fixation on adequate P supply makes P the principal limiting element of ecosystems. The biological importance of P, means that ecosystems have developed mechanisms by which P is taken up, recycled and retained efficiently (Cole *et al.* 1977; Attiwill and Adams 1993). In dystrophic or oligotrophic tropical forests, a significant portion of P is cycled biologically within the plant biomass, thus it is protected from conversion to soil P of low solubility. Lal *et al.* (2001) estimated that between 20% to 91% of the P demand of Indian dry forest trees was satisfied by re-translocation of P prior to leaf abscission. The extent of re-translocation reflects soil nutrient availability (Tiessen *et al.* 1994). Recycling through mineralization of organic P from plant residues also contributes to plant P requirements (Frossard *et al.* 2000). Annual recycling of P to soil in above- and below-ground plant residues represented 18–38% and 15–80% (mean 55%) of plant P uptake in temperate crop and forest ecosystems, respectively (Hanway and Olsen 1980; Pritchett and Fisher 1987). In dystrophic forests, much of the P uptake by roots may be directly from plant litter via the hyphae of mycorrhiza. The biological and biochemical processes of P cycling are more important in tropical soils than in most temperate environments, because of the combination of lower inorganic P availability and greater biological activity in the tropics. The high biological potential of the humid tropics is evident in the large biomass production and rapid turnover of organic matter, and also in the very intensive land use. Up to four crops can be planted and harvested per year on a single plot, generating high nutrient demand.

THE AGRICULTURAL IMPORTANCE OF P

Much of tropical agriculture is undercapitalized with respect to P fertility. Mailly *et al.* (1997) illustrate a P-budget typical for most low input agriculture over a six-year cultivation cycle in Java. Of the 130 kg P ha^{-1} accumulated in plants during the cycle, half were removed in harvested materials. Fertilization replenished only 45 kg P ha^{-1}. In addition, P stocks of arable fields are frequently depleted further by erosion. In a study on the P fluxes in low input agriculture in northeast Brazil, Menezes and Sampaio (2002) showed the largest P flow was associated with erosion from cultivated fields (6 kg ha^{-1} $year^{-1}$), followed by the erosive flow from (generally overstocked) pastures. By comparison to these erosive flows, the output of P from the farm by crop (2 kg ha^{-1} $year^{-1}$) and animal (0.2 kg ha^{-1} $year^{-1}$) products was minor.

Biological cycling of P is rarely sufficient to supply P to highly productive cropping systems. Continued inputs of P in the form of fertilizers are required to sustain high levels of production after the initial mineralization of soil organic matter, commonly associated with bringing land under cultivation. On North American grasslands, this initial release of P from inherited soil organic matter lasted for some 60 years of cultivation without P inputs, and resulted in 20–30% decreases in soil organic P (Tiessen *et al.* 1982). Following these first 60 years of cultivation, fertilizer use in the region increased many-fold.

Crop P fertilizer requirement varies with soil type, from <1 kg P ha^{-1} in relatively unweathered soils of arid environments to 200 or 300 kg P ha^{-1} in oxide rich tropical or volcanic soils. As a result, oxidic tropical and subtropical soils, which are often in regions with low purchasing power, account for 50% of the world inorganic P fertilizer requirements for crop production. Crop production on such soils without inorganic P fertilisation will degrade the (agro-) ecosystem. The important role of P in maintaining ecosystem quality is demonstrated by the need for large P inputs in restoration strategies that combine P fertilisation with the planting of N-fixing legumes for degraded lands in Southeast Asia. During the plant succession following P application, P is recycled in plant residues, which decompose to release P in quantities and at rates greater than those determined by inorganic P availability. As a result, plants with higher P demand can grow, thus re-creating an organic P supply cycle that can effectively compete with inorganic P sorption and prevent P losses by runoff and leaching. Soil P availability will remain a major constraint on food production. Management of P fertilizer inputs, together with an understanding of organic matter cycling, should allow for the development of more sustainable agriculture practices.

ECOSYSTEM P AND THE IMPACT OF HUMAN ACTIVITY

The distribution of total ecosystem P between soils and plants varies widely. In a grazed permanent pasture, herbage P amounted to only 1% of the total P content of topsoil (0–20 cm; Williams and Haynes 1992), whereas above- and below-ground plant parts of a temperate forest accounted for 38% of total P in the ecosystem (Hart *et al.* 2003). In many dystrophic tropical forests, the largest reservoir of nutrient elements is in the plant biomass. In the total (300 t ha^{-1}) aboveground biomass, of a P-limited Colombian tropical rainforest, Rodriguez-Jimenez (1988) measured 40 kg P. Such differences in P contents and elemental ratios reflect plant community adaptation to geochemical constraints.

Elevated P concentrations in the environment are often an indicator of (past) biological or human activity. Megalithic and Khalahari campsites and Terras Pretas do Indio in the Amazon all show a clearly elevated P content. Even the "industrial" fertilizer P, mined from rock phosphate with a P concentration of approximately 150,000 mg kg^{-1}, is ultimately derived from biological processes. It is the product of the sedimentation and accumulation of marine organisms.

Human impacts have substantially altered global P transformations and transfers. Phosphorus transfers in the environment are closely correlated with human occupation, as shown by the regression between P exports from watersheds and their population density: loss rates of 0.3 kg P km^{-2} at a population density of 0.1 km^{-2}, rising to 30 kg P km^{-2} at densities of 300 km^{-2} (Caraco 1995). In the Canadian province of British Columbia, diatom records in lake sediments show a significant P enrichment since 1850, the time of European settlement. The advent of inorganic P fertilizers and concentrated livestock production in some areas has greatly increased P loads in surface waters (Anderson 1997). Particularly in richer nations, inorganic P fertilizer use and concentrated livestock production have increased P loads and transfers to surface waters. The P balance for a lake watershed in Sweden, a wealthy nation with a sound environmental policy, suggests 62% of P inputs came from inorganic P fertilizer, 30% from manure, 5% from sewage, 2% from atmospheric deposition and 0.3% from the natural weathering of rocks (Ryding *et al.* 1990). Phosphorus outputs were, 93% in crop exports, 6% in erosion from arable land and 0.6% was leached to ground water. The total P inputs were three times greater than outputs; i.e. the watershed showed a net P accumulation.

The surplus of P in highly developed regions is in stark contrast to the nutrient deficiencies in many developing countries. While P reserves in soils of highly developed agricultural areas are increasing, and even reaching saturation, many tropical soils have a large P deficit, aggravated by high P fixation capacities that reduce fertilizer availability. Increased food demands from old, weathered, often tropical soils has reduced the fertility of these soils beyond the traditional capacity for regeneration (such as under shifting cultivation) and land degradation is evident. Phosphorus availability in these soils is very low and will not sustain growing populations without extra P inputs.

World trade movements amount to some 10 million tonnes P $year^{-1}$, with 81% of trade in fertilizer, rock or phosphoric acid, 15% in plant and 1% in livestock commodities (Beaton *et al.* 1995). Since trade flows are uneven between world regions, these figures imply an important P enrichment in regions with large and wealthy populations, which import both fertilizers and other P-containing commodities. An equalization of P availabilities around the globe would require large resources, investments in effective P recycling and an increased value-added activity in primary producer regions. Shipping livestock products rather than soybean for instance, would avoid some of the nutrient concentration in rich intensive livestock production regions. While the need for efficient use of P has prompted human populations to evolve elaborate management strategies for maintaining P fertility in agriculture, recycling P and limiting P pollution from wastes, improvements in management and shipping patterns are still needed if increasing populations are to manage their environments sustainably.

Regions that import large P surpluses endanger the biological integrity of their surface waters. Although most watersheds will show a net P retention, even with substantial P inputs, they are leaky, and rates of P export are higher after P application than natural levels. Phosphorus inputs into fresh waters can increase growth of algae and aquatic weeds and lead to oxygen shortages due to their decomposition. Remedial action in such cases has focused on the role of P, although N is also essential for the growth of aquatic biota. The focus on P is due to the difficulty in controlling the fixation of atmospheric N by blue-green algae (Sharpley and Rekolainen 1997). Some 10% of P export from land occurs by leaching and ground water transport, while 90% is transported by overland flow as sediment or dissolved P. Despite the small proportion of leached P, it has a greater effect on eutrophication of receiving waters because it is soluble and therefore easily available to biota. Phosphorus losses by over-land transport range from 0.1 to 10 kg ha^{-1} $year^{-1}$, or more on highly erosive sites.

On average, surface runoff waters carry 10 $\mu g\ L^{-1}$ of dissolved and 1,000 $\mu g\ L^{-1}$ of sediment P (Melak 1995). While P is transported down-stream, sedimentation and re-suspension occur. Phosphorus is released from the sediment when solution P is diminished. The concomitant movement and recycling of P results in a "spiraling" of P as it moves down rivers. Inland lakes are affected by past fertilizer and animal waste management. More dialogue between freshwater and land use practitioners on the transfers of nutrients between systems is needed to develop regional plans to prevent P loss.

Globally some 33 million tonnes P $year^{-1}$ are discharged into oceans, of which, more than half is carried by rivers and the remainder is coastal runoff (Howarth *et al.* 1995). Most of the input to oceans is ultimately buried in sediments, through processes lasting millions of years. Coastal seas receive both sediment-bound and dissolved P. The bio-available portion of P in fluvial transport is estimated to be approximately 2 million tonnes P $year^{-1}$ and is responsible for the high productivity of near-shore waters. The sensitivity of coastal regions means that prevention should be favored over damage management. Water of the high oceans contains only 0.01 mg P L^{-1}, and P is critically limiting productivity of these waters.

One problem with quantifying the role of P in the global environment is the difficulty in measuring the biologically active or available portion of P. Phosphate availability is a function of chemical equilibrium-controlled solubilities and sorption reactions, and of rate-limited biological processes. Most methods for determining available P attempt to quantify only the chemical solubility of P using different extractants, but few relate this to P supply rates to plants. Soil test methods extract a portion of soil P that is related to plant available P, as estimated by regression equations established over years of agronomic experimentation and testing of fertilizer responses. Results obtained with this approach are rarely transferable between crops or soil types. The approach does not work at all when perennial plants or natural ecosystems are examined, because measurable pools are often small, and biological P re-cycling determines P availability. Since available P is constantly replenished, it is strongly time dependent. This makes available P a functional concept rather than a measurable quantity.

REFERENCES

Anderson NJ (1997) Reconstructing historical phosphorus concentrations in rural lakes using diatom models. In: Tunney H, Carton OT, Brookes PC, Johnston AE (eds), Phosphorus Loss from Soil to Water. CAB International, Wallingford, pp 95–118

Attiwill PM, Adams MA (1993) Nutrient cycling in forests. New Phytol 124: 561–582

Beaton JD, Roberts TL, Halstead EH, Cowell LE (1995) Global transfers of P in fertilizer materials and agricultural commodities. In: Tiessen H (ed), Phosphorus in the Global Environment: Transfers, Cycles and Management SCOPE 54. Wiley, Chichester, pp 7–26

Caraco NF (1995) Influence of human populations on phosphorus transfers to aquatic systems: a regional case study using large rivers. In: Tiessen H (ed), Phosphorus in the Global Environment: Transfers, Cycles and Management SCOPE 54. Wiley, Chichester, pp 235–244

Cole CV, Innis GS, Stewart JWB (1977) Simulation of phosphorus cycling in semi-arid grassland. Ecology 58: 1–15

Frossard E, Condron LM, Oberson A, Sinaj S, Fardeau JC (2000) Processes governing phosphorus availability in temperate soils. J Environ Qual 29: 15–23

Hanway JJ, Olsen RA (1980) Phosphate nutrition of corn, soya beans, and small grains. In: Khasawneh FE, Sample EC, Kamprath EJ (eds), The Role of Phosphorus in Agriculture. ASA/CSSA/SSSA, Madison, WI, pp 681–692

Harrison AF (1987) Soil Organic Phosphorus - A Review of World Literature. CAB International, Wallingford

Hart PBS, Clinton PW, Allen RB, Nordmeyer AH, Evans G (2003) Biomass and macro-nutrients (above- and below-ground) in a New Zealand beech (Nothofagus) forest ecosystem: implications for carbon storage and sustainable forest management. Forest Ecol Manag 174: 281–294

Howarth RH, Jensen HS, Mariano R, Postma H (1995) Transport to and processing of P in near-shore and oceanic waters. In: Tiessen H (ed), Phosphorus in the Global Environment: Transfers, Cycles and Management SCOPE 54. Wiley, Chichester, pp 321–345

Lal CB, Annarurna C, Raghubanshi AS, Singh JS (2001) Foliar demand and resource economy of nutrients in dry tropical forest species. J Veg Sci 12: 5–14

Mailly D, Chritanty L, Kimmins JP (1997) Without bamboo, the land dies: nutrient cycling and biogeochemistry of a Javanese bamboo talun-kebun system. Forest Ecol Manag 91: 155–173

Melak JM (1995) Transport and transformations of P, fluvial and lacustrine ecosystems. In: Tiessen H (ed), Phosphorus in the Global Environment: Transfers, Cycles and Management SCOPE 54. Wiley, Chichester, pp 245–254

Menezes RSC, Sampaio EVSB (2002) Simulação dos fluxos e balanços de fósforo em uma unidade de produção agrícola familiar no semi-árido paraibano. In: Silveira LM, Petersen P, Sabourin E (eds), Agricultura familiar e agroecologia no semi-árido: avanços a partir do Agreste da Paraíba. Rio de Janeiro, RJ, pp 249–260

Pritchett WL, Fisher RF (1987) Nutrient cycling in forest ecosystems. In: Properties and Management of Forest Soils (2nd edition). Wiley, New York, pp 180–204

Rodriguez-Jimenez LVA (1988) Consideraciones sobre la biomasa, composicion quimica y dinamica del bosque pluvial tropical de colinas bajas, Bajo Calima, Buenaventura, Colombia. In: Serie Documentacion, Corporacion Nacional de Investigacion y Fomento Forestal. Bogota, Colombia

Ryding SO, Enell M, Wennberg L (1990) Swedish agricultural non-point source pollution: A summary of research and findings. Lake Reservoir Manag 6: 207–217

Sharpley AN, Rekolainen S (1997) Phosphorus in agriculture and its environmental implications. In: Tunney H, Carton OT, Brookes PC, Johnston AE (eds), Phosphorus Loss from Soil to Water. CAB International, Wallingford, pp 1–54

Tiessen H, Stewart JWB, Bettany JR (1982) Cultivation effect on the amounts and concentrations of carbon, nitrogen and phosphorus in grassland soils. Agron J 74: 831–835

Tiessen H, Chacon P, Cuevas E (1994) Phosphorus and nitrogen status in soils and vegetation along a toposequence of dystrophic rainforests on the upper Rio Negro. Oecologia 99: 145–150

Williams PH, Haynes RJ (1992) Balance sheet of phosphorus, sulphur, and potassium in a long-term grazed pasture supplied with superphosphate. Fert Res 31: 51–60

Chapter 2
CARBON/NITROGEN/PHOSPHORUS ALLOMETRIC RELATIONS ACROSS SPECIES

Karl J. Niklas

INTRODUCTION

This chapter reviews some of the ecological and evolutionary implications of carbon (C), nitrogen (N), and phosphorus (P) stoichiometry and the allometric relationships among these elements reported for terrestrial plant species because the patterns of C mass allocation and N:P-stoichiometry for different plant organ-types are of general interest to understanding a broad range of ecological and evolutionary phenomena (Aerts and Chapin 2000; Bazzaz and Grace 1997; Chapin *et al.* 1986; Grime 1979; Niklas and Enquist 2001, 2002; Westoby *et al.* 2002; Wright *et al.* 2004; Niklas *et al.* 2005, 2007; Kerkhoff *et al.* 2006). Much of the functional-trait variation observed among species differing in overall size can be attributed to differences in the amount of C, N or P allocated to the construction of leaves, stems, roots, and reproductive structures as well as to differences in overall body size (Grime 1979; Field and Mooney 1986; Tilman 1988; Bazzaz and Grace 1997; Jackson *et al.* 1997; Milberg and Lamont 1997; Weiher *et al.* 1999; Niklas and Enquist 2001, 2002; Enquist and Niklas 2002; Westoby *et al.* 2002; Wright *et al.* 2004; Niklas *et al.* 2005, 2007; Kerkhoff *et al.* 2006). Likewise, the P and N concentrations in plant tissues critically influence the material and energy cycles of whole ecosystems (Chapin *et al.* 1997; De Angelis 1980; Kerkhoff *et al.* 2005; Koerselman and Meuleman 1996; Silver 1994; Sterner and Elser 2002; Vitousek 1982; Vogt *et al.* 1986; Ågren and Bosatta 1996) and phylogenetic functional trait differences in the ability to acquire and use N or P are temperature-dependent, such that climatic shifts of sufficient magnitude (e.g., along latitudinal or altitudinal gradients) can have major affects on the C economy of terrestrial vegetation (Kerkhoff *et al.* 2005; Wright *et al.* 2005; Westoby and Wright 2006).

The allocation of N and P to leaves is of particular interest because leaves provide the principal means by which vascular plants capture sunlight. Both N and P play pivotal roles in photosynthesis and respiration, and their abundance also influences the consumption of leaves by herbivores (Cebrian 1999; Elser *et al.* 2000a,b; Mattson 1980). In addition, the availability of N and P is known to limit the growth of both terrestrial (Chapin *et al.* 1986; Ågren 1988; Güsewell 2004) and aquatic plants (e.g. Klausmeier *et al.* 2004; Karpinets *et al.* 2006). Further, coordinated patterns of variation in N and P have been observed across the leaves

P.J. White, J.P. Hammond (eds.) *The Ecophysiology of Plant-Phosphorus Interactions*,

of phylogentically and ecologically unrelated plant species. These patterns have received particular attention in the light of recent stoichiometric models that draw attention to the biochemical constraints imposed on plant growth by the allocation of N to tissue proteins (which are particularly N-rich) and the allocation of P to the ribosomal RNA "machinery" used to synthesize proteins (Sterner and Elser 2002; Ågren 2004; Güsewell 2004; Wright *et al.* 2004; Kerkhoff *et al.* 2005; Niklas and Cobb 2005; Niklas *et al.* 2005, 2006, 2007).

QUARTER-POWER SCALING "RULES" AND N:P-STOICHIOMETRY

Stoichiometric models are especially relevant to theories purporting to explain the apparent ubiquity of quarter-power scaling "laws" that span all levels of biological organization, from molecules to ecosystems, across prokaryotes and eukaryotes, and among plants and animals (Hemmingsen 1960; Peters 1983; Calder 1984, 1996; Schmidt-Nielsen 1984). For example, across animal species ranging in size from that of a mouse to an elephant, maximum life span in captivity, blood volume circulation, fast muscle contraction, and a host of other phenomena each scale closely to the 1/4 power of body mass (Lindstedt and Calder 1981). Perhaps the most famous of these rules is Kleiber's, which states that basal metabolic rates scale as the 3/4 power of body mass (Kleiber 1932, 1961) – a scaling relationship that finds its analog in the allometry of growth rate versus body mass across the polyphyletically and ecologically diverse unicellular algae and terrestrial plants (Banse 1976; Niklas 1994; Niklas and Enquist 2001).

Yet, the identification of an unambiguous mechanistic explanation for the origin of these scaling rules remains an open theoretical problem. Numerous theories have been advanced, but each has been viewed with considerable skepticism (e.g. Blaxter 1965; Blum 1977; Gray 1981; Heusner 1982; Feldman 1995; Economos 1982, 1983; Prothero 1986a). Among the most recent of these, is the theory of Geoffrey B. West, James H. Brown and Brian J. Enquist (denoted hereafter as the WBE theory) who assert that all quarter-power scaling rules (and their 1/4 multiples like 3/4) emerge from the interplay among the physical or geometric constraints resulting from three functional properties of every biological system (West *et al.* 1997, 1999, 2001). Specifically, their theory claims that for biological systems, all networks (1) are space-filling, hierarchical branching systems, (2) have terminal branch elements that are invariant in size, and (3) by virtue of natural selection, minimize the energy required to transport and deliver nutrients (and thus minimize either the time or distance nutrients are moved).

As so many theories before it, the WBE theory has been heavily criticized on empirical, theoretical, and even strictly mathematical grounds (e.g. Dodds *et al.* 2001; Darveau *et al.* 2002; Weibel 2002). Arguably, the first assumption (i.e. that biological delivery networks are "fractal" in nature) is consistent with the "self-similarity" typically observed when branched nutrient networks within multicellular organisms are dissected and numerically quantified. However, if the WBE theory is valid across all levels of biological organization, from that of molecules to ecosystems as claimed by its authors, fractal-like delivery networks must exist at each level, which is difficult to imagine for some levels of biological organization (e.g. molecules) and undocumented for others (e.g. organelles and ecosystems). Similar concerns exist for the two remaining assumptions of the WBE theory. It has yet to be established that capillaries, bronchioles, and terminal xylary elements are invariant in size or that they minimize the time and energy required to exchange mass or energy.

Despite these concerns (or perhaps because of them), the WBE theory has engendered a renaissance in the field of allometric theory and empirical enquiry – one in which alternative theories for the existence of quarter-power scaling rules continue to be sought. It is in this context that recent developments in modeling the effects of N and P allocation patterns on protein synthesis rates (and thus "growth") are particularly exciting. These models emerge from the perspective that, irrespective of phyletic affinity or ecological preference, the growth rate of any kind of organism is positively correlated with ribosome number and rate of activity and negatively correlated with protein concentration (Dobberfuhl 1999; Sterner and Elser 2002; Elser *et al.* 2003; Ågren 2004; Vrede *et al.* 2004). Conceptually, the amounts of ribosomes and proteins are thought of as respective measures of an organism's protein-production "machinery" and the "overhead" that must be produced per unit time to maintain a constant growth rate. N:P-stoichiometry is emphasized, because large fractions of an organism's N and P are allocated to the construction of proteins and rRNA, respectively. Thus, N:P-stoichiometry is predicted to correlate with growth rate at the level of cells, tissues and the whole organism (Dobberfuhl 1999; Sterner and Elser 2002; Vrede *et al.* 2004). Specifically, growth rates should correlate positively with increasing rRNA (and P) investments relative to protein (and N) investments.

This prediction is particularly relevant to three previously reported allometric relationships for plants (Niklas and Enquist 2001, 2002). First, annual growth rates in body mass across phyletically and ecologically diverse species appear to scale as the 3/4 power of body size. Second, growth rates scale linearly (isometrically) with the capacity to intercept sunlight. Third, total leaf N appears to scale as the 3/4 power of total leaf P, across and within some species (Niklas and Cobb 2005; Niklas *et al.* 2005, 2007). The goal of this paper is to review these relationships and to explore them empirically with the aid of a recently expanded database for nonwoody and woody plant species ranging across eleven orders of magnitude in total body size.

A STATISTICAL ASIDE

Biological scaling relationships, referred to as "power rules", comply mathematically with the formula

$$Y_o = \beta Y_a^{\alpha},$$

where Y_o and Y_a are the variables plotted on the ordinate and abscissa axes, respectively, β is the normalization constant, and α is the scaling exponent. In most, but not all cases, Y_a is some measure of mass (typically, but not invariably, expressed in units of carbon mass). When $\alpha = 1$, the formula $Y_o = \beta\, Y_a^{\alpha}$ describes an isometric relationship that plots as a straight line on both linear and logarithmic axes. When $\alpha \neq 1$, the formula $Y_o = \beta\, Y_a^{\alpha}$ describes an allometric relationship that plots as a linear function on logarithmic axes. Logarithmic transformation shows that log β and α are the Y_o-intercept ("elevation") and the slope ("scaling exponent") of the log-log linear allometric relationship, respectively, i.e.

$$\log Y_o = \log \beta + \alpha \log Y_a$$

The linearization of data by means of logarithmic transformation has become a conventional practice in allometric studies, in part because it minimizes the sum of squared residuals for the transformed as opposed to the original function. It should be noted, however, that regression parameters estimated in this way do not invariably provide the best fit of data to a regression model compared to minimizing the squared residuals for the actual function by using nonlinear regression protocols. Analyses of residuals are required to determine whether log-log linear or log-log nonlinear functions optimize the goodness of fit. This protocol does not appear to be a "standard practice", perhaps because most allometric theories assert (or insist on) the existence of numerically unique scaling exponents, which do not exist for log-log nonlinear relationships.

The objective of the vast majority of allometric studies is to determine the numerical values of log β and α. When a predictive relationship is sought, simple ordinary least squares regression (OLS) analysis can be used. When the objective is to establish a functional relationship between Y_o and Y_a, as is generally the case, OLS regression analysis is ill equipped for this purpose, in part because it is based on the assumption that Y_a is biologically independent of Y_o and that it is measured without error. Three regression methods have been suggested to overcome this limitation, i.e. Bartlett's three-group method, principal axis regression, and reduced major axis regression (Sokal and Rohlf 1980), which has been recently renamed as standardized major axis regression (Warton *et al.* 2006). Considerable controversy revolves around which of these methods is the most appropriate (Smith 1980; Harvey 1982; Prothero 1986*b*; Seim 1983; Rayner 1985; McArdle 1988, 2003; Jolicoeur 1990; Warton *et al.* 2006). This issue is not trivial, especially when the goal is to "test" when empirically determined scaling exponents agree statistically with those predicted by a particular theory, because the numerical values of α and log β depend on the regression techniques used and because different techniques can produce significantly different numerical values even for the same data set.

Space precludes a detailed discussion of the merits and detractions of each of the three regression methods. However, standardized major axis (SMA) regression analysis has emerged as a "standard" allometric technique over the past few years. Statistical software is available to perform SMA regression analyses, but access to this software is not critical, because OLS regression summary statistics provide all the necessary information to compute the numerical values of α and log β, and their corresponding 95% confidence intervals.

Specifically, these regression parameters can be computed using the formulas

$$\alpha_{\text{SMA}} = \frac{\alpha_{\text{OLS}}}{r}$$

and

$$\log\beta_{\text{SMA}} = \overline{\log Y_o} - \alpha_{\text{RMA}}\, \overline{\log Y_a},$$

where α_{SMA} is the (reduced major axis) scaling exponent, α_{OLS} is the OLS regression slope, r is the OLS correlation coefficient, log β_{SMA} is now called the allometric constant, and $\overline{\log Y}$ denotes the mean value of log Y. The corresponding 95% confidence intervals of these two regression parameters are computed using the formulas

$$\alpha_{\text{SMA}} \pm t_{N-2}\left(\frac{MSE}{SS_a}\right)^{1/2}$$

and

$$\log\beta_{\text{SMA}} \pm t_{N-2}\left[MSE\left(\frac{1}{\text{N}} + \frac{\overline{\log Y_a}^{2}}{SS_a}\right)\right]^{1/2},$$

where MSE is the OLS mean square error, SS_a is the OLS sums of squares for log Y_a, N is the sample size, and $t_{N-2} = 1.96$ when $N - 2 > 120$.

LIGHT, GROWTH, AND BODY SIZE

Two scaling relationships appear to cut across phyletically diverse unicellular algae and tree-sized embryophytes (Banse 1976; Niklas 1994, 2004; Niklas and Enquist 2001). Growth in dry C mass per individual per year ("annual growth", G_T) scales isometrically with respect to the capacity to intercept sunlight (quantified by pigment concentration per cell for unicellular algae, C_P, and by standing leaf mass for tree species, M_L), and annual growth scales as the 3/4 power of body mass (total cell or organism dry mass, M_T). Respectively, these scaling relationships are expressed by the isometric and allometric formulas

$$H = \beta_0\, G_T,$$

and

$$G_T = \beta_1 M_T^{3/4},$$

where H denotes C_P or M_L and allometric constants are distinguished from each other by different numerical subscripts. Combining these two scaling relationships leads to the prediction that the ability to harvest sunlight as gauged by C_P or M_L is proportional to the 3/4 power of total body mass, i.e.

$$H = \beta_2 M_T^{3/4},$$

where $\beta_2 = \beta_0 \beta_1$. These log-log linear scaling relationships are illustrated in Figures 2.1 and 2.2.

An isometric relationship between H and G_T makes intuitive sense. Even though the ability to "harvest sunlight" and its corresponding "energy use efficiency" are very different biophysical phenomena, it is not unreasonable to expect growth rates to correlate linearly with the ability to capture radiant energy. In contrast, it is not obvious why either annual growth rate or light-harvesting ability should scale as the 3/4 power of body mass. Early workers exploring the relationship between basal metabolic rates across animals differing in body size expected a 2/3 scaling exponent, because they assumed that the ability of cells or entire organisms to exchange mass or energy with the environment is dictated by body surface area (which scales

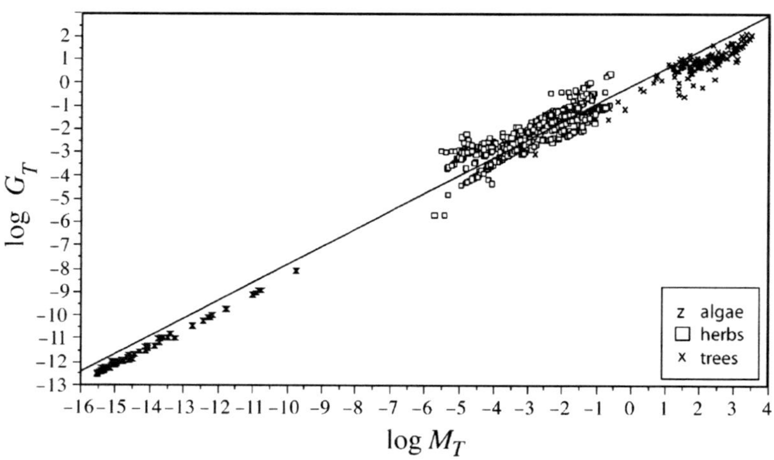

Fig. 2.1 Bivariate relationship between $\log_{10}$-tranformed data for annual growth in total dry mass per plant (G_T) and total body mass (M_T) for unicellular algae, herbaceous species, and trees (see insert for key to symbols). Original units: G_T = pg C per cell per day (algae) and kilogram dry mass per embryophyte per year; M_T = pg C per algal cell (algae) and kilogram dry mass (embryophytes). Solid line, standardized major axis regression curve (for all data); see Table 2.1 for regression statistical summary. (Data from Niklas and Enquist 2001.)

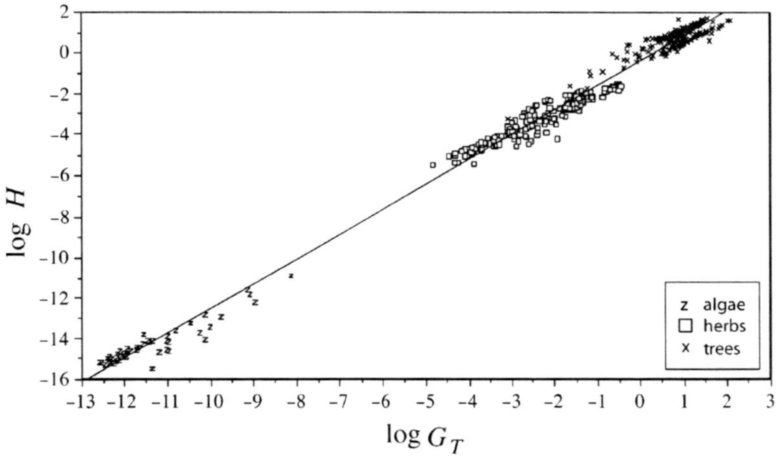

Fig. 2.2 Bivariate relationship between $\log_{10}$-tranformed data for total light-harvesting capacity (H) and annual growth in total dry mass per plant (G_T) and total body mass (M_T) for unicellular algae, herbaceous species, and trees (see insert for key to symbols). Original units: H = pg photosynthetic pigments per cell per day (algae) and kilogram dry leaf mass per plant per year; G_T = pg C per cell per day (algae) and kilogram dry mass per embryophyte per year. Solid line, standardized major axis regression curve (for all data); see Table 2.1 for regression statistical summary. (Data from Niklas and Enquist 2001.)

as the square of any linear reference dimension L) and that the demand for nutrients is correlated with body volume (which scales as the cube of L). Importantly, the 2/3 scaling relationship between surface area and volume holds true only for a series of geometrically identical objects that retain the same shape as they increase in size – two conditions that are repeatedly violated by unicellular and multicellular organisms, both ontogenetically and phylogenetically.

Regardless of the mechanistic explanation for why the three scaling relationships exist, each receives reasonably strong statistical support when "tested" against empirically observed trends for phyletically diverse unicellular algae and tree-sized dicots and conifers (Table 2.1). For these organisms, the 95% confidence intervals of the slope of the log-log linear relationship between light harvesting capability and annual growth approach or include unity. Likewise, the intervals of the slope of the log-log linear relationship between annual growth and total body mass include 0.75. Thus, the proportional relationships summarized by $H \alpha G_T \alpha M_T^{3/4}$ are reasonably accurate across unicellular algae and tree-sized plants.

In pointed contrast, the allometry of nonwoody plants (i.e. herbaceous species and one-year old dicot and conifer tree species) deviates from these predictions, because it is strongly isometric in terms of all three biological variables, i.e., $H \alpha G_T \alpha M_T$ (Table 2.1). Consequently, the 3/4 scaling "rule" is neither "invariant" nor "universal".

Table 2.1 Standardized major axis regression scaling exponents, allometric constants, and their respective 95% confidence intervals for $\log_{10}$-transformed annual growth rates G_T, light interception capabilities H, and total dry body mass M_T of unicellular algae, nonwoody plants, and woody plants (see Figures 2.1 and 2.2)

	α_{SMA} (95% CI)	$\log \beta_{SMA}$ (95% CI)	r^2
Unicellular algae (H gauged by cell pigment concentration C_P) (n = 68)			
$\log H$ vs. $\log G_T$	0.95 (0.87; 1.03)	−3.51 (−4.43; −2.60)	0.886
$\log G_T$ vs. $\log M_T$	0.75 (0.73; 0.76)	−0.91 (−1.10; −0.70)	0.995
$\log H$ vs. $\log M_T$	0.71 (0.64; 0.78)	−4.38 (−5.25; −3.50)	0.876
Nonwoody species (H gauged by standing leaf mass M_L) (n = 1,147)			
$\log H$ vs. $\log G_T$	1.01 (0.97; 1.06)	−0.91 (−1.01; −0.80)	0.903
$\log G_T$ vs. $\log M_T$	0.99 (0.95; 1.04)	0.51 (0.39; 0.63)	0.907
$\log H$ vs. $\log M_T$	1.01 (0.99; 1.03)	−0.39 (−0.46; −0.32)	0.975
Woody species (H gauged by standing leaf mass M_L) (n = 265)			
$\log H$ vs. $\log G_T$	0.90 (0.84; 0.97)	−0.11 (−0.19; −0.03)	0.790
$\log G_T$ vs. $\log M_T$	0.77 (0.71; 0.83)	−0.74 (−0.87; −0.61)	0.804
$\log H$ vs. $\log M_T$	0.70 (0.64; 0.75)	−0.78 (−0.91; −0.65)	0.766

LEAF N:P-STOICHIOMETRY

That growth does not invariably scale as the 3/4 power of body mass is evident from the analyses of data for nonwoody vascular plants presented in the previous section. However, the claim that annual growth across ecologically and phyletically diverse unicellular and multicellular photoautotrophic eukaryotes scales isometrically or nearly so with respect to light harvesting ability (see Niklas and Enquist 2001, 2002) is statistically robust (Table 2.1). In the case of unicellular photoautotrophs, H is measured in units of photosynthetic pigment concentrations per cell, C_P. However, for terrestrial embryophytes, H is measured in terms of standing dry leaf mass per plant, M_L. Thus, annual growth appears to be inexorably linked to the "machinery" of photosynthesis in some very basic way that cuts across otherwise sharply defined phyletic boundaries.

This linkage probably exists at numerous metabolic and structural levels, but the view advocated here is that it is sensitive to the manner in which N and P is allocated in light harvesting structures (e.g. moss phyllids, "microphylls" and "euphylls", and entire tree canopies). This perspective is based on the comparatively strong scaling relationships that exist between total leaf carbon mass (M_C) and total leaf N and P (M_N and M_P, respectively) – relationships that appear to obey their own quarter-power "rules" across and within those species that have been examined in sufficient detail.

For example, based on stoichiometric data collected from 131 herbaceous species, including C_3 and C_4 species, Niklas *et al.* (2005) report that leaf N content scales almost isometrically with respect to increasing leaf carbon content, whereas leaf P content scales as the 4/3 power of leaf carbon content (Figure 2.3a; Table 2.2).

For these species, it follows from $M_N \propto M_C$ and $M_P \propto M_C^{4/3}$ such that $M_N \propto M_P^{3/4}$, which should also hold true for N and P content expressed as percentages (Figure 2.3b; Table 2.2). Although stoichiometric analyses of plant conspecifics differing in size are sparse, the data that are available indicate that intraspecific trends may abide by the same "rules". For example, in a study of *Eranthis hyemalis* (a perennial member of the Ranunculaceae), Niklas and Cobb (2005) report scaling exponents

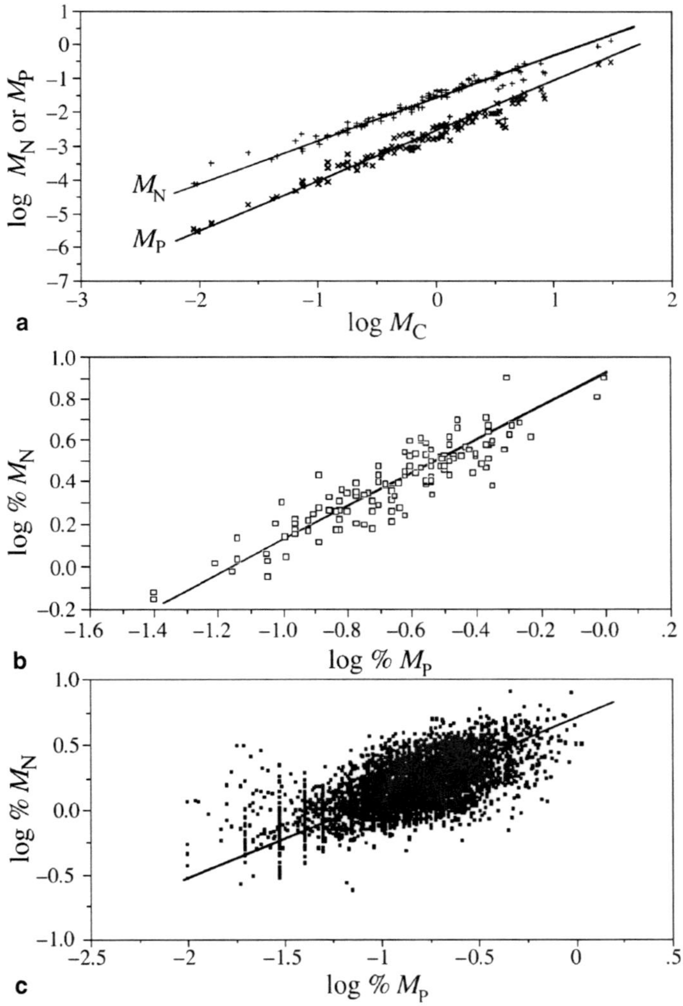

Fig. 2.3 Bivariate relationships among $\log_{10}$-tranformed data for total leaf nitrogen, phosphorus, and carbon content (M_N, M_P, and M_C, respectively) and percentage of leaf nitrogen and phosphorus content. **a.** M_N and M_P versus M_C for 131 species. (Data from Niklas *et al.* 2005.) **b.** % M_N versus % M_P for 131 species. (Data from Niklas *et al.* 2005.) **c.** % M_N versus % M_P for 7,445 species. (Data from Reich and Oleksyn 2005.) Original units: gram mass per leaf gram dry mass. Solid lines, standardized major axis regression curves; see Table 2.2 for regression statistical summaries

Table 2.2 Standardized major axis regression scaling exponents, allometric constants, and their respective 95% confidence intervals for $\log_{10}$-transformed data of total leaf nitrogen, phosphorus, and carbon content (M_N, M_P, and M_C, respectively)

	α_{SMA} (95% CI)	$\log \beta_{SMA}$ (95% CI)	r^2
Across herbaceous species (n = 131)			
$\log M_N$ vs. $\log M_C$	1.06 (0.95; 1.17)	−1.67 (−1.76; −1.58)	0.941
$\log M_P$ vs. $\log M_C$	1.37 (1.27; 1.48)	−2.61 (−2.70; −2.52)	0.968
$\log M_N$ vs. $\log M_P$	0.78 (0.72; 0.85)	−0.74 (−0.72; −0.76)	0.948
Eranthis hyemalis (n = 17)			
$\log M_N$ vs. $\log M_C$	1.00 (0.98; 1.03)	−1.33 (−1.39; −1.26)	0.996
$\log M_P$ vs. $\log M_C$	1.37 (1.32; 1.42)	0.77 (0.65; 0.90)	0.993
$\log M_N$ vs. $\log M_P$	0.73 (0.70; 0.76)	−1.89 (−1.97; −1.82)	0.996
Across Reich and Oleksyn (2004) data set (n = 7,445)			
$\log M_N$ vs. $\log M_P$	0.73 (0.71; 0.75)	1.08 (1.07; 1.08)	0.33

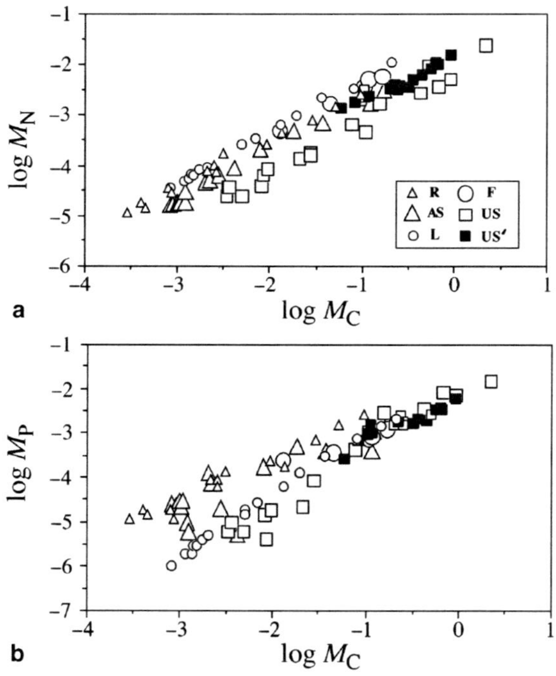

Fig. 2.4 Bivariate relationships among $\log_{10}$-tranformed data for nitrogen, phosphorus, and carbon content (M_N, M_P, and M_C, respectively) for organs and plant parts of *Eranthis hyemalis* (see insert for key to symbols: *R* = roots, *AS* = aerial stems, *L* = leaves, *F* = developing fruits and seeds, *US* = tubers from growing plants; *US'* = tubers from winterized plants). See Table 2.2 for regression statistical summaries. (Data from Niklas and Cobb 2005.)

for M_N, M_P, and M_C relationships that are statistically indistinguishable from the proportional relationships $M_N \propto M_C$ and $M_P \propto M_C^{4/3}$ and $M_N \propto M_P^{3/4}$ even at the level of organ-type (Table 2.2; Figure 2.4).

Whether these scaling relationships are "universal" properties of vascular plant biology remains problematic. Based on an extensive world-wide survey of leaf N and P composition, Wright *et al.* (2004) reported that leaf N content scales roughly as the 2/3 power of leaf P content. In contrast, using an expanded version of the leaf N and P data reported by Reich and Oleksyn (2004) consisting of 7,445 entries for individual species reflecting conspecifics differing in age, regression analysis of M_N versus M_P reveals a scaling exponent of 0.73 with 95% confidence intervals that include the numerical value of 3/4 but exclude that of 2/3 (Figure 2.3c; Table 2.2). This inconsistency may be the result of phyletic effects (i.e. biases introduced by differences in the taxonomic composition of the data sets used). However, regardless of the reason, it is clear that the relationship between leaf N and P content is allometric and governed by the generic formula

$$M_N = \beta_3 \, M_P^{\alpha < 1.0}$$

N:P-STOICHIOMETRY AND GROWTH MODELS

This "generic" formula has added significance when it is juxtaposed with stoichiometric models for predicting relative growth rates based on cell or tissue N and P content. Dobberfuhl (1999) first proposed that growth depends on total body N (N_T) and total body P (P_T) allocation to protein and ribosomal RNA (also see Sterner and Elser 2002; Ågren 2004; Vrede *et al.* 2004). This model conceptually relates relative growth rates to N:P-stoichiometry by envisioning proteins as the "overhead" required to achieve growth and rRNA as the protein-output "machinery" used to maintain or recycle this overhead. Dobberfuhl and others noted that, when an organism maintains a constant chemical composition, its relative growth rate μ can be mathematically expressed in terms of the amounts and rates of change of carbon (C), nitrogen (N), and phosphorus (P) content by the formula

$$\mu = \frac{1}{C}\left(\frac{dC}{dt}\right) = \frac{1}{N}\left(\frac{dN}{dt}\right) = \frac{1}{P}\left(\frac{dP}{dt}\right).$$

For any one of these essential substances, designated as *X*, μ can be approximated by the formulas

$$\mu = \frac{1}{X}\left(\frac{dX}{dt}\right) = \ln\left(\frac{X_2}{X_1}\right) \cdot \left(t_2 - t_1\right)^{-1},$$

where X_2 is the total concentration of substance X at time t_2 and X_1 is the total concentration of X at time t_1 (see Hunt 1990). If X is some measure of protein synthesis, the formulas for μ can be recast as

$$\mu = \ln\left[\frac{f_N \mathrm{N_T} + \left(\dfrac{k_s r_e F f_P \mathrm{P_T}}{m_r}\right)}{f_N \mathrm{N_T}}\right] \cdot t^{-1},$$

where f_N is the decimal fraction of $\mathrm{N_T}$ invested in proteins, k_s is the protein synthesis rate per ribosome, r_e is the protein retention efficiency, F is the decimal fraction of total RNA allocated to rRNA, f_P is the decimal fraction of $\mathrm{P_T}$ invested in RNA, m_r is the mass of an average ribosome, and t now denotes the time interval $t_2 - t_1$ (Dobberfuhl 1999; Vrede *et al.* 2004).

The relative growth rates of very different unicellular algae and small aquatic animals have been successfully predicted using this approach or ones similar to it (e.g. Nielsen *et al.* 1996; Klausmeier *et al.* 2004; Vrede *et al.* 2004) despite the assumptions that $\mathrm{N_T}$ and $\mathrm{P_T}$ allocation patterns are ontogenetically invariant, that balanced growth has been achieved, and the supposition that resources are not limiting. In addition, this approach has been integrated with allometric theory by noting that the ability to harvest light scales isometrically with respect to total annual plant growth across unicellular algae and vascular plant species (Table 2.1; Niklas and Enquist 2001).

As noted for vascular plants, this ability is gauged by standing leaf dry mass M_L. Consequently, the relative growth rate of leaves μ_L may provide a reliable gauge of the relative growth rate of the entire plant body. Accordingly, if the formula for μ is *generally* valid, μ_L should be governed by total leaf N and P such that

$$\mu_L = \ln\left[1 + \left(\frac{k_s r_e F}{m_r}\right)\left(\frac{f_\mathrm{P}}{f_\mathrm{N}}\right)\left(\frac{M_\mathrm{P}}{M_\mathrm{N}}\right)\right] \quad t^{-1}.$$

Combining this relationship with the observation that $M_\mathrm{N} = \beta_3\, M_\mathrm{P}^{\alpha}$ obtains a quantitative description of leaf relative growth rates in terms of the allometry of total leaf N and P:

$$\mu_L = \ln\left[1 + \left(\frac{k_s r_e F}{m_r}\right)\left(\frac{f_\mathrm{P}}{f_\mathrm{N}}\right)\left(\frac{M_\mathrm{P}^{\frac{1}{\alpha}-1}}{M_\mathrm{N}^{\frac{1}{\alpha}}}\right)\right] t^{-1}.$$

Note that this formula predicts that leaf relative growth rates will increase across species with either increasing leaf N or P allocations when $\alpha < 1.0$ (see Niklas *et al.* 2005).

TESTING THE MODEL

The relationship adduced in the previous section for μ_L has three significant attributes: (1) it incorporates an allometric relationship for leaf N and P allocation directly, (2) it relates the N:P-stoichiometry of leaves directly to relative growth rates (and thus prior allometric theory treating the relationship between leaf dry mass and total plant annual growth), and (3) it can be examined empirically based on observed leaf growth rates, thereby setting limits on the numerical values of α and other allometric or physiological parameters.

However, this relationship has an important distraction. It can be evaluated empirically only if the numerical values of all physiological variables are known. For bacteria and animals, the values of some of these variables have been experimentally determined, i.e., $k_s = 2.5 \times 10^{-11}$ μg protein ribosome^{-1} day^{-1}, $r_e = 0.60$, $F = 0.80$, and $m_r = 4.53 \times 10^{-12}$ μg rRNA ribosome^{-1} (Campana and Schwartz 1981; McKee and Knowles 1987; Mathers *et al.* 1993; Sadava 1993; Vrede *et al.* 2004). Assuming that these values are equally applicable to plants, it follows that $k_s\, r_e\, F/m_r = 2.648$. The value of f_N, however, is problematic for plants. Prior work indicates that between 16% and 27% of total leaf N is incorporated in RuBisCo (Evans 1989) and that, depending on whether ambient light conditions are high or low, between 15% and 60% of total leaf N is found in chloroplast thylakoids (pigment-protein complexes, electron transport constituents, reaction centers, components of the electron transport chain, particularly cytochrome b/f; and ferredoxin; Evans 1989). Therefore, based on published N allocation to RuBisCo and thylakoids, it is reasonable to suppose that $f_N \sim 0.55$ across otherwise diverse species.

Likewise, from prior analyses of 131 herbaceous species, the allometry of leaf N with respect to leaf P is reasonably well approximated by the formula $M_N = 0.18\, M_P^{3/4}$ (Figure 2.2; Table 2.2; Niklas *et al.* 2005). Using this scaling relationship gives a model for the relative growth rates of leaves that lacks the numerical value of only one parameter, the decimal fraction of total leaf P allocated to RNA:

$$\mu_L = \ln\left(1 + 47.37 f_P M_N^{1/3}\right)\cdot t^{-1} = \ln\left(1 + 26.75 f_P M_P^{1/4}\right)\cdot t^{-1}.$$

Assuming that f_P varies little or not at all across species, this formula predicts that μ_L will increase as a function of either increasing M_N or M_P (either or both of which increase with leaf C mass). It also predicts that μ_L should decrease as a function of increasing M_N/M_P.

Unfortunately, reliable estimates for f_P for vascular plant species are currently unavailable. However, the aforementioned model has been used to calculate μ_L using different values of f_P and its results have been compared to observed leaf growth rates to set limits on the range of f_P-values that may occur in leaves (Niklas *et al.* 2005; Niklas 2006). Specifically, across 131 ecologically and phyletically diverse herbaceous species, observed μ_L scaled as the 0.33 power of total leaf N ($r^2 = 0.72$), as the 0.25 power of total leaf P ($r^2 = 0.76$), and as the 0.22 power of total leaf carbon mass ($r^2 = 0.78$; Figure 2.5). Likewise, as predicted, μ_L decreased as the quotient M_{LN}/M_{LP} increased.

CAVEATS AND THE ALLOMETRY OF CURRENT-YEAR SHOOTS

Nevertheless, considerable variation exists when observed leaf growth rates are plotted as a function of leaf carbon mass (Figure 2.1). This feature can be attributed to a number of factors not addressed in the model, but that are nevertheless of great ecological importance. Among the more obvious of these are species-specific differences or ecotypic variation in protein synthesis rates or retention efficiencies (Mathers *et al.* 1993; Sadava 1993; Güsewell 2004), differences in the fractional allocation of leaf N to proteins as a consequence of leaf age or different ambient light intensities (Evans 1989; Ryser *et al.* 1997), morphological and anatomical differences in leaf construction (Nielsen *et al.* 1996), recruitment of N and P from older organs during early growth (Meyer and Tukey 1965), differences in leaf tissue ploidy and N-use efficiency (Brown 1978), and changes in N or P allocation to leaf components during leaf ontogeny.

Despite the remarkable success of the model in light of the undoubted influence of these and other physiological and ecological variables on growth, there are grounds for concern. As noted, one of the attributes of N:P-stoichiometric models is that they can be challenged empirically by comparing the numerical values of physiological variables used to predict observed growth rates with those that are actually reported in the literature. One of these variables is the decimal fraction of total body P allocated to RNA construction, denoted by f_P. Inspection of Figure 2.5 indicates that the relative growth rates observed for 131 plant species plot between those predicted by a model in which f_P is *a priori* set equal to 0.05 and 0.15. The fact that all observed leaf growth rates plot within the "corridor" defined by these two values suggests that between 5% and 15% of total leaf P is invested in the construction of RNA.

However, this range does not agree with f_P-values reported for other life forms such as bacteria, small aquatic heterotrophs, or unicellular algae (Rhee 1978; Elser *et al.* 2003). Across these species, the decimal fraction of total cell or body P committed to RNA ranges between 0.20 and 0.90 (with an average value of 0.50 for animals). Therefore, the range $0.05 \leq f_P \leq 0.15$ identified by the model to set boundaries on the growth rates of vascular plant leaves is unusually low.

This discrepancy may be the result of systematically underestimating plant growth rates because of their ability to store and annually recycle large pools of N and P. Specifically, the relative growth rates plotted in Figure 2.5 are based on the difference in leaf N and P concentrations measured early and late in the growing season (Niklas *et al.* 2005). This procedure may introduce a systematic bias because there is substantial evidence that vascular plants recruit N and P from older organs as new tissues and organs are produced during the early growth season (e.g.

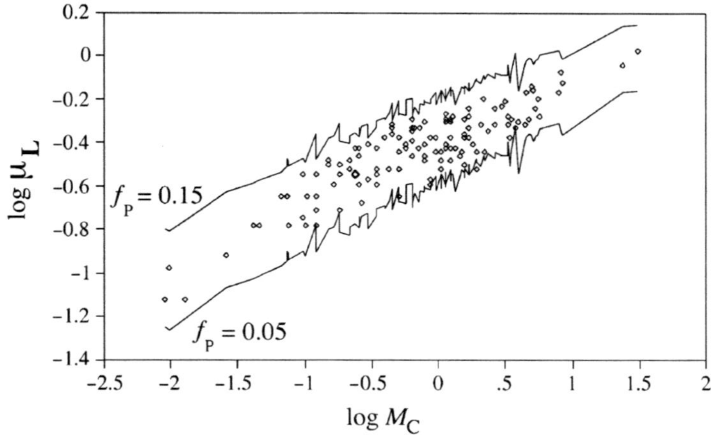

Fig. 2.5 Log_{10}-tranformed data for observed relative growth rates (μ_L) and those predicted by allometric theory plotted against log-transformed data for leaf carbon mass (M_C). Predicted values of μ_L assume that the decimal fraction of total leaf phosphorus content invested in RNA (f_P) equals 5% or 15% (predicted μ_L indicated by lines). (Data from Niklas *et al.* 2005.)

Mochizuki and Hanada 1958; Meyer and Tukey 1965; Taylor 1967; Taylor and May 1967; Niklas and Cobb 2005, 2006) and reabsorb substantial quantities of N and P toward the end of the growing season, e.g. an average of 50% of total P is reabsorbed before leaf senescence (50% of which comes from nucleic acid hydrolysis; Chapin and Kedrowski 1983; Aerts 1996). Therefore, tissue N and P concentrations in newly formed leaves may be significantly higher than those reached once constant growth rates are achieved (thus violating a basic assumption of the model), and leaf N and P concentrations may be on the decline even before visible signs of leaf senescence. If leaf relative growth rates are systematically underestimated for these (or other) reasons, the upper and lower f_P-levels identified by the model to fit the data would, likewise, be underestimated.

There is also good evidence that statistically significant differences exist for the N:P-stoichiometry of the current-year shoots of tree and herbaceous species. Niklas and Cobb (2006) report that M_N scales as the 1.17-power of dry leaf mass M_L and that M_P scales as the 1.09-power of M_L across current-year shoots of trees. The numerical values of these scaling exponents were found to be significantly different from those observed for herbaceous species (Table 2.3). Likewise, across tree shoots, M_N scaled as the 1.07-power of M_P, whereas, across herbaceous shoots, M_N scaled as the 0.77-power of M_P (Figure 2.6; Table 2.3). These data strongly caution against extrapolating N:P-stoichiometric scaling relationships across different plant size ranges (Reich *et al.* 2006).

Finally, there is evidence that the faction of M_N invested in photosynthetic tissues decreases with increasing M_L. For example, Takashima *et al.* (2004) studied the stoichiometry of two "evergreen" and two deciduous *Quercus* species and report that the fraction of M_N allocated to the photosynthetic machinery of the

Table 2.3 Standardized major axis (SMA) regression parameters for leaf mass and N:P–stoichiometry relationships. M_N = total leaf nitrogen content, M_P = total leaf phosphorus content; M_L = total leaf dry mass

	α_{SMA} (95% CI)	log β_{SMA} (95% CI)	r^2
Tree shoots (n = 112)			
log M_N vs. log M_L	1.17 (1.14; 1.20)	−1.75 (−1.78; −1.71)	0.949
log M_P vs. log M_L	1.09 (1.04; 1.14)	−2.77 (−2.82; −2.73)	0.885
log M_N vs. log M_P	1.07 (1.02; 1.12)	1.22 (1.07; 1.38)	0.887
Herbaceous shoots (n = 131)			
log M_N vs. log M_L	1.06 (1.02; 1.10)	−1.67 (−1.70; −1.64)	0.945
log M_P vs. log M_L	1.37 (1.34; 1.41)	−2.61 (−2.64; −2.58)	0.966
log M_N vs. log M_P	0.77 (0.74; 0.80)	0.34 (0.25; 0.43)	0.948

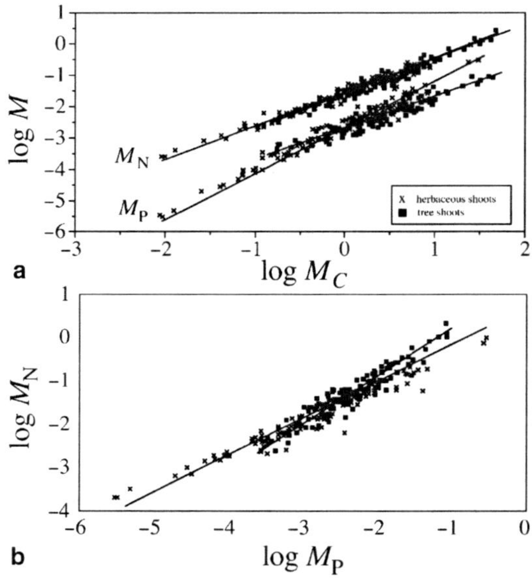

Fig. 2.6 Bivariate relationships among $\log_{10}$-tranformed data for nitrogen, phosphorus, and carbon content of all leaves from first-year shoots of herbaceous and tree species (M_N, M_P, and M_C, respectively; see insert for symbols). Solid lines, standardized major axis regression curves; see Table 2.3 for regression statistical summaries. **A**. M_N and M_P versus M_C. **B**. M_N versus M_P. (Data from Niklas and Cobb 2006.)

evergreen species is smaller than that allocated in the deciduous species, whereas the fraction allocated to cell wall proteins was correspondingly higher. Across these species, the fraction of M_N in cell wall proteins was also positively correlated with M_L per unit area, suggesting a trade-off exists in how N is partitioned between photosynthetic and cell wall structural proteins. This suggestion is consistent with the observation that M_N increases across tree species with increasing M_L more rapidly than across herbaceous species (whose leaves tend to be less structurally reinforced than those of tree species), whereas M_P increases more rapidly with respect to M_L across herbaceous species (whose leaves tend to possess larger fractions of living tissues; Niklas and Cobb 2005, 2006). These trends are also consistent with studies reporting non-uniform M_N distribution patterns in tree canopies (e.g. Hirose and Werger 1987; Werger and Hirose 1991; Sellers *et al.* 1992; Sands 1995; Anten *et al.* 2000) that accord with the theoretical prediction that optimal M_N distribution patterns should increase with increased light intensities (Field and Mooney 1986; Anten *et al.* 1995).

VARIATION AND FUTURE DIRECTIONS

That significant differences in leaf N:P-stoichiometry exist even among vascular plants is strongly suggested by the allometric and stoichiometric trends reviewed here and elsewhere (Broadley *et al.* 2004; Watanabe *et al.* 2007). As noted, the relationships among total leaf mass, annual growth rates, and total body mass differ between nonwoody and woody plants and they differ between the current-year shoots of herbaceous and woody species. Across non-woody plants, total leaf mass scales isometrically with respect to both annual growth rate and total body mass (i.e., $M_L \propto G_T \propto M_T$), whereas, across woody plants, leaf mass scales as the 3/4 power to total body mass (i.e., $M_L \propto G_T \propto M_T^{3/4}$). Yet, based on the data presented here and elsewhere, total leaf mass (in units of C mass) scales isometrically with respect to total leaf N and allometrically albeit roughly as the 3/4 power of total leaf P (i.e., $M_C \propto M_N \propto M_P^{3/4}$; Table 2.2). Assuming that these relationships are definitive, it follows that total annual growth rates should scale across all species as the 3/4 power of total leaf P (i.e. $G_T \propto M_P^{3/4}$). However, it also follows that total leaf P should increase as the 4/3 power of the total body mass for nonwoody plants (i.e. $M_P \propto M_T^{4/3}$) but scale isometrically with respect to the body mass for woody plants (i.e. $M_P \propto M_T$).

These predictions have yet to be explored empirically, but they are in accord with the observation that much of the total body mass of woody plants is composed of physiologically inert material (heartwood) that increases in volume fraction with each passing year. Indeed, we need to know much more about the allometry and stoichiometry of what may be called "necromass" – organic constituents that contribute to total body mass but that do not participate in metabolic activity or resource utilization, e.g. cell wall materials and secondary metabolites sequestered in the lumens of dead cells, which continue to accumulate throughout the lifetime of the multicellular individual. We also need to know much more about the allometry of annual biomass accumulation with respect to the N:P-stoichiometry of meristematic tissues, both for herbaceous non-woody and woody species.

It is also clear that the juxtaposition of allometric theory and observation with the potential insights gained from N:P-stoichiometric models is in its infancy. This approach clearly offers great promise (if for no other reason than that it helps to identify and quantify interdependencies across every level of biological organization, from molecules to ecosystems, and across bacteria to multicellular eukaryotes), but it is perhaps best viewed as a heuristic device with which to explore important conceptual issues. To be more effective, this juxtaposition would benefit greatly from more detailed measurements of how stoichiometric parameters vary ontogenetically and phylogenetically. In particular, more detailed data sets are needed for protein synthesis rates per ribosome, protein retention efficiencies, and the proportion of total P and N committed respectively to the construction of rRNA and nonstructural proteins. Mutants of unicellular algae, like those of *Chlamydomonas*, and parasitic plants with a "leaf-stem" construction, like *Monotropa*, should be used to "dissect" how total N and P of cells and tissues are allocated to the construction of different parts of the photosynthetic machinery. Perhaps in this way, we will be able to explain why tissue and organ N and dry mass scale as the 3/4 power of P and why so many other phenomena seem to obey similar quarter-power rules.

REFERENCES

Aerts R (1996) Nutrient resorption from senescing leaves of perennials: are there general patterns? J Ecol 84: 597–608

Aerts R, Chapin FS (2000) The mineral nutrition of wild plants revisited: a re-evaluation of processes and patterns. Adv Ecol Res 30: 1–67

Ågren GI (1988) Ideal nutrient productivities and nutrient proportions in plant growth. Plant Cell Environ 11: 613–620

Ågren GI (2004) The C:N:P stoichiometry of autotrophs - theory and observations. Ecol Lett 7: 185–191

Ågren GI, Bosatta E (1996) Theoretical Ecosystem Ecology: Understanding Nutrient Cycles. Cambridge University Press, Cambridge

Anten NPR, Schieving F, Werger MJA (1995) Patterns of light and nitrogen distribution in relation to whole canopy carbon gain in C-3 and C-4 monocotyledonous and dicotyledonous species. Oecologia 101: 504–513

Anten NPR, Hikosaka K, Hirose T (2000) Nitrogen utilisation and the photosynthetic system. In: Marshall B, Roberts JA (eds), Leaf Development and Canopy Growth. Sheffield Academic Press, Sheffield, pp 171–203

Banse K (1976) Rates of growth, respiration and photosynthesis of unicellular algae as related to cell size - a review. J Phycol 12: 135–140

Bazzaz FA, Grace J (1997) Plant Resource Allocation. Academic, San Diego, CA

Blaxter KL (ed) (1965) Proceedings of the 3rd Symposium on Energy Metabolism. Troon, Scotland, May 1964. Academic, New York

Blum JJ (1977) On the geometry of four-dimensions and the relationship between metabolism and body mass. J Theor Biol 64: 599–601

Broadley MR, Bowen HC, Cotterill HL, Hammond JP, Meacham MC, Mead A, White PJ (2004) Phylogenetic variation in the shoot mineral concentration of angiosperms. J Exp Bot 55: 321–336

Brown RH (1978) A difference in N use efficiency in C_3 and C_4 plants and its implications in adaptation and evolution. Crop Sci 18: 93–98

Calder WA III (1984) Size, Function, and Life History. Harvard University Press, Cambridge, MA

Calder WA III (1996) Size, Function, and Life History. Dover, New York

Campana T, Schwartz LM (1981) Section 9. RNA and associated enzymes. In: Schwartz LM, Azar MM (eds), Advanced Cell Biology. Van Nostrand Reinhold, New York, pp 877–944

Cebrian J (1999) Patterns in the fate of production in plant communities. Am Nat 154: 449–468

Chapin FS, Kedrowski RA (1983) Seasonal changes in nitrogen and phosphorus fractions and autumn retranslocation in evergreen and deciduous taiga trees. Ecology 64: 376–391

Chapin FS, Vitousek PM, Vancleve K (1986) The nature of nutrient limitation in plant-communities. Am Nat 127: 48–58
Chapin FS, Walker BH, Hobbs RJ, Hooper DU, Lawton JH, Sala OE, Tilman D (1997) Biotic control over the functioning of ecosystems. Science 277: 500–504
Darveau CA, Suarez RK, Andrews RD, Hochachka PW (2002) Allometric cascade as a unifying principle of body mass effects on metabolism. Nature 417: 166–170
De Angelis DL (1980) Energy-flow, nutrient cycling, and ecosystem resilience. Ecology 61: 764–771
Dobberfuhl DR (1999) Elemental Stoichiometry in Crustacean Zooplankton: Phylogenetic Patterns, Physiological Mechanisms, and Ecological Consequences. Dissertation, Arizona State University, Tempe, Arizona
Dodds, PS, Rothman DH, Weitz JS (2001) Re-examination of the "3/4 - law" of metabolism. J Theor Biol 209: 9–27
Economos AE (1982) On the origin of biological similarity. J Theor Biol 94: 25–60
Economos AE (1983) Elastic and/or geometric similarity in mammalian design. J Theor Biol 103: 167–172
Elser JJ, Fagan WF, Denno RF, Dobberfuhl DR, Folarin A, Huberty A, Interlandi S, Kilham SS, McCauley S, Schulz KL, Siemann EH, Sterner RW (2000a) Nutritional constraints in terrestrial and freshwater food webs. Nature 408: 578–580
Elser JJ, Sterner RW, Gorokhova E, Fagan WF, Markow TA, Cotner JB, Harrison JF Hobbie SE, Odell GM, Weider LJ (2000b) Biological stoichiometry from genes to ecosystems. Ecol Lett 3: 540–550
Elser JJ, Acharya K, Kyle M, Cotner J, Makino W, Markow T, Watts T, Hobbie W, Fagan W, Schade J, Sterner RW (2003) Growth rate-stoichiometry couplings in diverse biota. Ecol Lett 6: 936–943
Enquist BJ, Niklas KJ (2002) Global allocation rules for patterns of biomass partitioning in seed plants. Science 295: 1517–1520
Evans JR (1989) Photosynthesis and nitrogen relationships in leaves of C_3 plants. Oecologia 78: 9–19
Feldman HA (1995) On the allometric mass exponent, when it exists. J Theor Biol 172: 187–197
Field C, Mooney HA (1986) The photosynthesis-nitrogen relationship in wild plants. In: Givnish TJ (ed), On the Economy of Plant Form and Function. Cambridge University Press, Cambridge, pp 25–55
Gray BF (1981) On the "surface law" and basal metabolic rate. J Theor Biol 93: 757–767
Grime JP (1979) Plant Strategies and Vegetation Processes. Wiley, Chichester
Güsewell S (2004) N:P ratios in terrestrial plants: variation and functional significance. New Phytol 164: 243–266
Harvey PH (1982) Rethinking allometry. J Theor Biol 95: 37–41
Hemmingsen AM (1960) Energy metabolism as related to body size and respiratory surfaces, and its evolution. Reports of the Steno Memorial Hospital and Nordisk Insulin Laboratorium 9: 6–110
Heusner A (1982) Energy metabolism and body size. I. Is the 0.75 mass exponent of Kleiber a statistical artifact? Resp Physiol 48: 1–12
Hirose T, Werger MJA (1987) Maximizing daily canopy photosynthesis with respect to the leaf nitrogen allocation pattern in the canopy. Oecologia 72: 520–526
Hunt R (1990) Basic Growth Analysis. Unwin Hyman, London
Jackson RB, Mooney HA, Schulze ED (1997) A global budget for fine root biomass surface area and nutrient contents. Proc Natl Acad Sci USA 94: 7362–7366
Jolicoeur P (1990) Bivariate allometry: interval estimation of the sloped of the ordinary and standardized normal major axes and structural relationship. J Theor Biol 144: 275–285
Karpinets TV, Greenwood DJ, Sams CE, Ammons JT (2006) RNA:protein ratio of the unicellular organism as a characteristic of phosphorus and nitrogen stoichiometry of the cellular requirement of ribosomes for protein synthesis. BMC Biol 4: 30
Kerkhoff AJ, Enquist BJ, Elser JJ, Fagan WF (2005) Plant allometry, stoichiometry and the temperature-dependence of primary productivity. Global Ecol Biogeogr 14: 585–598

Kerkhoff AJ, Fagan WF, Elser JJ, Enquist BJ (2006) Phylogenetic and growth form variation in the scaling of nitrogen and phosphorus in the seed plants. Am Nat 168: E103–E122
Klausmeier CA, Litchman E, Daufresne T, Levin SA (2004) Optimal nitrogen-to-phosphorus stoichiometry of phytoplankton. Nature 429: 171–174
Kleiber M (1932) Body size and metabolic rate. Physiol Rev 27: 511–541
Kleiber M (1961) The Fire of Life. An Introduction to Animal Energetics. Wiley, New York
Koerselman W, Meuleman AFM (1996) The vegetation N:P ratio: a new tool to detect the nature of nutrient limitation. J Appl Ecol 33: 1441–1450
Lindstedt SL, Calder WA III (1981) Body size, physiological time, and longevity of homeothermic animals. Quart Rev Biol 56: 1–16
Mathers EM, Houlihan DF, McCarthy ID, Burren LJ (1993) Rates of growth and protein-synthesis correlated with nucleic-acid content in fry of rainbow trout, *Oncorhynchus mykiss* - effects of age and temperature. J Fish Biol 43: 245–263
Mattson WJ (1980) Herbivory in relation to plant nitrogen-content. Ann Rev Ecol Syst 11: 119–161
McArdle BH (1988) The structural relationship: regression in biology. Can J Zool 66: 2329–2339
McArdle BH (2003) Lines, models, and errors: regression in the field. Limnol Oceanogr 48: 1363–1366
McKee MJ, Knowles CO (1987) Levels of protein, RNA, DNA, glycogen and lipid during growth and development of *Daphnia magna* Straus (Crustacea: Cladocera). Freshwater Biol 18: 342–351
Meyer MM, Tukey HB Jr (1965) Nitrogen, phosphorus, and potassium plant reserves and the spring growth of Taxus and Forsythia. Proc Am Soc Hort Sci 87: 537–544
Milberg P, Lamont BB (1997) Seed/cotyledon size and nutrient content play a major role in early performance of species on nutrient-poor soils. New Phytol 137: 665–672
Mochizuki T, Hanada S (1958) The effect of nitrogen on the formation of the anisophylly on the terminal shoots of apple trees. Soil Plant Food 4: 68–74
Nielsen SL, Enríquez S, Duarte CM, Sand-Jensen S (1996) Scaling maximum growth rates across photosynthetic organisms. Funct Ecol 10: 167–175
Niklas KJ (1994) Size-dependent variations in plant growth rates and the "3/4-power rule". Am J Bot 81: 134–144
Niklas KJ (2004) Plant allometry: Is there a grand unifying theory? Biol Rev 79: 871–889
Niklas KJ (2006) Plant allometry, leaf nitrogen and phosphorus stoichiometry and interspecific trends in annual growth rates. Ann Bot 97: 155–163
Niklas KJ, Cobb ED (2005) N, P, and C stoichiometry of *Eranthis hyemalis* (L). Salib. (Ranunculaceae) and the allometry of plant growth. Am J Bot 92: 1263–1268
Niklas KJ, Cobb ED (2006) Biomass partitioning and leaf N,P-stoichiometry: comparisons between tree and herbaceous current-year shoots. Plant Cell Environ 29: 2030–2042
Niklas KJ, Enquist BJ (2001) Invariant scaling relations for interspecific plant biomass production rates and body size. Proc Natl Acad Sci USA 98: 2922–2927
Niklas KJ, Enquist BJ (2002) On the vegetative biomass partitioning of seed plant leaves, stems, and roots. Am Nat 159: 1517–1520
Niklas KJ, Owens T, Reich PB, Cobb ED (2005) Nitrogen/phosphorus leaf stoichiometry and the scaling of plant growth. Ecol Lett 8: 636–642
Niklas KJ, Cobb ED, Niinemets Ü, Reich PB, Sellin A, Shipley B, Wright IJ (2007) "Diminishing returns" in the scaling of functional leaf traits across and within species groups. Proc Natl Acad Sci USA 104: 8894–8896
Peters RH (1983) The Ecological Implications of Body Size. Cambridge University Press, Cambridge
Prothero J (1986a) Scaling of energy-metabolism in unicellular organisms - a reanalysis. Comp Biochem Physiol A: Physiol 83: 243–248
Prothero J (1986b) Methodological aspects of scaling in biology. J Theor Biol 118: 259–286
Rayner JMV (1985) Linear relations in biomechanics: the statistics of scaling functions. J Zool 206: 415–439

Reich PB, Oleksyn J (2004) Global patterns of plant leaf N and P in relation to temperature and latitude. Proc Natl Acad Sci USA 101: 11001–11006
Reich PB, Tjoelker MG, Machado JL, Oleksyn J (2006) Universal scaling of respiratory metabolism, size and nitrogen in plants. Nature 439: 457–461
Rhee G-Y (1978) Effects of N:P atomic ratios and nitrate limitation on algal growth, cell composition, and nitrate uptake. Limnol Oceanogr 23: 10–25
Ryser P, Verduyn B, Lambers H (1997) Phosphorus allocation and utilization in three grass species with contrasting response to N and P supply. New Phytol 137: 293–302
Sadava D (1993) Cell Biology: Organelles, Structure and Function. Jones and Bartlett, Boston, MA
Sands PJ (1995) Modelling canopy production. I. Optimal distribution of photosynthetic resources. Aust J Plant Physiol 22: 603–614
Schmidt-Nielsen K (1984) Scaling: Why Is Animal Size So Important? Cambridge University Press, Cambridge
Seim E (1983) On rethinking allometry: which regression model to use? J Theor Biol 104: 161–168
Sellers PJ, Berry JA, Collatz GJ, Field CB, Mooney HA (1992) Canopy reflectance photosynthesis and respiration. III. A reanalysis using improved leaf models and a new canopy integration scheme. Remote Sens Environ 42: 187–216
Silver WL (1994) Is nutrient availability related to plant nutrient use in humid tropical forests. Oecologia 98: 336–343
Sokal RR, Rohlf (1980) Biometry. Freeman, New York
Smith RJ (1980) Rethinking allometry. J Theor Biol 87: 97–111
Sterner RW, Elser JJ (2002) Ecological Stoichiometry: The Biology of Elements from Molecules to the Biosphere. Princeton University Press, Princeton, NJ
Takashima T, Hikosaka K, Hirose T (2004) Photosynthesis or persistence: nitrogen allocation in leaves of evergreen and deciduous *Quercus* species. Plant Cell Environ 27: 1047–1054
Taylor BK (1967) The nitrogen nutrition of the peach tree. I. Seasonal changes in nitrogenous constituents in mature trees. Aust J Biol Sci 20: 379–387
Taylor BK, May LH (1967) The nitrogen nutrition of the peach tree. II. Storage and mobilization of nitrogen in young trees. Aust J Biol Sci 20: 389–411
Tilman D (1988) Plant strategies and the dynamics and structure of plant communities. Monographs in Population Biology. Volume 26. Princeton University Press, Princeton, NJ
Vitousek P (1982) Nutrient cycling and nutrient use efficiency. Am Nat 119: 553–572
Vogt KA, Grier CC, Vogt DJ (1986) Production, turnover, and nutrient dynamics of aboveground and belowground detritus of world forests. Adv Ecol Res 15: 303–377
Vrede T, Dobberfuhl DR, Kooijman SALM, Elser JJ (2004) Fundamental connections among organism C:N:P stoichiometry, macromolecular composition and growth. Ecology 85: 1217–1229
Warton D, Wright IJ, Falster DS, Westoby M (2006) Bivariate line-fitting methods in allometry. Biol Rev 81: 259–291
Watanabe T, Broadley MR, Jansen S, White PJ, Takada J, Satake K, Takamatsu T, Tuah SJ, Osaki M (2007) Evolutionary control of leaf element composition in plants. New Phytol 174: 516–523
Weibel ER (2002) Physiology - The pitfalls of power laws. Nature 417: 131–132
Weiher E, vanderWerf A, Thompson K, Roderick M, Garnier E, Eriksson O (1999) Challenging Theophrastus: a common core list of plant traits for functional ecology. J Veg Sci 10: 609–620
Werger MJA, Hirose T (1991) Leaf nitrogen distribution and whole canopy photosynthetic carbon gain in herbaceous stands. Vegetatio 97: 11–20
West GB, Brown JH, Enquist BJ (1997) A general model for the origin of allometric scaling laws in biology. Science 276: 122–126
West GB, Brown JH, Enquist BJ (1999) The fourth dimension of life: fractal geometry and allometric scaling of organisms. Science 284: 167–169

West GB, Brown JH, Enquist BJ (2001) A general model for ontogenetic growth. Nature 413: 628–631

Westoby M, Wright IJ (2006) Land-plant ecology on the basis of functional traits. Trends Ecol Evol 21: 261–268

Westoby M, Falster DS, Moles AT, Vesk PA, Wright IJ (2002) Plant ecological strategies: some leading dimensions of variation between species. Ann Rev Ecol Syst 33: 125–159

Wright IJ, Reich PB, Westoby M, Ackerly DD, Baruch Z, Bongers F, Cavender-Bares J, Chapin T, Cornelissen JHC, Diemer M, Flexas J, Garnier E, Groom PK, Gulias J, Hikosaka K, Lamont BB, Lee T, Lee W, Lusk C, Midgley JJ, Navas M-L, Niinemets Ü, Oleksyn J, Osada N, Poorter H, Poot P, Prior L, Pyankov VI, Roumet C, Thomas SC, Tjoelker MG, Veneklaas EJ, Villar R (2004) The worldwide leaf economics spectrum. Nature 428: 821–827

Wright IJ, Reich PB, Cornelissen JHC, Falster DS, Groom PK, Hikosaka K, Lee W, Lusk CH, Niinemets Ü, Oleksyn J, Osada N, Poorter H, Warton DI, Westoby M (2005) Modulation of leaf economic traits and trait relationships by climate. Global Ecol Biogeogr 14: 411–421

Chapter 3
PHOSPHORUS AND AQUATIC PLANTS

Gabrielle Thiébaut

THE ROLE OF PHOSPHORUS IN AQUATIC ECOSYSTEMS

Aquatic systems receive the bulk of their nutrient supply from stream inflow. In stream communities, and also in lakes with a stream outflow, the export of nutrients in outgoing stream water is a major factor in nutrient budgets of aquatic communities. By contrast, in lakes without an outflow, nutrient accumulation in permanent sediments is often the major export pathway. Only a small fraction of available nutrients is incorporated into the biological interactions of stream communities (Winterbourn and Townsend 1991). In streams and rivers, the majority of nutrients flow on, as particles or dissolved in the water, to be discharged into a lake or the sea. Nevertheless, some nutrients do cycle from inorganic forms in freshwater, to inorganic forms in animals or plants, to inorganic forms in water, and so on. Because of the transport downstream, the displacement of nutrients may be best represented as a spiral (Elwood *et al.* 1983), where rapid phases of inorganic nutrient displacement alternate with periods when the nutrients are locked in biomass (e.g. in aquatic plants). Aquatic plants may obtain nitrogen (N) and phosphorus (P) from the sediment and then release these elements into the water. These plants function as a source for nutrients, by trapping fine organic and inorganic particles, enhancing mineralization of organic matter through oxidation of the sediments, and altering the localized environment, thus enabling P release through reducing conditions and increased pH and temperature. Oxygen translocation to the roots of plants has the effect of oxidizing the immediate sediment environment, and this may limit P availability (Moore *et al.* 1994; Wigand *et al.* 1997). Aquatic plants can also have a significant impact on a system's light environment and nutrient budget (Reckhow and Chapra 1999).

Considerable attention has been devoted to investigating P dynamics in lacustrine ecosystems (Lofgren and Bostrom 1989) and microcosm experiments (Pelton *et al.* 1998) that do not accurately reflect natural conditions. Although the nutrient enrichment of flowing waters is of great concern (Smith *et al.* 1999), the P dynamics of stream plants have received little attention to date (Madsen and Adams 1988; Chambers *et al.* 1989; Robach *et al.* 1995; Pelton *et al.* 1998; Clarke and Wharton 2001a,b; Thiébaut and Muller 2003; Baldy *et al.* 2007). Nutrient availability

P.J. White, J.P. Hammond (eds.) *The Ecophysiology of Plant-Phosphorus Interactions*,

(primarily P and, when P is in excess, N) plays an important role in controlling the abundance and development of aquatic plants (e.g. Sculthorpe 1967; Abernethy 1994; Carr and Chambers 1998).

NUTRIENT REQUIREMENTS OF AQUATIC PLANTS: ADAPTATIONS-COMPETITIVENESS

Relative trophic status of water and sediment (Best and Mantai 1979; Rattray *et al.* 1991) and particularly the amount of soluble reactive phosphorus (SRP; Denny 1972; Pelton *et al.* 1998) are important factors in the nutrition of aquatic plants. For many years, controversy has existed regarding the relative roles of roots versus shoots, and sediment versus open water in the nutrition of submerged aquatic plants.

Role of sediment versus open water in the nutrition of aquatic plants

The vast majority of work on the nutrition of aquatic plants has been done in lentic mineral waters. The conclusion of these extensive studies is that lacustrine sediments are generally more important sources of P than lake water (Barko and Smart 1980, 1981; Raven 1981; Rattray *et al.* 1991; Barko *et al.* 1991; Rattray 1995). Most studies on lacustrine plant species have been conducted in laboratory chambers (Denny 1972; De Marte and Hartman 1974; Bole and Allan 1978; Best and Mantai 1979; Barko and Smart 1980; Gabrielson *et al.* 1984; Smart and Barko 1985; Best *et al.* 1996). These stagnant conditions are not appropriate for plants whose ecology is dependent on water flow. In contrast to lakes, the sediment pore water of running waters is permanently flowing.

Water and sediments exhibit a high degree of variability within rivers and between different rivers (Figures 3.1 and 3.2). However, when little sediment P is available, uptake of water P by plants can be observed (Chambers *et al.* 1989; Pelton *et al.* 1998). Madsen and Cedergreen (2002) showed that in nutrient rich Danish streams, aquatic plants were able to satisfy their nutrient demands by leaf uptake alone, in spite of the low water/pore water SRP ratio. However, since these authors de-rooted the plants for their experiments, they could not exclude the possibility that nutrients were taken up by roots of intact plants. Significant positive correlations were found between SRP in the water column and total P in plant tissues for *Elodea nuttallii* Planchon St John, *Elodea canadensis* Michaux, *Ranunculus peltatus* Schrank and *Callitriche platycarpa* Kütz (Table 3.1; Thiébaut and Muller 2003). Although Robach *et al.* (1995) did not observe a significant correlation between total plant P and total sediment P, positive correlations between total sediment P and total tissue P have been observed for *R. peltatus*, *E. nuttallii*, *Callitriche obtusangula* Le Gall. and *C. platycarpa* (Table 3.1; Thiébaut and Muller 2003).

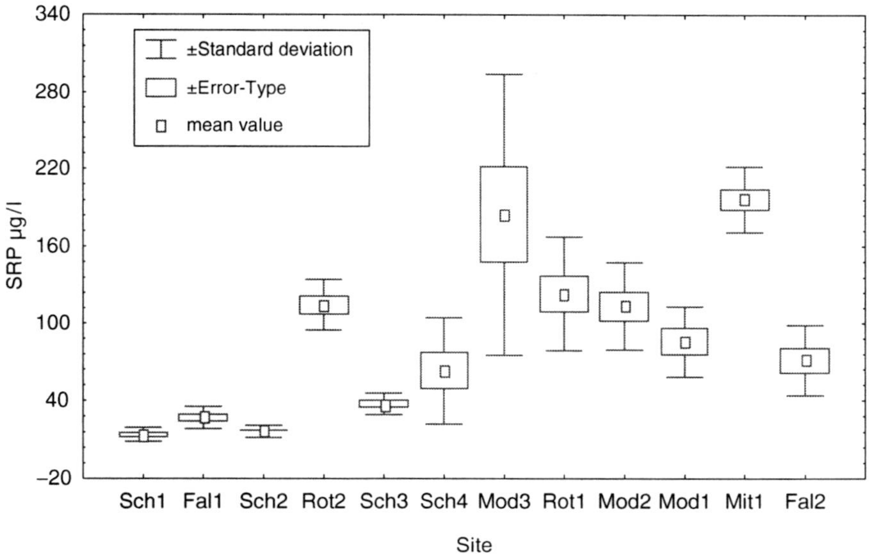

Fig. 3.1 Boxplots of soluble reactive phosphorus in water versus sites (mean values). Values were obtained from nine samples from each of the sites. n = 12 sites. (Thiébaut and Muller 2003.)

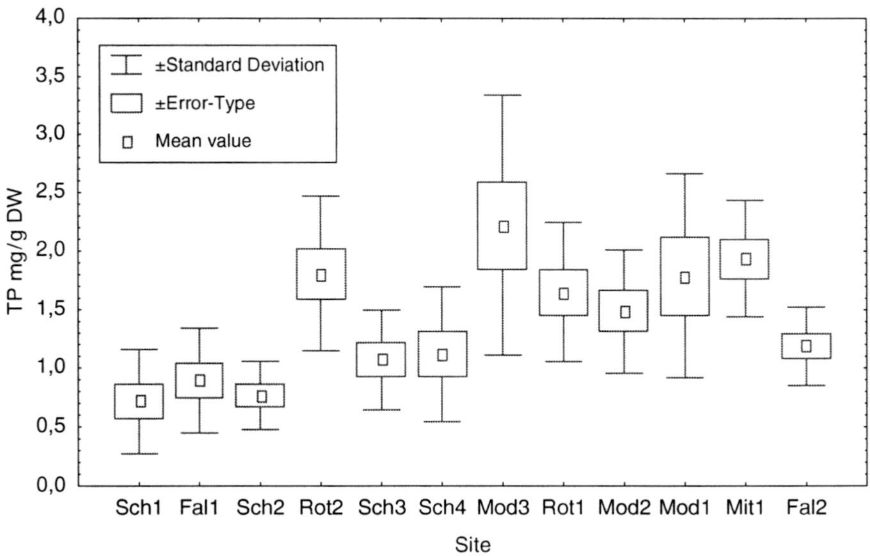

Fig. 3.2 Boxplots of sediment total phosphorus versus sites (mean values). Values were obtained from nine samples from each site. n = 12 sites. (Thiébaut and Muller 2003.)

Table 3.1 Relationships between water SRP and plant P concentrations and between sediment total P and plant P concentrations (mean values). r^2: Pearson correlation coefficient, ns: non significant, $^*p < 0.1$, $^{**}p < 0.05$; $^{***}p < 0.001$. (Thiébaut and Muller 2003.)

Species	Water SRP	Sediment TP
C. hamulata	ns	ns
C. platycarpa	$r^2 = 0.72^{**}$	$r^2 = 0.84^{***}$
C. obtusangula	ns	$r^2 = 0.66^*$
E. nuttallii	$r^2 = 0.69^*$	$r^2 = 0.75^*$
E. canadensis	$r^2 = 0.95^*$	ns
R. peltatus	$r^2 = 0.77^*$	$r^2 = 0.79^*$

Role of root versus shoot in the nutrition of aquatic plants

Plant morphology, and particularly root to shoot ratio, lead to preferential use of different sources of nutrients. For example, detached floating species, such as *Lemna*, must obtain nutrients from the water column, because there is no contact with the substrate. Sediment P can be the main nutrient source for plants with large root systems (Denny 1972). However, some results are contradictory. For example, *Berula erecta* (Huds.) Coville has a large root system and a strictly submerged growth form and *C. obtusangula* has few roots and floating and submerged shoots. Baldy *et al.* (2007) reported positive correlations between the P concentration of shoots and roots of *B. erecta* and water P concentration but not sediment P concentration, but did not find any correlations between shoot P concentration of *C. obtusangula* and water or sediment P concentration. However, significant correlations between shoot P concentration of *C. obtusangula* and sediment P concentration have been reported (Thiébaut and Muller 2003).

The P uptake mechanism is not well known in the case of aquatic species. Phosphorus is taken up by plant cells in the form of phosphate (Pi) using transporter proteins, such as $H_2PO_4^-/H^+$ symporters in the plasma membrane (Schachtman *et al.* 1998; Smith *et al.* 2003). Phosphate can be translocated from shoots to roots or from roots to shoots, as has been demonstrated in laboratory studies (De Marte and Hartman 1974; Eugelink 1998). While shoot absorption is still a subject of debate, root uptake of Pi is commonly accepted as the mode aquatic plants use to acquire Pi (Bole and Allan 1978; Best and Mantai 1979; Barko and Smart 1981; Barko *et al.* 1988). Even plants with limited root systems have been shown to take up Pi via their roots (Barko and James 1998). The sediment appears to be a nutrient source for rooted plants mainly when nutrients are not sufficiently available in the water (Bole and Allan 1978; Barko and Smart 1981). Therefore, the importance of sediment nutrients to aquatic plants will depend upon both species (morphological type) and the environment (trophic status, flow, shade, competition, pollution, management).

The most important function of roots in many submerged species may be anchorage and not nutrient acquisition. This is in agreement with the observation that many species of fast-flowing habitats (e.g. *Ranunculus* species) have shallow roots, which curl around large sediment particles and do not penetrate deeply into the sediment. Evidence that both roots and shoots are able to obtain nutrients (Agami and Waisel 1986), and that Pi translocation in two *Elodea* species was greater from shoot to root than from root to shoot (Eugelink 1998), suggest that nutrient acquisition by roots may be a secondary function in some species.

The relationships between aquatic plants and the trophic status of rivers is complex, partly because rooted plants can absorb nutrients from both the sediment and the water column, and partly because of the effects of a wide range of environmental variables (Clarke and Wharton 2001a,b).

Role of ecological factors in the phosphorus nutrition of aquatic plants

Ecological factors, such as substratum (Clarke and Wharton 2001a,b) and water velocity (Royle and King 1991; Carr and Chambers 1998), can modify P availability and thus plant nutrition. However, limited information is available on the influence of water velocity on Pi uptake by plants. Water velocity may cause a decrease in Pi uptake by impeding nutrient adsorption (Boeger 1992) or it may increase Pi uptake by increasing the amount available in the water or sediment (Jarvie *et al.* 2002). The few studies focusing on the role of water velocity on P dynamics have been conducted in aquatic plant communities (Dawson 1988), or on sediment chemistry (Prairie and Kalff 1988; Chambers *et al.* 1992; Chambers and Prepas 1994). These studies reported a relatively long-term effect of hydrological regimes on P dynamics, with positive or negative correlations between dissolved P and water discharge. These correlations varied in magnitude and direction among streams. With the exception of the study by Baldy *et al.* (2007), there is no literature available on the role of water velocity on P storage by plant species in lotic ecosystems. These authors established that *C. obtusangula* stores P more efficiently in its shoots in conditions of low water velocity than in conditions high water velocity. Moreover, as Boeger (1992) observed, the water velocity effect may be mediated by the substratum. *B. erecta* grows preferentially on coarse substrate (gravel) with high water velocities, in contrast with *C. obtusangula*, which grows on finer sediments (sand, mud) with low water velocities.

Phosphorus storage in aquatic plants

Although P assimilation and storage in aquatic plants depends on the vegetation type and their growth characteristics (Reddy *et al.* 1999), aquatic plants form a continuum across a unique pattern of change in nutrient concentrations. This continuum exists despite considerable differences in the plants' architectural, evolutionary and life histories, and the growth conditions encountered in their habitats (Duarte 1992).

Tissue P concentration differs in different parts of submerged angiosperms (Rorslett *et al.* 1985) and in plants collected during different seasons (Gerloff and Krombholz 1966; Carpenter and Adams 1977; Bole and Allan 1978; Carignan and Kalff 1980; Rorslett *et al.* 1985; Madsen and Adams 1988; Royle and King 1991; Robach *et al.* 1995; Eugelink 1998; Thiébaut and Muller 2003; Garbey *et al.* 2004a). Tissue P concentration also varies according to the plant species. Tissue P concentration varied according to year and season in *C. platycarpa*, *C. hamulata* and *E. nuttallii*, and tissue P storage depended on the trophic level of sediment and water, and varied by species (Figure 3.3). The rank order of tissue P concentration was *C. obtusangula* > *R. peltatus* ≥ *E. nuttallii* ≥ *E. canadensis* > *C. platycarpa* > *C. hamulata* (Thiébaut and Muller 2003). The mean tissue P concentrations in *E. nuttallii* and *E. canadensis*, are comparable to the tissue P concentrations obtained by Barko and Smart (1980) for two other hydrocharitaceae species *Hydrilla verticillata* (L.f) Royle and *Egeria densa*

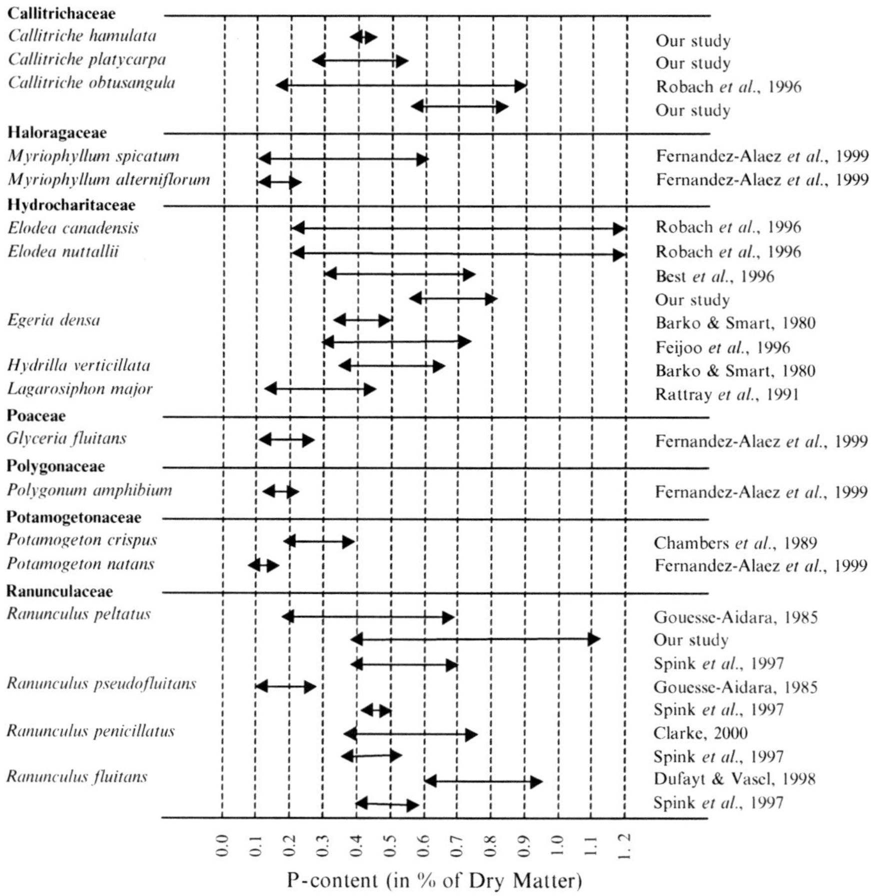

Fig. 3.3 A comparison of the tissue P concentration in macrophytes, expressed as percentage of dry matter. (Using data from our study (Garbey *et al.* 2004a) and from the literature.)

Planchon, but higher than the tissue P concentrations measured by Dawson (1976) for another *Ranunculus* species. Gerloff and Krombholz (1966) suggested that the distribution patterns of plants with tissue P concentrations above critical concentrations (1.3 mg P g^{-1} dry plant weight), were not limited by nutrient availability. When the growth rate was at its maximum in the spring for *R. peltatus* and *Elodea* species, there was sufficient P to satisfy their nutrient requirements. In contrast, competition for P could occur between invasive species and *C. platycarpa* in nutrient-poor sites.

Adaptation to temporally fluctuating resource availability

The flow rates and amount of P in plants varies continuously throughout the year. This is illustrated in Figure 3.4, which shows a seasonal effect, and is evident from the literature (Madsen and Adams 1988; Rorslett *et al.* 1985; Robach *et al.* 1995). Seasonal variation could be a consequence of annual growth and decomposition patterns of the plant species, as was shown for *Potamogeton pectinatus* L. (Howard-Williams and Allanson 1981), *Ranunculus penicillatus* (Dum.) Bab. (Dawson 1976), *R. peltatus* (Garbey *et al.* 2004b), and *E. nuttallii* (Kunii 1984). The temporal variability in Pi uptake by aquatic plants could be explained by phenology and morphology. This variability reflected the fluctuating nutrient availability at the sites.

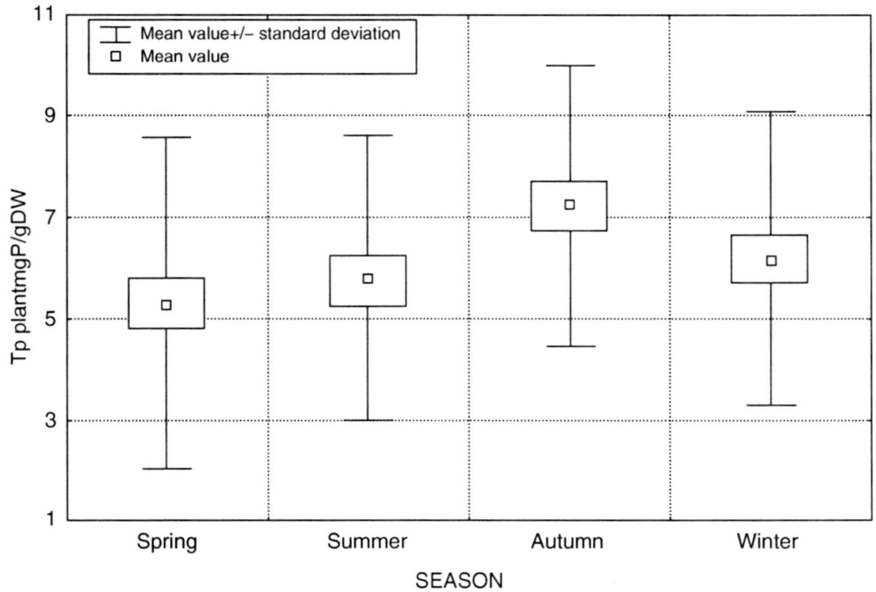

Fig. 3.4 Seasonal variability in tissue P concentration of aquatic vascular plants (Thièbaut 2005)

Aquatic plant tissue P concentrations were the lowest in spring, and the highest in autumn (Thiébaut 2005). New shoots developing in autumn probably played an important role in accumulating P for the next year's initial growth for *E. nuttallii*. The success of *Elodea* species might be correlated with their ability to remain in a vegetative condition throughout the winter season. The increase in biomass in spring progressively restricts flow and thus reduces velocity and turbulence. Reduced turbulence increases the boundary layer of static water around the plant, and the uptake of nutrients is then reduced to the diffusion rate across this zone (Dawson 1976). Reproductive growth also occurs in spring *Elodea* species and *R. peltatus*. *Elodea* species, as well as *R. peltatus*, are adapted to seasonal P fluctuations whereas *C. platycarpa*, a slower growing species, displayed low storage ability. High spatial and/or temporal fluctuations in nutrient availability favored the spread of these aquatic plants and supports the theory of invasibility of Davis *et al.* (2000).

However, plant growth and nutrient availability are not always directly correlated (Kern-Hansen and Dawson 1978; Canfield and Hoyer 1988; Feijoo *et al.* 1996). Indeed, if plant development results from the mobilization of energy reserves and structural materials accumulated in different storage organs, nutrient-uptake is likely to exceed the rate of utilization in growth, allowing reserves to accumulate (Grime 1988; Chapin and Van Cleve 1989). This "luxury" uptake may benefit the plant later, if water P concentrations diminish. Species distributions may be partly influenced by their P usage; their ecological range being a function of their ability to adapt to low or fluctuating water P concentrations. Therefore, competitiveness for nutrients has to integrate both the plant's ability to use that nutrient for rapid growth, and its ability to extract and retain nutrients for when nutrient availability is low. This ability might be described more accurately by physiological traits such as nutrient storage in plant tissues (e.g. Garbey *et al.* 2004a).

SPECIES DISTRIBUTION AND WATER PHOSPHATE OR SEDIMENT PHOSPHORUS

Nutrient availability in the sediment and water column is known to affect the community composition of submerged plants. For example, the occurrence of the exotic *Elodea* species in the streams of Northern Vosges corresponded with an increase in nutrient concentrations (Thiébaut *et al.* 2004).

Within the aquatic plant community there is a spectrum of tolerances to nutrient enrichment (Table 3.2). Many of the more abundant species are fairly "cosmopolitan", occurring over a wide range of water Pi concentrations. Within this minimum-maximum tolerance range, however, individual species preferences are apparent.

Some species, although fairly cosmopolitan, were found either rarely or at low abundances at low water Pi concentrations, and were found more frequently or at greater abundances as the water Pi concentration increased (e.g. *E. nuttallii* in Northern Vosges streams; Figure 3.5).

Table 3.2 Species distribution and water nutrient status

Species	Nutrient status
Acorus calamus L.	Eutrophic
Alisma lanceolatum With	Eutrophic
Alisma plantago-aquatica L.	Eutrophic
Apium inundatum L.	Oligotrophic
Apium nodiflorum (L.) Lag.	Mesotrophic to eutrophic
Berula erecta (Huds.) Coville	Oligotrophic
Butomus umbellatus L.	Eutrophic
Callitriche hamulata Kütz. ex Koch	Oligotrophic to eutrophic
Callitriche obtusangula Le Gall	Eutrophic
Callitriche platycarpa Kütz.	Oligotrophic to eutrophic
Callitriche stagnalis Scop.	Oligotrophic to mesotrophic
Callitriche truncata Guss. ssp. *occidentalis*	Mesotrophic
Carex rostrata Stokes	Oligotrophic
Catabrosa aquatica (L.) Beauv.	Oligotrophic to mesotrophic
Ceratophyllum demersum L.	Eutrophic
Ceratophyllum submersum L.	Eutrophic
Eleocharis palustris (L.) Roemer & Schultes	Oligotrophic to mesotrophic
Elodea canadensis Michx	Mesotrophic
Elodea nuttallii (Planchon) St John	Mesotrophic to eutrophic
Glyceria fluitans R. Br.	Oligotrophic to eutrophic
Groenlandia densa (L.) Fourr.	Oligotrophic to mesotrophic
Helodes palustris Spach	Oligotrophic
Hippuris vulgaris L.	Oligotrophic to mesotrophic
Hottonia palustris L.	Oligotrophic to mesotrophic
Hydrocharis morsus-ranae L.	Mesotrophic
Hydrocotyle vulgaris L. fo aq.	Oligotrophic
Iris pseudacorus L.	Oligotrophic to eutrophic
Juncus bulbosus L.	Oligotrophic
Juncus subnodulosus Schrank aq. fo.	Oligotrophic
Lemna gibba L.	Eutrophic
Lemna minor L.	Mesotrophic to eutrophic
Lemna trisulca L.	Mesotrophic
Littorella uniflora (L.) Ascherson	Oligotrophic
Luronium natans (L.) Rafin.	Oligotrophic
Lycopus europaeus L.	Oligotrophic to eutrophic
Mentha aquatica L.	Oligotrophic to eutrophic
Menyanthes trifoliata L.	Oligotrophic
Montia fontana L. agg.	Oligotrophic
Myosotis gr. *palustris* (*M. scorpioides* L.)	Mesotrophic
Myriophyllum alterniflorum DC.	Mesotrophic
Myriophyllum spicatum L.	Eutrophic
Myriophyllum verticillatum L.	Mesotrophic
Najas marina L.	Eutrophic
Najas minor L.	Eutrophic
Nasturtium officinale R. Br. agg.	Mesotrophic to eutrophic
Nuphar lutea (L.) Sibth. & Sm.	Eutrophic
Nymphaea alba L.	Oligotrophic to mesotrophic
Nymphoides peltata (S. G. Gmelin) O. Kuntze	Mesotrophic
Oenanthe aquatica (L.) Poiret	Oligotrophic to mesotrophic
Oenanthe crocata L.	Oligotrophic to mesotrophic
Oenanthe fluviatilis (Bab.) Coleman	Mesotrophic to eutrophic

(continued)

Table 3.2 (continued)

Species	Nutrient status
Phalaris arundinacea L.	Oligotrophic to eutrophic
Phragmites australis (Cav.) Trin. ex Steudel	Eutrophic
Polygonum amphibium L.	Eutrophic
Polygonum hydropiper L. aq. fo.	Eutrophic
Potamogeton acutifolius Link	Mesotrophic
Potamogeton alpinus Balbis	Oligotrophic to mesotrophic
Potamogeton berchtoldii Fieber	Eutrophic
Potamogeton coloratus Hornem.	Oligotrophic
Potamogeton compressus L.	Eutrophic
Potamogeton crispus L.	Eutrophic
Potamogeton friesii Rupr.	Mesotrophic
Potamogeton gramineus L.	Mesotrophic
Potamogeton lucens L.	Eutrophic
Potamogeton natans L.	Oligotrophic to mesotrophic
Potamogeton nodosus Poiret	Eutrophic
Potamogeton obtusifolius Mert. & Koch	Mesotrophic
Potamogeton panormitanus Biv.	Mesotrophic to eutrophic
Potamogeton pectinatus L.	Eutrophic
Potamogeton perfoliatus L.	Mesotrophic to eutrophic
Potamogeton polygonifolius Pourret	Oligotrophic
Potamogeton praelongus Wulfen	Oligotrophic to mesotrophic
Potamogeton trichoïdes Cham. & Schelcht	Eutrophic
Ranunculus aquatilis L.	Oligotrophic to mesotrophic
Ranunculus circinatus Sibth.	Mesotrophic
Ranunculus fluitans Lam.	Oligotrophic to eutrophic
Ranunculus hederaceus L.	Oligotrophic to mesotrophic
Ranunculus hololeucos Lloyd	Oligotrophic
Ranunculus omiophyllus Ten.	Oligotrophic
Ranunculus peltatus Schrank.	Oligotrophic to eutrophic
R. penicillatus ssp. *penicillatus*	Oligotrophic to eutrophic
Ranunculus trichophyllus Chaix	Oligotrophic to mesotrophic
Rorippa amphibia (L.) Besser	Oligotrophic to eutrophic
Sagittaria sagittifolia L.	Eutrophic
Scirpus fluitans L.	Oligotrophic
Scirpus lacustris L.	Eutrophic
Scirpus sylvaticus L.	Oligotrophic to eutrophic
Sparganium angustifolium Michaux	Oligotrophic
Sparganium emersum Rehmann sh. l.	Oligotrophic to eutrophic
Sparganium erectum L.	Oligotrophic to eutrophic
Sparganium minimum Wallr	Oligotrophic
Spirodela polyrhiza (L.) Schleiden	Eutrophic
Trapa natans L.	Mesotrophic
Typha angustifolia L.	Eutrophic
Typha latifolia L.	Eutrophic
Vallisneria spiralis L.	Eutrophic
Veronica anagallis-aquatica L.	Oligotrophic to mesotrophic
Veronica beccabunga L.	Oligotrophic to eutrophic
Wolffia arhiza (L.) Horkel & Wimmer	Eutrophic
Zannichellia palustris L.	Eutrophic

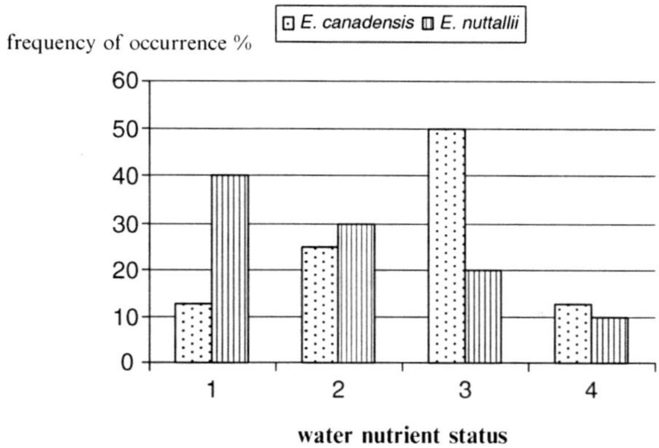

Fig. 3.5 Frequency of occurrence of *Elodea* species, expressed as percentages, according to the nutrient status of water in the Northern Vosges streams (North Eastern France). Distribution of *Elodea* species in four groups: group 1 (oligotrophic status): $< 40 \mu g\ P\ l^{-1}$; group 2: $40 < x < 60 \mu g\ P\ l^{-1}$; group 3: $60 < x < 100 \mu g\ P\ l^{-1}$; group 4 (eutrophic status): $> 100 \mu g\ P\ l^{-1}$

However, aquatic plant species have different trophic requirements (Holmes and Newbold 1984) and respond differently to sediment nutrient supply. Denny (1972), suggests that across a number of rivers there should be evidence of different species responses to sediment P. Some species will be associated with low concentrations of sediment P and others will be associated with higher concentrations of sediment P. Therefore, there may be species that do not respond to sediment nutrients and will therefore show no particular preference for sediments of a particular nutrient status. The species associated with high sediment P concentrations (*Nuphar lutea* L. Sibth. & Sm., *Potamogeton natans* L., *Sagitaria sagittifolia* L.) are characteristic of silt and clay sediments and sluggish flows (Clarke and Wharton 2001a). In contrast, *Myriophyllum spicatum* L. appears to be associated with sediments with relatively low P and with high water nutrient levels.

THE USE OF AQUATIC PLANTS

The use of aquatic plants for nutrient removal from wastewaters

Some floating aquatic plants are used in constructed wetlands, mainly in tropical countries, due to their capacity to absorb and store large quantities of nutrients, and their rapid growth rates. Their high productivity and nutrient removal capability have created substantial interest in their use for wastewater treatment and resource recovery. However, most studies have focused on the treatment of municipal wastewater and have tested the efficiency of constructed wetlands for the treatment of effluents. Relatively few studies have been reported on the use of floating plants in animal manure-based wastewater treatment or as an alternative approach for fish farm management. The application of aquatic plants for treatment of effluents has mainly involved constructed wetlands. Constructed wetlands have an appeal as a farm waste management practice because they are low cost, use simple technology, and ideally require little maintenance or management after construction (Hammer 1992). The resulting gains in vegetative biomass can also provide economic returns when harvested.

Among the floating aquatic plants, water hyacinth (*Eichhornia crassipes*), a species which has spread worldwide, has been evaluated on a large scale for nutrient removal from wastewaters (Reddy and Smith 1987). Studies have shown that among the floating plants, water hyacinth (*Eichhornia crassipes* (Mart.) Solms), water lettuce (*Pistia stratiotes* L.), and pennywort (*Hydrocotyle umbellate* L.) were more productive than small-leaf plants such as salvinia (*Salvinia rotundifolia* Willd.), azolla (*Azolla caroliniana* Willd.) and duckweed (*Lemna minor* L.; Reddy *et al.* 1983). In terms of nutrient reduction, chemical oxygen demand (COD), solids, and salinity, water hyacinth performed better than water lettuce and pennywort in diluted anaerobically digested flushed dairy manure wastewater (Sooknah and Wilkie 2004).

The use of aquatic plants in biological assessment of water quality

The concept of using living organisms to identify, monitor and assess pollution is well established. There are several advantages to using aquatic plants as the basis for bioindication: plants are stationary; there are relatively few species within any one region; many are rooted and thus reflect both water and sediment quality; and they are relatively long-lived and therefore can integrate seasonal or disturbance factors (Carbiener *et al.* 1990). Additionally, plants may encompass a broad taxonomic base (Kelly and Whitton 1998), increasing the likelihood of detecting a variety of pollution effects. The disadvantages of using aquatic plants as monitoring tools include: marked seasonal variations in community composition and species abundance; sparse vegetation in many freshwater systems due to abiotic factors; and the fact that the ecology of many aquatic plant species and their response to pollution is not well documented. There is currently considerable interest in utilizing plant-based bioindication and biomonitoring techniques to identify, assess, and inform management policy on the anthropogenic nutrient enrichment (eutrophication) of freshwater systems.

Water quality monitoring using aquatic plants has been developed in Europe over the past two decades (Harding 1981; Holmes and Newbold 1984; Carbiener *et al.* 1990; Holmes 1995; Thiébaut and Muller 1999; Dawson *et al.* 1999;

Schneider and Melzer 2003; Haury *et al.* 2006). For the operational monitoring demanded by the European Union Water Framework Directive, three trophic indices can be used as metrics to describe aquatic plant communities in running waters. The hypothesis is that biological indices express the trophic status of a river in terms of the response of the aquatic plant communities to nutrient status.

The mean trophic rank (MTR)

At present, the most widely employed method utilizing aquatic plants in the trophic assessment of rivers is the Mean Trophic Rank (MTR; Holmes 1995). The UK Environment Agency has recently commissioned the development of the MTR scheme (Holmes *et al.*, 1999) as a bioindication tool to aid in the implementation and monitoring of the European Union Urban Waste Water Treatment Directive (EU UWWTD; Dawson *et al.* 1999).

The "MTR" is a biotic index, specifically for the purposes of biological monitoring under the EU UWWTD. It is successfully used for monitoring changing nutrient concentrations, principally at sewage treatment works. It is based on the presence and abundance of aquatic plants and uses a simple scoring system to derive a single index to describe the trophic status of a site. Although MTR has been shown to differentiate sites upstream and downstream of nutrient inputs on a catchment scale, relationships between MTR and water quality can be obscured by site factors such as land-use, shade and localized disturbance (Demars and Harper 1998).

The trophic index of macrophytes[1] (TIM)

The "Trophic Index of Macrophytes" (TIM) is a tool tested in Germany for indicating the trophic state of running waters (Schneider and Melzer 2003). The method is based on both the P concentrations in the water body and the sediment pore water. The TIM is useful for detecting differences in the trophic state of running waters. It may be an important tool in finding the causes of degradation according to the European Union Water Framework Directive and in long-term monitoring of improvements in water quality (Schneider and Melzer 2003).

---

[1] The term "aquatic macrophytes" is commonly used for all macroscopic forms of aquatic vegetation; it includes macroscopic algae, bryophytes, some pteridophytes and many flowering plants (angiosperms).

The "Biological index of macrophytes in rivers" (IBMR)

The French trophic index "Biological Index of Macrophytes in Rivers" (IBMR) is a biotic index used by the French Water Agencies as a macrophyte-based method to assess the trophic status of rivers (Haury *et al.* 2006). Species are assigned a score according to their tolerance to eutrophication (the higher the score, the less the tolerance) and a mean score for the site is then calculated, weighted by the abundance of the individual species. The French trophic index appears to be a useful means for assessing eutrophication of running waters (Figure 3.6).

Limitations of trophic indices

As simple indices may provide useful ecosystem management tools to assess rapidly the possible detrimental effects of ecosystem pollution, several studies have evaluated their suitability for monitoring water quality (Thiébaut *et al.* 2002; Thiébaut 2006). The trophic indices appeared to be more pertinent tools for monitoring the trophic status of the streams. Aquatic plant indices may not be adequate tools for discovering the causes of habitat and water quality degradations (industrial pollution, flood

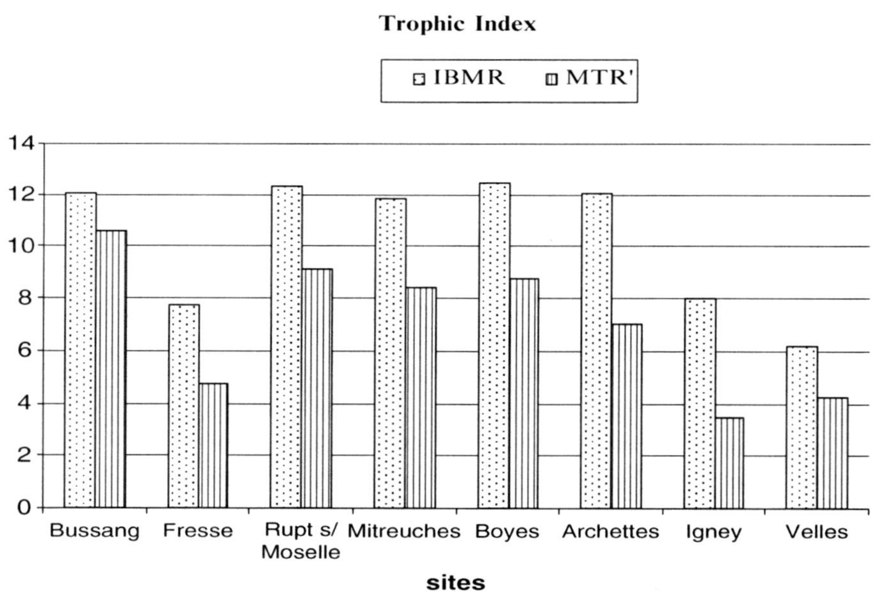

Fig. 3.6 Comparison of the trophic status of the Moselle river (NE France) from upstream (Bussang) to downstream (Velle). The trophic status, based on aquatic macrophyte communities, was calculated using the Biological Index of Macrophytes in Rivers (IBMR) and the modified Mean Trophic Rank (MTR) index. The values obtained with the modified MTR index corresponded to the values of the MTR index divided by five, allowing for comparison with IBMR

disturbances). Some deficiencies and limitations to the current methodology have been discussed by Thiébaut *et al.* (2002). However, the effectiveness of the MTR and IBMR methods is limited at the full catchment scale by the low numbers of indicator species in headwater streams (Demars and Harper 1998; Haury *et al.* 2006). Furthermore, a number of other factors might affect aquatic plant species in streams undergoing water pollution that are not evaluated by trophic indices.

CONCLUSIONS AND SUMMARY

The input of large amounts of P from human activities plays an important role in controlling the presence and abundance of aquatic plants. Relative trophic status of water and sediment, and particularly the amount of SRP, are important factors in aquatic plant nutrition. For many years controversy has existed regarding the role of roots versus shoots and sediment versus open water in the nutrition of submerged aquatic plants. Laboratory and *in situ* experiments have indicated that rooted submerged plants obtain nutrients from both sediment and water. Environmental (e.g. water velocity) and biological factors play a role in determining plant P nutrition. However, the source of nutrients for submerged, rooted plants has not been clearly defined despite considerable effort over almost 40 years. Over the course of a year, the flow rates and P in plants varies continuously. Phosphorus storage in tissues depends on the plant species, on the season, and on the trophic status of the site. Within the aquatic community, there is a spectrum of tolerances to nutrient enrichment. Many of the more abundant species are fairly cosmopolitan. Some floating aquatic plants are used in constructed wetlands due to their rapid growth rates and their capacity to absorb and store large quantities of nutrients. They are also used in biological assessment of water quality.

REFERENCES

Abernethy VJ (1994) Functional ecology of euhydrophyte communities of European riverine wetland ecosystems. Dissertation, University of Glasgow, Scotland

Agami M, Waisel Y (1986) The ecophysiology of roots of submerged vascular plants. Physiol Veg 24: 607–624

Baldy V, Trémolières M, Andrieu M, Belliard J (2007) Changes in phosphorus content of two aquatic plants according to water velocity, trophic status and time period in hardwater streams. Hydrobiologia 575: 343–351

Barko JW, Smart RM (1980) Mobilization of sediment phosphorus by submerged freshwater macrophytes. Freshwater Biol 10: 229–238

Barko JW, Smart RM (1981) Sediment-based nutrition of submersed macrophytes. Aquat Bot 10: 339–352

Barko JW, James WF (1998) Effects of submerged aquatic macrophytes on nutrient dynamics, sedimentation, and resuspension. In: Sondergaard S, Sondergaard, C (eds), The Structuring

Role of Submerged Macrophytes in Lakes. Springer, Berlin/Heidelberg/New York, pp 197–214
Barko JW, Smart RM, McFarland DG, Chen RL (1988) Interrelationships between the growth of *Hydrilla verticillata* (L.f.) Royle and sediment nutrient availability. Aquat Bot 32: 205–216
Barko JW, Gunnison D, Carpenter SR (1991) Sediment interactions with submersed macrophyte growth and community dynamics. Aquat Bot 41: 41–65
Best MD, Mantai KE (1979) Growth of *Myriophyllum*: sediment or lake water as the source of nitrogen and phosphorus. Ecology 59: 1075–1080
Best EPH, Woltman H, Jacobs FHH (1996) Sediment-related growth limitation of *Elodea nuttallii* as indicated by a fertilization experiment. Freshwater Biol 36: 33–44
Boeger R (1992) The influence of substratum and water velocity on growth of *Ranunculus aquatilis* L. (Ranunculaceae). Aquat Bot 42: 351–359
Bole JB, Allan JR (1978) Uptake of phosphorus from sediment by aquatic plants, *Myriophyllum spicatum* and *Hydrilla verticillata*. Water Res 12: 353–358
Canfield DEJ, Hoyer MV (1988) Influence of nutrient enrichment and light availability on the abundance of aquatic macrophytes in Florida streams. Can J Fish Aquat Sci 45: 1467–1472
Carbiener, R, Trémolières M, Mercier JL, Ortscheit A (1990) Aquatic macrophyte communities as bioindicators of eutrophication in calcareous oligosaprobe stream waters (Upper Rhine plain, Alsace). Vegetatio 86: 71–88
Carignan R, Kalff J (1980) Phosphorus source for aquatic weeds: water or sediment. Science 207: 987–988
Carpenter SR, Adams SA (1977) The macrophyte tissue nutrient pool of hardwater eutrophic lake: implications for macrophyte harvesting. Aquat Bot 3: 239–255
Carr GM, Chambers PA, (1998) Macrophyte growth and sediment phosphorus and nitrogen in a Canadian prairie river. Freshwater Biol 39: 525–536
Chambers PA, Prepas EE (1994) Nutrient dynamics in riverbeds: the impact of sewage effluent and aquatic macrophytes. Water Res 28: 453–464
Chambers PA, Prepas EE, Bothwell ML, Hamilton HR (1989) Roots versus shoots in nutrient uptake by aquatic macrophytes in flowing waters. Can J Fish Aquat Sci 46: 435–439
Chambers PA, Prepas EE, Gibson K (1992) Temporal and spatial dynamics in riverbed chemistry: the influence of flow and sediment composition. Can J Fish Aquat Sci 49: 2128–2140
Chapin FS, Van Cleve K (1989) Approaches to studying nutrient uptake, use and loss in plants. In: Pearcy RW, Mooney HA, Ehleringer JR, Rundel PW (eds), Physiological Ecology. Chapman & Hall, New York, pp 185–207
Clarke SJ, Wharton G (2001a) Using macrophytes for the environmental assessment of rivers: the role of sediment nutrients. R&D Technical Report E1-S01/TR, 90 p
Clarke SJ, Wharton G (2001b) Sediment nutrient characteristics and aquatic macrophytes in lowland English rivers. Sci Total Environ 266: 103–112
Davis MA, Grime JP, Thompson K (2000) Fluctuating resources in plant communities: a general theory of invasibility. J Ecol 88: 528–534
Dawson FH (1976) The annual production of aquatic macrophyte *Ranunculus penicillatus* var. calcareous (R.W. Butcher) CDK Cook. Aquat Bot 2: 51–73
Dawson FH (1988) Water flow and the vegetation of running waters. In: Symoens JJ (ed), Vegetation of Inland Waters. Kluwer, The Netherlands, 283–308
Dawson FH, Newman JR, Gravelle MJ, Rouen KJ, Henville P (1999) Assessment of the trophic status of rivers using macrophytes. Evaluation of the Mean Trophic Rank. R&D Technical Report E39, Environment Agency, Bristol, UK, 178 p
Demars BOL, Harper DM (1998) The aquatic macrophytes of an English lowland river system: assessing response to nutrient enrichment. Hydrobiologia 384: 75–88
De Marte JA, Hartman RT (1974) Studies on absorption of ^{32}P, ^{59}Fe and ^{45}Ca by water-milfoil (*Myriophyllum excalbescens* Fernald). Ecology 55: 188–194
Denny P (1972) Sites of nutrient absorption in aquatic macrophytes. J Ecol 60: 819–929
Duarte C (1992) Nutrient concentration of aquatic plants: patterns across species. Limnol Oceanogr 37: 882–889

Elwood JW, Newbold JD, O'Neill RV, van Winckle W (1983) Resource spiralling: an operational paradigm for analyzing lotic ecosystems. In: Fontaine TD, Bartell SM (eds), Dynamics of Lotic Ecosystems. Ann Arbor Science, Ann Arbor, MI, pp 3–28

Eugelink AH (1998) Phosphorus uptake and active growth of *Elodea canadensis* Michx. and *Elodea nuttallii* (Planch.) St. John. Water Sci Technol 37: 59–65

Feijoo CS, Momo FR, Bonetto CA, Tur NM (1996) Factors influencing biomass and nutrient content of the submersed macrophyte *Egeria densa* Planch. in a pampasic stream. Hydrobiologia 341: 21–26

Gabrielson JO, Perkins MA, Welch EB (1984) The uptake, translocation and release of phosphorus by *Elodea densa*. Hydrobiologia 3: 43–48

Garbey C, Murphy KJ, Thiébaut G, Muller S (2004a) Variation in P-content in aquatic plant tissues offers an efficient tool for determining plant growth strategies along a resource gradient. Freshwater Biol 49: 346–356

Garbey C, Thiébaut G, Muller S (2004b) Morphological plasticity of a spreading aquatic macrophyte, *Ranunculus peltatus*, in response to environmental variables. Plant Ecol 173: 125–137

Gerloff GC, Krombholz PH (1966) Tissue analysis as a measure of nutrient availability for the growth of angiosperm aquatic plants. Limnol Oceanogr 11: 529–537

Grime JP (1988) The CRS model of primary plant strategies: origins, implications, and tests. In: Gottlieb LD, Jain SK (eds), Plant Evolutionary Biology. Chapman & Hall, London, pp 371–393

Hammer DA (1992) Designing constructed wetlands systems to treat agricultural non point source pollution. Ecol Eng 1: 49–82

Harding JPC (1981) Macrophytes as monitors of river quality in the southern NWWA area, Report No. TS-BS-81–2. North West Water, Rivers Division, Scientists Dept. Technical Support Group

Haury J, Peltre M-C, Trémolières M, Barbe J, Thiébaut G, Bernez I, Daniel H, Chatenet P, Haan-Archipof G, Muller S, Dutartre A, Laplace-Treyture C, Cazaubon A, Lambert-Servien E (2006) A new method to assess water trophy and organic pollution: the Macrophyte Biological Index for Rivers (IBMR): its application to different types of river and pollution. Hydrobiologia 570: 153–158

Holmes NTH (1995) Macrophytes for Water and Other River Quality Assessments. National Rivers Authority, Anglian Region, Peterborough

Holmes NTH, Newbold C (1984) River plant communities - reflectors of water and substrate chemistry. Focus Nat. Conserv. 9: 535–539

Holmes NTH, Newman JR, Chadd S, Rouen KJ, Saint L, Dawson FH (1999) Mean Trophic Rank: A User's Manual. R&D Technical Report E38. Environmental Agency, Bristol, UK

Howard-Williams C, Allanson BR (1981) Phosphorus cycling in a dense *Potamogeton pectinatus* L. bed. Oecologia 49: 56–66

Jarvie HP, Neal C, Williams RJ, Neal, M, Wickham HD, Hill LK, Wade AJ, Warwick A, White J (2002) Phosphorus sources, speciation and dynamics in the lowland eutrophic River Kennet, UK. Sci Total Environ 282–283: 175–203

Kelly MG, Whitton BA (1998) Biological monitoring of eutrophication in rivers. Hydrobiologia 384: 55–67

Kern-Hansen U, Dawson FH (1978) The standing crop of aquatic plants of lowland streams in Denmark and the inter-relationships of nutrients in plant, sediment and water. In: Proceedings, EWRS 5th International Symposium On Aquatic Weeds, pp 143–150

Kunii H (1984) Seasonal growth and profile structure development of *Elodea nuttallii* (Planch.) St John in pond Ojaga-Ike, Japan. Aquat Bot 18: 239–247

Lofgren S, Bostrom B (1989) Interstitial water concentrations of phosphorus, iron and manganese in a shallow, eutrophic Swedish lake - implications for phosphorus cycling. Water Res 23: 1115–1125

Madsen JD, Adams MS (1988) The nutrient dynamics of a submersed macrophyte community in a stream ecosystem dominated by *Potamogeton pectinatus*. J Freshwater Ecol 4: 541–550

Madsen TV, Cedergreen N (2002) Sources of nutrients to rooted submerged macrophytes growing in a nutrient-rich stream. Freshwater Biol 46: 283–291

Moore BC, Lafer JE, Funk WH (1994) Influence of aquatic macrophytes on phosphorus and sediment porewater chemistry in a freshwater wetland. Aquat Bot 49: 137–148

Pelton DK, Levine SN, Braner M (1998) Measurements of phosphorus uptake by macrophytes and epiphytes from the La Platte River (VT) using ^{32}P in streams microcosms. Freshwater Biol 39: 285–299

Prairie YT, Kalff J (1988) Dissolved phosphorus dynamics in headwater streams. Can J Fish Aquat Sci 45: 200–209

Rattray MR (1995) The relationship between P, Fe and Mn, uptakes by submersed rooted angiosperms. Hydrobiologia 308: 107–120

Rattray MR, Howard-Williams C, Brown JM (1991) Sediment and water as sources of nitrogen and phosphorus for submerged rooted aquatic macrophytes. Aquat Bot 40: 225–237

Raven JA (1981) Nutritional strategies of submerged benthic plants: the acquisition of C, N and P by rhizophytes and haptophytes. New Phytol 88: 1–30

Reckhow KH, Chapra SC (1999) Modelling excessive nutrient loading in the environment. Environ Pollut 100: 197–207

Reddy KR, Sutton DL, Bowes, G (1983) Freshwater aquatic plant biomass production in Florida. Soil Crop Sci Soc Fla Proc 42: 28–40

Reddy KR, Smith WH (eds) (1987) Aquatic Plants for Water Treatment and Resource Recovery. Magnolia Publishing, Orlando, FL

Reddy KR, Kadlec RH, Flaig E, Gale PM (1999) Phosphorus retention in streams and wetlands: a review. Crit Rev Environ Sci Technol 29: 83–146

Robach F, Hajnsek I, Eglin I, Trémolières M (1995) Phosphorus sources for aquatic macrophytes in running waters: water or sediment? Acta Bot Gall 142: 719–731

Rorslett B, Berge D, Johansen SW (1985) Mass invasion of *Elodea canadensis* in a mesotrophic, South Norwegian lake - impact on water quality. Verh Internat Verein Limnol 22: 2920–2926

Royle RN, King RJ (1991) Aquatic macrophytes in Lake Liddle, New south Wales: biomass, nitrogen and phosphorus status, and changing distribution from 1981 to 1987. Aquat Bot 41: 281–298

Schachtman DP, Reid RJ, Ayling SM (1998) Phosphorus uptake by plants: from soil to cell. Plant Physiol 116: 447–453

Schneider S, Melzer A (2003) The Trophic Index of Macrophytes (TIM) - a new tool for indicating the trophic state of running waters. Int Rev Hydrobiol 88: 49–67

Sculthorpe CD (1967) The Biology of Aquatic Vascular Plants. Edward Arnold, London

Smart RM, Barko JW (1985) Laboratory culture of submersed freshwater macrophytes on natural sediments. Aquat Bot 21: 251–263

Smith FW, Mudge SR, Rae AL, Glassop D (2003) Phosphate transport in plants. Plant Soil 248: 71–83

Smith VH, Tilman GD, Nekola JC (1999) Eutrophication: impact of excess nutrient inputs on freshwater, marine, and terrestrial ecosystems. Environ Pollut 100: 179–196

Sooknah RD, Wilkie AC (2004) Nutrient removal by floating aquatic macrophytes cultured in anaerobically digested flushed dairy manure wastewater. Ecol Eng 22: 27–42

Thiébaut G (2005) Does competition for phosphate supply explain the invasion pattern of *Elodea* species? Water Res 39: 3385–3393

Thiébaut G (2006) Aquatic macrophyte approach to assess the impact of disturbances on the diversity of the ecosystem and on river quality. Int Rev Hydrobiol 91: 483–497

Thiébaut G, Muller S (1999) A macrophyte communities sequence as an indicator of eutrophication and acidification levels in weakly mineralised streams in north-eastern France. Hydrobiologia 410: 17–24

Thiébaut G, Muller S (2003) Linking phosphorus pools of water, sediment and macrophytes in running waters. Ann Limnol 39: 307–316

Thiébaut G, Guerold F, Muller S (2002) Are trophic and diversity indices based on macrophyte communities pertinent tools to monitor water quality? Water Res 36: 3602–3610

Thiébaut G, Garbey C, Muller S (2004) Suivi biologique par les macrophytes aquatiques de la qualité des cours d'eau de la Réserve Biosphère des Vosges du Nord (N-Est de la France). Terre et Vie 59: 123–133

Wigand C, Stevenson JC, Cornell JC (1997) Effects of different submersed macrophytes on sediment biogeochemistry. Aquat Bot 56: 233–244

Winterbourn MJ, Townsend CR (1991) Streams and rivers: one-way flow systems. In: Barnes RSK, Mann KH (eds), Fundamentals of Aquatic Ecology. Blackwell Scientific, Oxford, pp 230–244

Chapter 4
PHOSPHORUS NUTRITION OF TERRESTRIAL PLANTS

Philip J. White and John P. Hammond

INTRODUCTION

Phosphorus (P) is essential for plant growth and fecundity. It is an integral component of genetic, metabolic, structural and regulatory molecules, in many of which it cannot be substituted by any other elements. Tissue P concentrations in well fertilized plants approximate 0.4–1.5% of the dry matter (Broadley *et al.* 2004), most of which is present as nucleic acids and nucleotides, phosphorylated intermediates of energy metabolism, membrane phospholipids and, in some tissues (principally seeds), as inositol phosphates. Some P also occurs in phosphoproteins and as inorganic phosphate (Pi) and pyrophosphate (PPi). It has been estimated that small metabolites, nucleic acids and phospholipids contribute approximately equally to leaf P content in P-replete plants (Figure 4.1; Marschner 1995; Dörmann and Benning 2002). Tissue P concentrations show no systematic differences between angiosperm species grown in P-replete conditions, but strong positive correlations occur between shoot P and shoot organic-N concentrations (Broadley *et al.* 2004). When plants are sampled from their natural environment, shoot N:P mass ratios vary between about 5:1 and 40:1 (e.g. Garten 1976; Thompson *et al.* 1997; Elser *et al.* 2000a; Tessier and Raynal 2003; Güsewell 2004; McGroddy *et al.* 2004; Güsewell *et al.* 2005; Han *et al.* 2005; Niklas *et al.* 2005; Wassen *et al.* 2005; Wright *et al.* 2005; Kerkhoff *et al.* 2006) and leaf N appears to scale as the 3/4 power of leaf P (Niklas *et al.* 2005; Niklas 2008). Ratios of 10:1 approximate the maximum critical organic-N:P ratios reported for a range of crop plants (Greenwood *et al.* 1980; Güsewell 2004). In general, leaf N:P ratios below 13.5 suggest N-limited plant growth, whilst leaf N:P ratios above 16 suggest P-limited plant growth (Aerts and Chapin 2000; Güsewell and Koerselman 2002; Tessier and Raynal 2003). Stoichiometric relationships between leaf N and leaf P appear to be a consequence of the requirements of N for proteins and of P for nucleic acids, membranes and metabolism (Elser *et al.* 2000b; Niklas 2008). Plant relative growth rate (RGR) is positively correlated with rRNA concentration and negatively correlated with protein concentration (Ågren 1988; Elser *et al.* 2000b; Niklas 2008). Thus, shoots of fast-growing herbaceous species characteristic of nutrient-rich, disturbed habitats tend to have

P.J. White, J.P. Hammond (eds.) *The Ecophysiology of Plant-Phosphorus Interactions*,

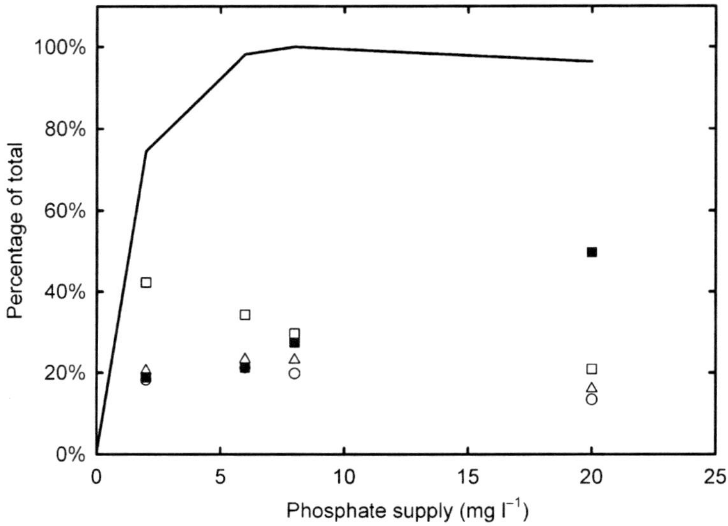

Fig. 4.1 The effect of P supply on leaf dry weight (line), expressed as a percentage of the maximum, and the percentage contributions of small metabolites (open triangles), nucleic acids (open squares), phospholipids (open circles) and inorganic phosphorus (filled squares) to the total leaf P content. (Data taken from Marschner 1995.)

higher concentrations of N and P, but lower N:P ratios, than shoots of slow-growing species characteristic of infertile habitats (Grime *et al.* 1997; Thompson *et al.* 1997; Grime 2001; Güsewell 2004; Niklas *et al.* 2005). Similarly, shoot N:P ratios increase during ontogeny as plant RGR declines (Güsewell 2004; Niklas 2008) and, between tissues, structural tissues have higher N:P ratios than metabolically active ones (Kerkhoff *et al.* 2006). This is consistent with the large P requirements for the growth of young tissues and the absolute cellular requirement for protein.

A lack of available P rapidly reduces plant growth rates. However, tissue P requirements and responses to P availability vary markedly between terrestrial plant species and among genotypes of a particular species (e.g. Bradshaw *et al.* 1960; Loneragan and Asher 1967; Rorison 1968; Ozanne *et al.* 1969; Greenwood *et al.* 1980, 2005, 2006; Coltman *et al.* 1986; Johnston *et al.* 1986; Alt 1987; Fageria *et al.* 1988; Föhse *et al.* 1988; Gunawardena *et al.* 1993; Gourley *et al.* 1994; Yan *et al.* 1995a,b, 2006; Beebe *et al.* 1997; Fageria and Baligar 1997, 1999; Li *et al.* 1997; Tian *et al.* 1998; Bolland *et al.* 1999; Narang *et al.* 2000; Sanginga *et al.* 2000; Baligar *et al.* 2001; Gaume *et al.* 2001; Górny and Sodkiewicz 2001; Liu *et al.* 2001; Osborne and Rengel 2002; Güsewell *et al.* 2003; Trehan and Sharma 2003; Blackshaw *et al.* 2004; Gahoonia and Nielsen 2004a,b; Zhao *et al.* 2004; Zhu and Lynch 2004; Hipps *et al.* 2005; Ozturk *et al.* 2005; White *et al.* 2005a,b; Wissuwa 2005; Oracka and Łapiñski 2006; Su *et al.* 2006; Marschner *et al.* 2007; Tesfaye *et al.* 2007; George and Richardson 2008), suggesting significant genetic variation in the ability of plants to acclimate to reduced P availability through the conservative use of P in tissues and/or an increased capability for P acquisition. Even mild P deficiency alters cellular biochemistry, biomass allocation and root morphology to match P acquisition with plant P requirements. Typical responses of plants to P starvation include the remobilization, reduction or replacement of P in inessential cellular compounds, the exudation of metabolites and enzymes into the rhizosphere to increase P availability and changes in root morphology and/or associations with microorganisms to acquire P more effectively from the soil. In this chapter, physiological responses of plants to the vagaries of P availability will be reviewed and set in the context of their consequences for plant growth and survival in natural and agricultural ecosystems.

PHOSPHORUS-CONTAINING COMPOUNDS IN PLANTS

Phosphorus is present in many chemical forms in plant cells (Marschner 1995). Some cellular compounds containing P are present at low concentrations or can be diminished and/or replaced with little consequence. It is the P-containing compounds that have unique cellular roles and those that are required in high concentrations by plant cells that define the absolute P requirement of plants. The acclimatory responses of plants to P starvation are directed towards maintaining essential cellular functions, either by utilizing plant P efficiently or by increasing P acquisition by the root system.

Nucleic acids

Phosphorus is an essential component of DNA and RNA, in which phosphodiester bridges link the deoxyribonucleotides or ribonucleotides. The requirement for DNA and RNA is greatest in tissues undergoing rapid cell division and/or cell expansion (Ågren 1988; Elser *et al.* 2000b; Niklas 2008). The plant cannot dispense with DNA or RNA and although DNA and RNA concentrations in plant cells can be reduced during P starvation this has a significant affect on plant growth rate (Raven 2008). In addition, P is required as ADP in photosynthesis and respiration, as ATP for energy transfer reactions in, for example, nucleic acid synthesis, metabolism, cytoskeletal rearrangements and membrane transport, as GTP for energy transfer reactions during nucleic acid biosynthesis, as NADPH in biosynthetic reactions and as signaling molecules such as GTP and cAMP. It is possible for a cell to reduce some dependence on ATP by rerouting biochemical pathways and utilizing PPi as an energy substrate (Figure 4.2; Plaxton and Carswell 1999; Hammond *et al.* 2004; Hammond and White 2008a), but a finite requirement for ATP cannot be avoided.

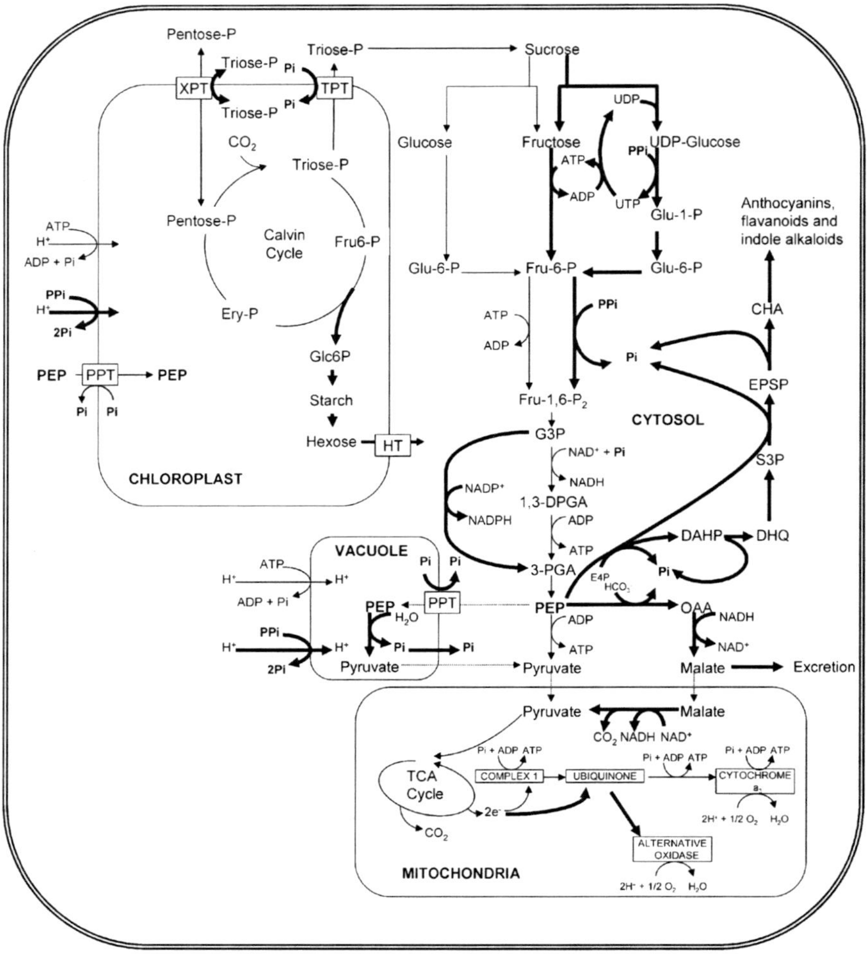

Fig. 4.2 Alternative metabolic processes for cytosolic glycolysis, mitochondrial electron transport, chloroplast processes and tonoplast H^+ pumping (bold arrows) that may enable plants to survive under P limiting conditions. Abbreviations for compounds are as follows; Glu-1-P, glucose 1-phosphate; Glu-6-P, glucose 6-phosphate; Fru-6-P, fructose 6-phosphate; Fru-1,6-P2, fructose 1,6-bisphosphate; G3P, glyceraldehyde-3-phosphate; 1,3-DPGA, 1,3-dephosphoglycerate; OAA, oxaloacetate; E4P, erythrose 4-phosphate; S3P, shikimate-3-phosphate; HT, hexose translocator; PEP, phosphoenolpyruvate; PPT, phosphoenolpyruvate/phosphate translocator; Pi, inorganic phosphate; TPT, triose phosphate/phosphate translocator; TrioseP, triose phosphates; XPT, xylulose 5-phosphate/phosphate translocator. (Figure redrawn from Plaxton and Carswell 1999 and Flügge *et al.* 2003 by Hammond and White 2008.)

Phosphorylated metabolites

A considerable quantity of cellular P occurs in the many phosphorylated intermediates of metabolic pathways. Phosphorylated compounds occur, for example, in the Calvin cycle, in the photorespiratory pathway, in glycolysis, in the pentose phosphate pathway, in nitrogen and sulfur assimilation, in the pathways of amino acid and nucleotide metabolism, and in pathways leading to the synthesis of polyphenols and lignin (Coruzzi and Last 2000; Dennis and Blakeley 2000; Malkin and Niyogi 2000; Siedow and Day 2000). In addition, where integrated metabolic transformations occur in different cellular compartments, it is often phosphorylated compounds that are transported across membranes. There is some flexibility in these metabolic pathways and, when plants lack sufficient P, alternative pathways requiring lower concentrations of phosphorylated intermediates are adopted (Figure 4.2; Plaxton and Carswell 1999; Flügge *et al.* 2003; Vance *et al.* 2003; Hammond *et al.* 2004; Hammond and White 2008a).

Phospholipids

In cell membranes, P occurs in phospholipids (phosphatidyl serine, phosphatidyl ethanolamine, phosphatidyl choline, phosphatidyl inositol and diphosphatidylglycerol), and the intermediate compounds of their biosynthesis (Somerville *et al.* 2000). In addition to their structural roles, phospholipids serve as substrates for the production of biochemical signals, such as inositol trisphosphate (IP_3), diacylglycerol, lysophosphatidyl choline, jasmonate and free headgroups (inositol, choline, ethanolamine, serine). Membrane lipids are required in abundance by photosynthetic tissues and tissue undergoing rapid cell division and/or cell expansion. The thylakoid membrane of the chloroplast is predominantly composed of sulphoquinovosyldiacylglycerol (SQDG), digalatosyldiacyglycerol (DGDG) and monogalatosyldiacyglycerol (MGDG), which is the most abundant lipid on the planet. By using these lipids in chloroplast membranes, plants reduce their requirements for phospholipids. Furthermore, when plants are starved of P, the relative abundance of SQDG, DGDG and MGDG increases in plant membranes, thereby contributing to tissue P economy (Essigmann *et al.* 1998; Härtel *et al.* 2000; Dörmann and Benning 2002; Andersson *et al.* 2003, 2005; Jouhet *et al.* 2004; Benning and Ohta 2005; Kobayashi *et al.* 2006; Li *et al.* 2006). Interestingly, these lipids are also found in the peribacteroid membranes surrounding rhizobial symbionts in legumes, where, again, their relative abundance increases during P starvation (Gaude *et al.* 2004).

Inorganic and storage P

In P-replete plants, over 85% of the cellular Pi is located in the vacuole (Marschner 1995). However, vacuolar Pi concentrations decrease rapidly when plants lack sufficient P, to maintain cytoplasmic Pi concentration ($[Pi]_{cyt}$) in the range 3–20 mM (Lee *et al.* 1990; Schachtman *et al.* 1998; Mimura 1999). If the P supplied to P-replete plants is reduced to the minimal amount required for optimal plant growth only the P in the inorganic fraction decreases substantially, which reflects the mobilization of surplus Pi from the vacuole (Figure 4.1; Marschner 1995). However, when the P supply to plants is decreased from an optimal to a suboptimal level, the P associated with nucleic acids, lipids, small metabolites and inorganic fractions all decrease. In contrast to most other tissues, Pi concentrations in seed are low and phytate (IP_6) is a dominant P fraction. In P-replete plants 50–90% of the total P in seeds occurs as phytate, but this value declines with decreasing P supply (Bieleski 1973; Marschner 1995; Mengel and Kirkby 2001). Remarkably, it is estimated that the amount of P contained in seed phytate is as much as 50% of all phosphate fertilizer applied annually on a global scale (Lott *et al.* 2000).

In addition, a small, but important, amount of P occurs in proteins. Phosphorylation and dephosphorylation of serine residues affects the activity of many enzymes (Raven 2008). Appropriate regulation of these enzymes is often critical for cellular homeostasis and, therefore, the P associated with this function is indispensable.

SYMPTOMS OF PHOSPHORUS IMBALANCE IN PLANTS

The symptoms of P deficiency in plants reflect the roles of P in plant cells. Phosphorus deficiency results in a diminutive or spindly habit, acute leaf angles, suppression of tillering, prolonged dormancy, early senescence and decreased size and number of flowers and buds (Bould *et al.* 1983; Bergmann 1992; Marschner 1995; Mengel and Kirkby 2001). Symptoms of P deficiency occur first in older leaves. The development of dark green or blue-green foliage is among the first symptoms of P deficiency. Red, purple or brown pigments develop in leaves, especially along veins. This is a consequence of anthocyanin production, which is induced by increased leaf sucrose concentrations (Müller *et al.* 2005; Teng *et al.* 2005; Amtmann *et al.* 2006; Solfanelli *et al.* 2006) and is thought to protect nucleic acids from UV damage and chloroplasts from photoinhibitory damage caused by P-limited photosynthesis (Hoch *et al.* 2001). Severe P deficiency results in chloroplast abnormalities, such as a reduction in the number of grana and their morphology (Bould *et al.* 1983). There is a gradual reduction in rates of cell division, cell expansion, photosynthesis and respiration, and changes in the abundance of C, N and S metabolites and concentrations of plant growth regulator substances during P starvation (Bould *et al.* 1983; Marschner 1995).

In agriculture, P-deficiencies of crops are usually treated by the addition of P-fertilizers to the soil. This has the added advantage of increasing soil P reserves for future crops. Foliar sprays of ammonium or potassium phosphate can be used, but may cause damage to the leaves (Bould *et al.* 1983). A lack of phytoavailable Zn in the soil can also cause P toxicity in crops (Loneragan *et al.* 1982). In nature, P rarely accumulates to toxic concentrations in plant tissues, except occasionally in species adapted to soils with excessively low P availability (Shane *et al.* 2004). However, in the laboratory, when roots of P-deficient plants are transferred to solutions containing high Pi concentrations, P may accumulate to toxic levels in shoots (e.g. Green *et al.* 1973; Clarkson and Scattergood 1982; Cogliatti and Clarkson 1983). This is a consequence of an innate inability to rapidly downregulate the high Pi uptake capacity of roots of P-deficient plants.

PHOSPHORUS AVAILABILITY TO PLANTS

Phosphorus is the 11th most abundant element in the earth's crust, and its concentration in soils generally lies between 100 and 3,000 mg P kg^{-1} soil, or 200–6,000 kg P ha^{-1} (Hedley *et al.* 1995; Mengel 1997).

In the soil, P is present as free Pi in the soil solution, labile Pi bound to soil particles, especially clays, as insoluble inorganic salts, such as calcium (Ca) phosphate in alkaline soils or aluminum (Al) and iron (Fe) phosphates in acidic soils, as complex organic compounds in the soil organic material, which may constitute 30–60% of the P in the topsoil, and the P in living soil biomass, which comprise about 5% of soil P (Mengel 1997; Hinsinger 2001; Oberson and Joner 2005; Turner 2007; Kirkby and Johnston 2008). These P sources must be solubilized or degraded (mineralized) to release soluble Pi for plant nutrition. The rates by which P is interconverted between these P fractions varies widely (Barber 1995; Mengel 1997), and both the amounts of P in each fraction and the rates of their interconversion are influenced by vegetation, amounts and chemical constituents of any Pi-fertilizers applied, the total P concentration in soil, soil structure, organic matter content and mineralogy, soil pH (Pi availability is highest between pH 6.5 and 7.5), temperature, soil moisture, and the abundance and identity of the soil micro-organisms present.

Nevertheless, P availability frequently limits plant growth in both natural and agricultural ecosystems (Epstein 1972; Chapin *et al.* 1986; Ågren 1988; Vance *et al.* 2003; Güsewell 2004). The simple reason for this is that both plant roots, and their associated mycorrhizal fungi, can only acquire P as orthophosphate (Schachtman *et al.* 1998; White 2003), which is present at extremely low concentrations (<10 μM) in the soil solution due to the low solubility products of inorganic P salts (Bieleski 1973; Barber 1995; Hedley *et al.* 1995; Marschner 1995). As a consequence, the diffusion of Pi through the soil solution is slow, and plant roots with their associated mycorrhizal fungi must occupy the soil volume at high density to acquire Pi at a sufficient rate for maximal growth (Bieleski 1973; Barber 1995; Marschner 1995). In addition, plant available Pi in the rhizosphere soil solution is rapidly depleted, and the replenishment of Pi in the rhizosphere soil solution from soil P sources is slow (Bieleski 1973; Barber 1995). For these reasons, conventional agriculture applies Pi-fertilizers to increase Pi concentrations in the rhizosphere to maximize crop P uptake and growth. Unfortunately, the reserves of commercially exploitable Pi rock are currently estimated to last less than 150 years (Mengel 1997; Steen 1998; Vance *et al.* 2003; Cohen 2007), so alternative strategies for P-fertilisation of crops may be required in the future.

PLANT STRATEGIES TO INCREASE THE ACQUISITION OF P

A variety of strategies are employed by plants to mobilize and acquire Pi from the soil (Vance *et al.* 2003; Hammond *et al.* 2004; Ticconi and Abel 2004; Raghothama and Karthikeyan 2005; Rengel and Marschner 2005; White *et al.* 2005a;

Lambers *et al.* 2006; Jain *et al.* 2007b). In general, these utilize the excess carbon assimilated when plant growth is limited by factors other than photosynthesis (Mengel and Kirkby 2001; Morgan *et al.* 2005; Hermans *et al.* 2006; Hammond and White 2008a). In response to P deficiency: (1) Plant roots acidify the rhizosphere and secrete low-molecular-weight organic anions and phosphatase enzymes into the soil to mobilize Pi from inorganic and organic P sources. (2) Plants invest a greater proportion of their biomass in their root system. (3) The morphology of the root system is altered, not only to explore the soil volume more effectively but also to exploit any localized patches of high Pi availability. (4) There is a general increase in the capacity of plant roots to take up Pi, accelerating the rate of Pi uptake from the soil solution. (5) Most plants foster symbiotic relationships with mycorrhizal fungi to increase their ability to explore the soil volume and mobilize P from remote inorganic and organic sources. None of these strategies are mutually exclusive and plants often employ several simultaneously to avert P-deficiency.

Rhizosphere modification

Roots of P-deficient plants often release protons (H^+) to acidify the rhizosphere (Marschner 1995; Hinsinger 2001). In addition, they secrete low-molecular-mass organic anions, such as carboxylates and piscidic acid (Jones 1998; López-Bucio *et al.* 2000b; Hocking 2001; Ryan *et al.* 2001; Dakora and Phillips 2002; Jones *et al.* 2003; Delhaize *et al.* 2007). However, the effectiveness of these compounds in releasing Pi from soil minerals differs greatly (generally citrate > oxalate > malate = tartrate > acetate > succinate = lactate, but this order is dependent upon soil type; Jones 1998; Hinsinger 2001; Jones *et al.* 2003) and plant species differ in both the identity and quantity of the low molecular-mass organic acids they exude from their roots (e.g. Van ura and Hovadík 1965; Ohwaki and Hirata 1992; Dinkelaker *et al.* 1995; Jones 1998; Neumann and Römheld 1999; López-Bucio *et al.* 2000b; Hinsinger 2001; Dakora and Phillips 2002; Dechassa and Schenk 2004; Jain *et al.* 2007b), which may be related to their phylogeny, ecology and/or their ability to form mycorrhizal associations. Differences between genotypes of particular species in their ability to exude organic acids and to access different forms of mineral P have also been reported (Subbarao *et al.* 1997; Narang *et al.* 2000; Gaume *et al.* 2001; Ishikawa *et al.* 2002; Liao *et al.* 2006; Pearse *et al.* 2007, 2008). Although organic anions released by roots are rapidly decomposed in the soil (Jones 1998; Jones *et al.* 2003), plant roots secrete these compounds in locations where the abundance of microorganisms is low, such as the root apex or within clusters of lateral roots (Dinkelaker *et al.* 1995; Jones 1998; Gaume *et al.* 2001; Vance *et al.* 2003; Thornton *et al.* 2004; Lambers *et al.* 2006; Liao *et al.* 2006; Paterson *et al.* 2006). It is commonly observed that the release of carbon compounds increases microbial biomass and activity in the rhizosphere, but its consequences for microbial community structure and function are less well understood (Morgan *et al.* 2005).

Changes in microbial abundance and/or community structure can either promote or reduce Pi availability to plants, both directly and indirectly (Jones 1998; Barea *et al.* 2005; Morgan *et al.* 2005; Rengel and Marschner 2005; Marschner 2008).

Plant roots also secrete enzymes into the rhizosphere to release Pi from organic P compounds in the soil. These enzymes include acid phosphatases and phytases that hydrolyze organic phosphomonoesters, which are the dominant form of organic P in the soil, and apyrases and RNases that hydrolyze phosphodiesters (Thomas *et al.* 1999; Haran *et al.* 2000; Miller *et al.* 2001; Coello 2002; Wasaki *et al.* 2003; Tomscha *et al.* 2004; Zimmermann *et al.* 2004; Jain *et al.* 2007b). Although there is significant genotypic variation in the phosphatase activities secreted by roots both between and within plant species (Tadano and Sasaki 1991; Asmar *et al.* 1995; Li *et al.* 1997; Liu *et al.* 2001; Gaume *et al.* 2001), these are not always correlated with their ability to acquire P or grow in many soils (George *et al.* 2008). This may reflect the complementarity of the many compensatory mechanisms plants employ to acclimate to low P availability. It is noteworthy that rhizosphere microorganisms also release phosphomonoesterases and phosphodiesterases that contribute to P cycling in the soil and/or induce the release of these enzymes by plants (George and Richardson 2008; Marschner 2008).

Altered biomass allocation and root system modification

In response to P deficiency plants allocate more of their biomass to the root system, thereby increasing root growth rates and the volume of soil the root system can explore (Vance *et al.* 2003; Hutchings and John 2004; White *et al.* 2005a; Hermans *et al.* 2006; Hammond and White 2008a). To exploit the local effects of secreting organic anions and enzymes into the rhizosphere, plants increase their root length density in regions of higher P availability. In response to P deficiency, plants preferentially produce roots in the topsoil, since P is often concentrated close to the soil surface (Barber 1995; Lynch and Brown 2001; Rubio *et al.* 2003; Liao *et al.* 2004; Ho *et al.* 2005; Zhu *et al.* 2005), proliferate lateral roots in P-rich patches (Drew 1975; Robinson 1994; Hodge 2004; Hutchings and John 2004) and increase the length and density of root hairs to enlarge the effective surface area of the root system, thereby increasing the volume of soil explored for minimal biomass investment (Jungk 2001; Zhang *et al.* 2003). All these acclimatory responses increase P acquisition and plant growth, and there is considerable genetic variation both between and within plant species in these traits (e.g. O'Toole and Bland 1987; Sattelmacher *et al.* 1990; Klepper 1992; Oyanagi 1994; Barber 1995; Bonser *et al.* 1996; Manske *et al.* 2000; Lynch and Brown 2001; Stalham and Allen 2001; López-Bucio *et al.* 2002; Chevalier *et al.* 2003; Rubio *et al.* 2003; Gahoonia and Nielsen 2004a,b; Liao *et al.* 2004; Yan *et al.* 2004; Zhu and Lynch 2004; Ho *et al.* 2005; Malamy 2005; White *et al.* 2005a,b; Wissuwa 2005; Zhu *et al.* 2005, 2006; Reymond *et al.* 2006).

Increased Pi uptake capacity

The Pi uptake capacity of plant root cells is also increased in P deficient plants (Epstein 1972; Lee *et al.* 1990; Schachtman *et al.* 1998; Smith FW *et al.* 2003; Raghothama and Karthikeyan 2005; Bucher 2007; Jain *et al.* 2007b) even in the mature, suberised parts of the root (Clarkson *et al.* 1978; Rubio *et al.* 2004), but this is not considered to be a major factor affecting P acquisition efficiency (Barber 1995; Horst *et al.* 2001; White *et al.* 2005a; Lambers *et al.* 2006). Differences in root:shoot biomass ratio, root growth rate, root hair production and root system morphology generally account for most variation in P acquisition efficiency between and within plant species (Ozanne *et al.* 1969; Fageria *et al.* 1988; Föhse *et al.* 1988, 1991; Baligar *et al.* 2001; Lynch and Brown 2001; Dechassa *et al.* 2003; Wissuwa 2003; Zhu and Lynch 2004; White *et al.* 2005a; Zhu *et al.* 2005), although the exudation of organic acids contributes to the exceptionally high P acquisition efficiency of some crops, such as brassica (Hoffland 1992; Dechassa *et al.* 2003) and white lupin (López-Bucio *et al.* 2000b).

Improved symbiotic associations

To increase their exploration of the soil, most land plants form associations with mycorrhizal fungi (Harrison 1999; Karandashov and Bucher 2005; Bucher 2007; Smith and Read 2007). This association can benefit both partners, with the fungi receiving C from the plants and the plants receiving P and other mineral elements from the fungi. It is estimated that between 4% and 20% of net photosynthate is transferred from plants to their fungal partners (Johnson *et al.* 1997; Morgan *et al.* 2005). In return, the fungal partner acquires the mineral elements for the symbiosis. The fungal hyphae enlarge the volume of soil explored, increase the surface area for Pi uptake, extend into soil pores too small for roots to enter and, in some cases, hydrolyze organic P compounds that plants cannot (Bieleski 1973; Harrison 1999; Karandashov and Bucher 2005; Morgan *et al.* 2005; Bucher 2007). Consequently, roots of mycorrhizal plants can acquire between three to five times more Pi than those of non-mycorrhizal plants when grown in low P soils (Bieleski 1973; Marschner 1995; Johnson *et al.* 1997; Smith SE *et al.* 2003; Smith and Read 2007). By contrast, when Pi is readily available to plants, the C costs of mycorrhizal associations are not compensated for by improved P nutrition and a reduced mycorrhizal colonization of roots is often observed (Johnson *et al.* 1997; Graham 2000; Morgan *et al.* 2005). It is, perhaps, also noteworthy that nodulation and nodule growth in legumes are increased as plant P status improves, and this beneficial symbiosis stimulates plant growth enormously (Marschner 1995; Vádez *et al.* 1999; Schulze *et al.* 2006; Raven 2008).

CO-ORDINATING PLANT RESPONSES TO VARIATIONS IN P SUPPLY

Plants respond both to tissue P status, enabling the efficient use of C, N, S and P resources within the plant, and to local variations in soil Pi availability, enabling the proliferation of roots in Pi rich patches (Figure 4.3; White *et al.* 2005a; Amtmann *et al.* 2006; Hammond and White 2008a). Many of the responses of plants to P starvation appear to be initiated, or modulated, by a decrease in the delivery of Pi to the shoot (Figure 4.3[1]; Jeschke *et al.* 1997; Mimura 1999) and the consequent reduction in the Pi available for shoot metabolism. This often results in an immediate reduction in shoot growth rate before root growth is affected (Clarkson and Scattergood 1982; Cogliatti and Clarkson 1983). A reduction in $[Pi]_{cyt}$ impacts directly on photosynthesis, glycolysis and respiration (Plaxton and Carswell 1999; Hammond *et al.* 2004; Hammond and White 2008a), and changes in carbohydrate metabolism are reinforced by transcriptional reprogramming (Hammond *et al.* 2003, 2005; Wu *et al.* 2003; Misson *et al.* 2005; Hermans *et al.* 2006; Wasaki *et al.* 2006; Morcuende *et al.* 2007; Müller *et al.* 2007). This results in organic acids, starch and sucrose accumulating in leaves of P starved plants (Figure 4.3[2]; Rao *et al.* 1990; Cakmak *et al.* 1994; Ciereszko and Barbachowska 2000; Müller *et al.* 2004, 2005, 2007; Wissuwa *et al.* 2005; Hermans *et al.* 2006; Morcuende *et al.* 2007). Metabolism is rerouted by employing reactions that do not require Pi or adenylates

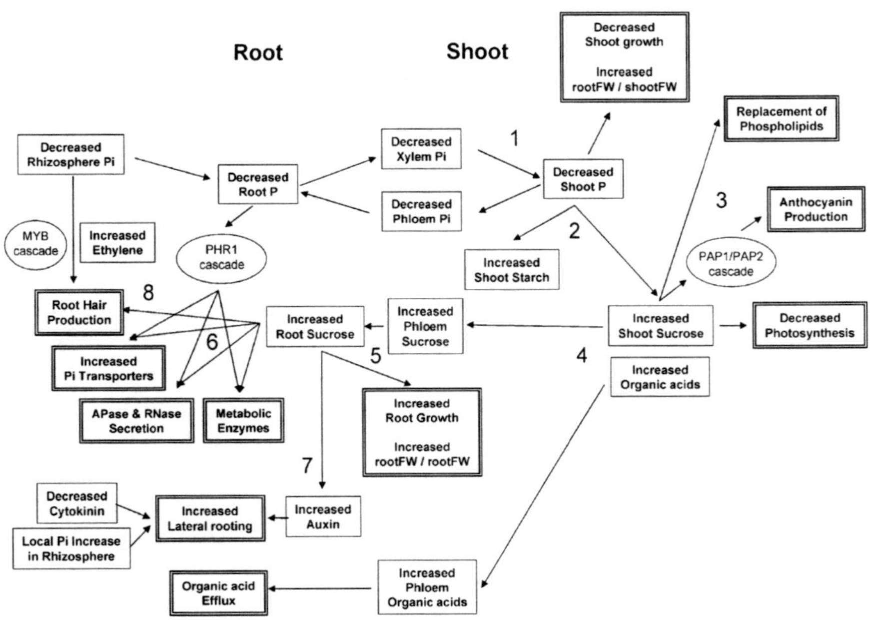

Fig. 4.3 Hypothetical signaling cascades initiating acclimatory responses to P starvation. (Based on reviews by White *et al.* 2005a; Amtmann *et al.* 2006; and Hammond and White 2008a.)

(Plaxton and Carswell 1999; Vance *et al.* 2003; Hammond *et al.* 2004; Hammond and White 2008a) and, under severe P-deficiency, intracellular phosphatases and nucleases are induced that remobilize P from cellular metabolites and nucleic acids (Bariola *et al.* 1994; Berger *et al.* 1995; Bosse and Köck 1998; Brinch-Pedersen *et al.* 2002; Petters *et al.* 2002; Hammond *et al.* 2003; Wasaki *et al.* 2006; Morcuende *et al.* 2007; Müller *et al.* 2005, 2007). A general decrease in tissue RNA concentration is also observed (Hewitt *et al.* 2005). Increased leaf sucrose concentrations lead indirectly to (i) a reduction of photosynthesis through the downregulation of many photosystem subunits and small subunits of RuBisCo (Paul and Pellny 2003; Lloyd and Zakhleniuk 2004; Amtmann *et al.* 2006; Hermans *et al.* 2006; Rook *et al.* 2006), (ii) an increase in leaf sulfolipid and galactolipid concentrations through the upregulation of genes involved in their biosynthesis (Essigmann *et al.* 1998; Dörmann and Benning 2002; Yu *et al.* 2002; Andersson *et al.* 2003, 2005; Hammond *et al.* 2003; Frentzen 2004; Benning and Ohta 2005; Franco-Zorrilla *et al.* 2005; Misson *et al.* 2005; Cruz-Ramírez *et al.* 2006; Kobayashi *et al.* 2006; Li *et al.* 2006), and (iii) the production of anthocyanins through a transcriptional cascade involving the transcription factors TTG1-TT8/EGL3-PAP1/PAP2 (Figure 4.3[3]; Lloyd and Zakhleniuk 2004; Teng *et al.* 2005; Amtmann *et al.* 2006; Solfanelli *et al.* 2006). An increased leaf sucrose concentration also results in the upregulation of transporters delivering organic acids and sucrose to the phloem, which facilitates the movement of these compounds to the root (Figure 4.3[4]; Gaume *et al.* 2001; Hermans *et al.* 2006).

The preferential allocation of C to the root system, and the resulting increased root:shoot biomass ratio, appears to be a direct consequence of altered shoot metabolism and is mediated by increased translocation of sucrose to the root (Figure 4.3[5]; Hermans *et al.* 2006; Hammond and White 2008a). In addition, the sucrose delivered to the root acts as a systemic signal (indicating low shoot P status) that can initiate changes in gene expression to alter root biochemistry and remodel root morphology (Liu *et al.* 2005; Amtmann *et al.* 2006; Hermans *et al.* 2006; Hernández *et al.* 2007; Karthikeyan *et al.* 2007; Tesfaye *et al.* 2007; Hammond and White 2008a). Increased root sucrose concentrations appear to upregulate genes encoding riboregulators, Pi transporters, RNases, phosphatases and metabolic enzymes in combination with the PHR1 transcriptional cascade (Figure 4.3[6]), whilst its effects on lateral rooting occur through modulation of auxin transport (Figure 4.3[7]) and those on root hair development are contingent upon changes in auxin transport and the local production of ethylene (Figure 4.3[8]).

The PHR1 protein is a MYB transcription factor that binds to an imperfect-palindromic sequence (P1BS; GNATATNC) present in the promoter regions of many genes whose expression responds to P starvation (PSR genes). These include genes encoding transcription factors, protein kinases, Pi transporters, RNases, phosphatases, metabolic enzymes and enzymes involved in the synthesis of sulfolipids and galactolipids (Figure 4.3; Rubio *et al.* 2001; Hammond *et al.* 2004; Franco-Zorrilla *et al.* 2004; Schünmann *et al.* 2004; Misson *et al.* 2005; Jain *et al.* 2007b). The expression of *PHR1* appears to be constitutive, but the PHR1 protein is targeted by a small ubiquitin-like modifier (SUMO) E3 ligase (SIZ1), whose expression is increased by P starvation

(Miura *et al.* 2005). Since the Arabidopsis *siz1* mutant constitutively exhibits phenotypic characteristics of P-deficient plants, it is hypothesized that SIZ1 acts as a repressor of plant responses to P starvation (Miura *et al.* 2005). One target of the PHR1 protein appears to be the microRNA family, miR399 (Bari *et al.* 2006; Chiou 2007). The expression of miR399 is specifically and rapidly up-regulated by P starvation (Fujii *et al.* 2005; Bari *et al.* 2006; Chiou *et al.* 2006). The target gene for miR399 is an ubiquitin E2 conjugating enzyme, also identified as the gene responsible for the *pho2* mutant phenotype (*AtUBC24*; At2g33770; Sunkar and Zhu 2004; Fujii *et al.* 2005; Aung *et al.* 2006; Bari *et al.* 2006; Chiou *et al.* 2006) and the expression of *AtUBC24* is downregulated during P starvation (Fujii *et al.* 2005; Bari *et al.* 2006; Chiou *et al.* 2006). It is thought that *AtUBC24* is a negative regulator of the expression of a subset of P starvation responsive genes, possibly through other intermediary transcription factors (Chiou 2007). Interestingly, there is some sequence similarity between miR399 and the TPSI1/Mt4/At4 family of non-coding transcripts, which allows them to bind to miR399 (Shin *et al.* 2006; Chiou 2007; Franco-Zorrilla *et al.* 2007). The expression of the TPSI1/Mt4/At4 family is induced rapidly and specifically in response to P starvation (Liu *et al.* 1997; Burleigh and Harrison 1999; Martín *et al.* 2000; Hou *et al.* 2005; Shin *et al.* 2006), and these non-coding transcripts sequester miR399 and serve to attenuate the miR399-mediated transcriptional responses to P starvation (Franco-Zorrilla *et al.* 2007). The recent characterization of the *At4* T-DNA knockout mutant suggests that it has a role in the internal redistribution of P from the shoots to the roots (Shin *et al.* 2006). It has a similar phenotype to the *pho2* mutant, which accumulates more P in leaves than wildtype plants (Delhaize and Randall 1995).

Recently, it has become apparent that most alterations in root morphology in response to P starvation arise from the interplay of local and systemic signals. Changes in the concentration, transport and/or sensitivity to auxin, ethylene, cytokinin and sucrose have all been implicated in the remodeling of root morphology in P-deficient plants (Martín *et al.* 2000; Forde and Lorenzo 2001; López-Bucio *et al.* 2002, 2003, 2005; Al-Ghazi *et al.* 2003; Casimiro *et al.* 2003; Casson and Lindsey 2003; Ma *et al.* 2003; Vance *et al.* 2003; Hammond *et al.* 2004; Ticconi and Abel 2004; Franco-Zorrilla *et al.* 2005; Malamy 2005; Nacry *et al.* 2005; White *et al.* 2005a; Amtmann *et al.* 2006; Jain *et al.* 2007a; Karthikeyan *et al.* 2007; Hammond and White 2008a), and the observed changes in concentrations of plant growth regulators are consistent with changes in the expression of genes known to be regulated by, or involved in the regulation of, auxin, ethylene, cytokinin and sucrose in roots of P-deficient plants (Al-Ghazi *et al.* 2003; Casson and Lindsey 2003; Uhde-Stone *et al.* 2003; Wu *et al.* 2003; Misson *et al.* 2005; Hermans *et al.* 2006; Hernández *et al.* 2007).

Contact of the root cap of P-starved plants with media lacking Pi appears to be necessary and sufficient to reduce meristematic activity in primary roots and slow their growth, in a response mediated by multicopper oxidases (Ticconi *et al.* 2004; López-Bucio *et al.* 2005; Sánchez-Calderón *et al.* 2005, 2006; Svistoonoff *et al.* 2007; Jain *et al.* 2007a). The proliferation of lateral roots of P starved plants in regions of increased Pi availability is also contingent upon growth of the primary root apex through these regions (Drew 1975; Robinson 1994; López-Bucio *et al.* 2003) but appears to be initiated by changes in auxin transport (López-Bucio *et al.* 2002, 2003, 2005; Al-Ghazi *et al.* 2003; Casimiro *et al.* 2003; Casson and Lindsey 2003; Malamy 2005; Nacry *et al.* 2005; Jain *et al.* 2007a), with greater sucrose availability increasing the responsiveness to auxin (Nacry *et al.* 2005; Jain *et al.* 2007a). It is also promoted by the reduction in cytokinin concentrations in roots of P-deficient plants (Martín *et al.* 2000; López-Bucio *et al.* 2002; Franco-Zorrilla *et al.* 2002), but the changes in cytokinin signaling during P starvation appear to be a secondary consequence of crosstalk between sugar and local P-signaling cascades (Franco-Zorrilla *et al.* 2005). This phenomenon is comparable to the proliferation of specialized cluster roots in regions of local Pi enrichment observed in diverse non-mycorrhizal plant species when they lack sufficient P (Dinkelaker *et al.* 1995; Shane *et al.* 2003; Shen *et al.* 2005; Lamont 2003; Lambers *et al.* 2006). The initiation and elongation of root hairs, once thought to be root cell autonomous, and regulated solely through local interactions between increasing auxin and ethylene concentrations (Bates and Lynch 1996; Jungk 2001; Casson and Lindsey 2003; Ma *et al.* 2003; Ticconi and Abel 2004; Zhang *et al.* 2003; He *et al.* 2005; Amtmann *et al.* 2006), now also appears to be modulated by plant P status through sucrose supply to the roots, since the roots of P starved plants have more and longer root hairs when supplied with sucrose (Jain *et al.* 2007a). Finally, the topsoil-foraging phenotype of P-deficient plants appears to be modulated primarily by the sensitivity of root gravitropism to ethylene, which increases with P starvation (Basu *et al.* 2007)

THE INFLUENCE OF P NUTRITION ON THE ECOLOGY OF TERRESTRIAL PLANTS

The most limiting, or most toxic, mineral element in an environment is likely to determine its ecology. In terrestrial ecosystems, recent anthropogenic inputs have raised N availability and most environments are now limited by the availability of P (Chapin *et al.* 1986; Güsewell 2004; Wassen *et al.* 2005). Thus, P availability will determine both primary production and species diversity of these ecosystems. In wild plants, as in agricultural plants, competitive advantage is gained by effective P acquisition and efficient utilization of P for growth and reproduction, and wild plants show the full range of responses to low P availability described in the previous sections. One proxy for the competitive ability of a species growing on P-limited soils is a high tissue N:P ratio, although slow growing species, perennials and legumes can provide exceptions (Güsewell 2004; Niklas 2008). The remobilization of P from senescing to developing leaves (Aerts 1996; Aerts and Chapin 2000; McGroddy *et al.* 2004; Güsewell 2005), and the storage of P between growth seasons (Güsewell *et al.* 2003; Güsewell 2004), are important factors for the P economy of wild plants, as is the ability to maintain the P-demands of symbiotic N-fixation in N-limited environments (Raven 2008). Supplying reproductive tissues with P is essential. Seeds have higher N and P concentrations and lower N:P ratios than vegetative tissues, and wild plants typically allocate between 15% and 60% of their P to reproduction depending upon P availability in their environment (Fenner 1986; Güsewell 2004). Tissue N:P ratio also determines the vulnerability of plants to herbivores, decomposers and pathogens.

In soils with extremely low P availability, such as those of Western Australia and South Africa, plants must be capable of acquiring P from less abundant and less readily-available P sources. This often requires special adaptations, and the flora of these regions is dominated by non-mycorrhizal plants that utilize tissue P efficiently and produce lateral or cluster roots that secrete citrate in local patches of P-rich soil (Dinkelaker *et al.* 1995; Lamont 2003; Lambers *et al.* 2006). Such plants include *Lupinus* and *Kennedia* (Fabales), Cyperaceae and Juncaceae (Poales), and Proteaceae (Proteales). Similarly, species that can solubilize P effectively in clayey, acid or alkaline soils often dominate the vegetation of these areas. Thus, it is noteworthy that calcicole and calcifuge plants differ in their efflux of organic acids (Ström 1997; Jones 1998). The ability to form mycorrhizal associations does not appear to be a specific adaptation to low P availability, since it was an obligate requirement for plants to colonize the land and most plant species maintain these associations (Karandashov and Bucher 2005; Smith and Read 2007). However, the costs and benefits to plants of this symbiosis depend critically on P availability and plant P requirements, and it may be disadvantageous under some circumstances (Johnson *et al.* 1997; Graham 2000; Morgan *et al.* 2005). In less extreme environments, slower-growing stress-tolerant species with low tissue P requirements and high N:P ratios often dominate when P is limiting (Thompson *et al.* 1997; Grime *et al.* 1997; Aerts and Chapin 2000; Grime 2001; Güsewell 2004). Since graminoids generally have lower tissue P concentrations and higher tissue N:P ratios than forbs, mixed pastures on soils with a low P availability are generally dominated by grasses that can also acquire Pi effectively (Güsewell 2004). It is thought that the diversity of plant species is highest in P-limited soils since P is relatively immobile in the soil and different plants show contrasting foraging strategies and/or acquire P from different sources, thereby minimizing competition (Janssens *et al.* 1998; Güsewell 2004; McCrea *et al.* 2004; Güsewell *et al.* 2005). In fact, a recent survey of temperate Eurasia suggests that P-limitation favors the persistence of endangered plant species (Wassen *et al.* 2005). However, this is not always observed. For example, shifts in the dominance of clonal graminoids can reduce the diversity of plant species on soils with low P availability through exclusion (Güsewell 2004).

Species composition changes rapidly after mineral fertilization and the long-term effects of a single fertilization event can persist for many decades, through its combined effects on plant and microbial community structures (Güsewell 2004). In the first year, the effects of fertilisation are often determined by the responses of species that dominated the original plant community that grow rapidly or are better able to exploit the timing or method of fertilisation. In subsequent years, subordinate or new species with different nutrient requirements may increase in abundance and reach dominance. Intriguingly, a heterogeneous distribution of mineral availability can increase the total biomass of plant communities more than a homogeneous supply (Hutchings and John 2004). It is expected that P-fertilisation will promote the growth of faster-growing plants with higher P requirements for growth and lower tissue N:P ratios (e.g. Bradshaw *et al.* 1960; Mamolos *et al.* 1995; Thompson *et al.* 1997; Elser *et al.* 2000b; Güsewell *et al.* 2003). These are primarily ruderal plant species, which are often annual species or short-lived perennials with high reproductive allocation and, consequently, greater P requirements, but also include competitive plant species (Thompson *et al.* 1997; Güsewell 2004; Han *et al.* 2005). An increase in the abundance of forbs and bryophytes, relative to grasses, is often observed following P fertilization (Güsewell 2004). In addition, since P-fertilisation promotes the growth of legumes and nitrogenase activity in their nodules (Smith 1992; Stöcklin *et al.* 1998), P-fertilisation of a N-limited environment can alter plant communities dramatically by increasing N availability.

PHOSPHORUS MANAGEMENT FOR SUSTAINABLE AGRICULTURE

Given that P limits agricultural productivity, and that Pi fertilizers are a finite resource, agronomists and plant breeders must work together to reduce the Pi-fertilizer inputs to agriculture without compromising yield or quality. This might be effected by improved agronomic strategies, greater use of alternative P-fertilizers, such as manures, animal wastes and recovered phosphates (Raven 2008), and through the development of crop genotypes that are more efficient in acquiring P from the soil and/or in utilizing P more economically in their tissues (Gahoonia and Nielsen 2004a; White *et al.* 2005a; Lambers *et al.* 2006).

In many developed countries, generous applications of P fertilizers in the past have led to an increase in soil P reserves, especially in arable areas, and many crops now show little response to P-fertilizer applications (Mengel and Kirkby 2001; Kirkby and Johnston 2008). In these circumstances, recommended fertilizer applications are often based on replacing P losses to the environment plus the P offtake by crops (e.g. Defra 2000). To reduce P fertilizer inputs to these crops, management practices should optimize the timing, placement and formulation of P-fertilizers to reduce P losses to the environment. This can be assisted by the use of decision support systems (Mengel 1997; Heathwaite *et al.* 2003; Fixen 2005; Zhang *et al.* 2007; Hammond and White 2008b), modern fertilizer placement techniques (Bryson 2005; Gregory and George 2005; White *et al.* 2005b) and the use of slow release fertilizers (Perrott and Kear 2000). These techniques can be complemented by growing crops that acquire P effectively from agricultural soils. For example, since P becomes concentrated in the surface/ploughed layers of agricultural soils, crop genotypes with a topsoil foraging phenotype would maximize P acquisition, but could make them susceptible to other edaphic or climatic stresses, such as drought. To reduce the P off-take by crops, the plant P requirements for optimal yields must be minimized. In regions where low soil P availability compromises crop production, which apparently exceed 5.7 billion hectares of potential agricultural land worldwide, improved agronomic practice and plant breeding should aim to increase P availability and acquisition. Liming can improve P availability in Pi-fixing soils with low pH (Mengel and Kirkby 2001). In some situations, the incorporation of soil organic matter into soils will improve the use of soil and fertilizer P, since organic compounds compete with Pi for binding sites on clays (Mengel 1997; Horst *et al.* 2001). Crop traits that improve P availability and acquisition include the development of an extensive root system, the release of organic acids and enzymes into the rhizosphere and the cultivation of beneficial associations with soil micro-organisms. Again, a reduced plant P requirement for optimal yields may be beneficial.

The P-use efficiency (PUE) of plants has been defined in many ways (Gourley *et al.* 1994; Baligar *et al.* 2001; Greenwood *et al.* 2005; Gregory and George 2005; White *et al.* 2005a). The following four definitions are used most often: (1) the increase in yield per unit P in the soil, which is often referred to as agronomic P efficiency (APE), (2) the amount of P in a plant divided by its root biomass, which is referred to as P acquisition efficiency (PAE), (3) the amount of P in a plant divided by the amount of P in the soil, which is referred to as P uptake efficiency (PUpE), and (4) crop yield divided by amount of P in the plant, which is referred to as tissue, or physiological, P utilization efficiency (PUtE). It will be apparent that APE is the product of PUpE and PUtE. For this reason, genetic strategies to increase the yield of crops on low P soils have focused on improving P acquisition by roots and tissue P utilization efficiency.

Earlier in this chapter, it was observed that greater root:shoot biomass ratios, root growth rates and root hair production, together with the proliferation of lateral roots in regions of local P availability and the exudation of organic acids and enzymes into the rhizosphere are the traits required for effective P mobilization and acquisition by plants. There is considerable genetic variation in all these traits in most crops, which might be selected through conventional breeding programs. In addition, it has been suggested that knowledge of the mechanisms whereby plants sense and respond to P availability in soils could facilitate selection, breeding and GM approaches to improve crop production on soils with low P availability (Vance *et al.* 2003; Hammond *et al.* 2004; White *et al.* 2005a; Jain *et al.* 2007b). The transcriptional cascades controlling appropriate facets of root morphology and/or the release of organic acids and enzymes into the rhizosphere could be targeted. Transgenic plants secreting more and/or different organic acids and hydrolytic enzymes into the rhizosphere have been engineered (Koyama *et al.* 2000; López-Bucio *et al.* 2000a,b; Richardson *et al.* 2001; Mudge *et al.* 2003; Zimmermann *et al.* 2003; George *et al.* 2004, 2005a,b; Xiao *et al.* 2005, 2006; George and Richardson 2008). However, although these genetic manipulations have shown promising results, they have not always been successful in promoting plant growth in natural soils (Delhaize *et al.* 2001; George *et al.* 2004, 2005a,b; George and Richardson 2008).

There is considerable variation both between and within crop species in the critical tissue P concentration required for maximum growth (Fageria *et al.* 1988; Fageria and Baligar 1997, 1999; Baligar *et al.* 2001; Osborne and Rengel 2002; Bentsink *et al.* 2003; Trehan and Sharma 2003; Zhu and Lynch 2004; Ozturk *et al.* 2005; White *et al.* 2005a,b). However, differences in the response of yield to P fertilisation do not appear to be correlated with PUtE (Greenwood *et al.* 1980; Alt 1987; Föhse *et al.* 1988; Fageria and Baligar 1999; Ozturk *et al.* 2005). Thus, selection for greater PUtE does not appear to be an effective strategy for developing crops that yield well on soils with low P availability. However, genotypes of crops that yield well and have lower tissue P concentrations can be used to reduce P-fertilizer inputs to soils on that require only maintenance P fertilisation.

Acknowledgements Work at SCRI was supported by the Scottish Executive Environment and Rural Affairs Department (SEERAD). Other work of PJW and JPH on phosphorus nutrition is supported by grants HH3504SPO, HH3507SFV, HH3509SFV and WQ0119 from the UK Department for Environment, Food and Rural Affairs (Defra).

REFERENCES

Aerts R (1996) Nutrient resorption from senescing leaves of perennials: are there general patterns? J Ecol 84: 597–608

Aerts R, Chapin FS (2000) The mineral nutrition of wild plants revisited: a re-evaluation of processes and patterns. Adv Ecol Res 30: 1–67

Ågren GI (1988) Ideal nutrient productivities and nutrient proportions in plant growth. Plant Cell Environ 11: 613–620

Al-Ghazi Y, Muller B, Pinloche S, Tranbarger TJ, Nacry P, Rossignol M, Tardieu F, Doumas P (2003) Temporal responses of *Arabidopsis* root architecture to phosphate starvation: evidence for the involvement of auxin signalling. Plant Cell Environ 26: 1053–1066

Alt D (1987) Influence of P- and K-fertilization on yield of different vegetable species. J Plant Nutr 10: 1429–1435

Amtmann A, Hammond JP, Armengaud P, White PJ (2006) Nutrient sensing and signalling in plants: potassium and phosphorus. Adv Bot Res 43: 209–257

Andersson MX, Stridh MH, Larsson KE, Liljenberg C, Sandelius AS (2003) Phosphate-deficient oat replaces a major portion of the plasma membrane phospholipids with the galactolipid digalactosyldiacylglycerol. FEBS Lett 537: 128–132

Andersson MX, Larsson KE, Tjellström H, Liljenberg C, Sandelius AS (2005) Phosphate-limited oat. The plasma membrane and the tonoplast as major targets for phospholipid-to-glycolipid replacement and stimulation of phospholipases in the plasma membrane. J Biol Chem 280: 27578–27586

Asmar F, Gahoonia TS, Nielsen NE (1995) Barley genotypes differ in activity of soluble extracellular phosphatase and depletion of organic phosphorus in the rhizosphere soil. Plant Soil 172: 117–122

Aung K, Lin SI, Wu CC, Huang YT, Su CL, Chiou TJ (2006) *pho2*, a phosphate overaccumulator, is caused by a nonsense mutation in a MicroRNA399 target gene. Plant Physiol 141: 1000–1011

Baligar VC, Fageria NK, He ZL (2001) Nutrient use efficiency in plants. Commun Soil Sci Plant Anal 32: 921–950

Barber SA (1995) Soil Nutrient Bioavailability: A Mechanistic Approach. Wiley, New York

Barea JM, Pozo MJ, Azcón R, Azcón-Aguilar C (2005) Microbial co-operation in the rhizosphere. J Exp Bot 56: 1761–1778

Bari R, Pant BD, Stitt M, Scheible WR (2006) PHO2, MicroRNA399, and PHR1 define a phosphate-signaling pathway in plants. Plant Physiol 141: 988–999

Bariola PA, Howard CJ, Taylor CB, Verburg MT, Jaglan VD, Green PJ (1994) The *Arabidopsis* ribonuclease gene *RNS1* is tightly controlled in response to phosphate limitation. Plant J 6: 673–685

Basu P, Zhang YJ, Lynch JP, Brown KM (2007) Ethylene modulates genetic, positional, and nutritional regulation of root plagiogravitropism. Funct Plant Biol 34: 41–51

Bates TR, Lynch JP (1996) Stimulation of root hair elongation in *Arabidopsis thaliana* by low phosphorus availability. Plant Cell Environ 19: 529–538

Beebe S, Lynch J, Galwey N, Tohme J, Ochoa I (1997) A geographical approach to identify phosphorus-efficient genotypes among landraces and wild ancestors of common bean. Euphytica 95: 325–336

Benning C, Ohta H (2005) Three enzyme systems for galactoglycerolipid biosynthesis are coordinately regulated in plants. J Biol Chem 280: 2397–2400

Bentsink L, Yuan K, Koorneef M, Vreugdenhil D (2003) The genetics of phytate and phosphate accumulation in seeds and leaves of *Arabidopsis thaliana*, using natural variation. Theor Appl Genet 106: 1234–1243

Berger S, Bell E, Sadka A, Mullet JE (1995) *Arabidopsis thaliana Atvsp* is homologous to soybean *VspA* and *VspB*, genes encoding vegetative storage protein acid phosphatases, and is regulated similarly by methyl jasmonate, wounding, sugars, light and phosphate. Plant Mol Biol 27: 933–942

Bergmann W (1992) Nutritional Disorders of Plants. Visual and Analytical Diagnosis. Gustav Fischer, Jena, Germany

Bieleski RL (1973) Phosphate pools, phosphate transport, and phosphate availability. Ann Rev Plant Physiol 24: 225–252

Blackshaw RE, Brandt RN, Janzen HH, Entz T (2004) Weed species response to phosphorus fertilization. Weed Sci 42: 406–412

Bolland MDA, Siddique KHM, Loss SP, Baker MJ (1999) Comparing responses of grain legumes, wheat and canola to applications of superphosphate. Nutr Cycl Agroecosyst 53: 157–175

Bonser AM, Lynch J, Snapp S (1996) Effect of phosphorus deficiency on growth angle of basal roots in *Phaseolus vulgaris*. New Phytol 132: 281–288

Bosse D, Köck M (1998) Influence of phosphate starvation on phosphohydrolases during development of tomato seedlings. Plant Cell Environ 21: 325–332

Bould C, Hewitt EJ, Needham P (1983) Diagnosis of Mineral Disorders in Plants. Volume 1: Principles. HMSO, London

Bradshaw AD, Chadwick MJ, Jowett D, Lodge RW, Snaydon RW (1960) Experimental investigations into the mineral nutrition of several grass species. Part III: phosphate level. J Ecol 48: 631–637

Brinch-Pedersen H, Sorensen LD, Holm PB (2002) Engineering crop plants: getting a handle on phosphate. Trends Plant Sci 7: 118–125

Broadley MR, Bowen HC, Cotterill HL, Hammond JP, Meacham MC, Mead A, White PJ (2004) Phylogenetic variation in the shoot mineral concentration of angiosperms. J Exp Bot 55: 321–336

Bryson R (2005) Proceedings of the International Fertiliser Society 577. Improvements in Farm and Nutrient Management Through Precision Farming. IFS, York

Bucher M (2007) Functional biology of plant phosphate uptake at root and mycorrhiza interfaces. New Phytol 173: 11–26

Burleigh SH, Harrison MJ (1999) The down-regulation of *Mt4*-like genes by phosphate fertilization occurs systemically and involves phosphate translocation to the shoots. Plant Physiol 119: 241–248

Cakmak I, Hengeler C, Marschner H (1994) Partitioning of shoot and root dry matter and carbohydrates in bean plants suffering from phosphorus, potassium and magnesium deficiency. J Exp Bot 45: 1245–1250

Casimiro I, Beekman T, Graham N, Bhalerao R, Zhang H, Casero P, Sandberg G, Bennett MJ (2003) Dissecting *Arabidopsis* lateral root development. Trends Plant Sci 8: 165–171

Casson SA, Lindsey K (2003) Genes and signalling in root development. New Phytol 158: 11–38

Chapin FS, Vitousek PM, Vancleve K (1986) The nature of nutrient limitation in plant-communities. Am Nat 127: 48–58

Chevalier F, Pata M, Nacry P, Doumas P, Rossignol M (2003) Effects of phosphate availability on the root system architecture: large scale analysis of the natural variation between *Arabidopsis* accessions. Plant Cell Environ 26: 1839–1850

Chiou TJ (2007) The role of microRNAs in sensing nutrient stress. Plant Cell Environ 30: 323–332

Chiou TJ, Aung K, Lin SI, Wu CC, Chiang SF, Su CI (2006) Regulation of phosphate homeostasis by microRNA in *Arabidopsis*. Plant Cell 18: 412–421

Ciereszko I, Barbachowska A (2000) Sucrose metabolism in leaves and roots of bean (*Phaseolus vulgaris* L.) during phosphate deficiency. J Plant Physiol 156: 640–644

Clarkson DT, Scattergood CB (1982) Growth and phosphate transport in barley and tomato plants during the development of, and recovery from, phosphate stress. J Exp Bot 33: 865–875

Clarkson DT, Sanderson J, Scattergood CB (1978) Influence of phosphate-stress on phosphate absorption and translocation by various parts of the root system of *Hordeum vulgare* L. (barley). Planta 139: 47–53

Coello P (2002) Purification and characterization of secreted acid phosphatase in phosphorus-deficient *Arabidopsis thaliana*. Physiol Plant 116: 293–298

Cogliatti DH, Clarkson DT (1983) Physiological changes in, and phosphate uptake by potato plants during development of, and recovery from phosphate deficiency. Physiol Plant 58: 287–294

Cohen D (2007) Earth's natural wealth: an audit. New Scientist 2605: 35–41

Coltman RR, Gerloff GC, Gabelman WH (1986) Equivalent stress comparisons among tomato strains differentially tolerant to phosphorus deficiency. J Am Soc Hort Sci 111: 422–426

Coruzzi G, Last R (2000) Amino acids. In: Buchanan BB, Gruissem W, Jones RL (eds), Biochemistry & Molecular Biology of Plants. ASPP, Rockville, MD, pp 358–419

Cruz-Ramírez A, Oropeza-Aburto A, Razo-Hernández F, Ramírez-Chávez E, Herrera-Estrella L (2006) Phospholipase DZ2 plays an important role in extraplastidic galactolipid biosynthesis and phosphate recycling in Arabidopsis roots. Proc Natl Acad Sci USA 103: 6765–6770

Dakora FD, Phillips DA (2002) Root exudates as mediators of mineral acquisition in low-nutrient environments. Plant Soil 245: 35–47

Dechassa N, Schenk MK (2004) Exudation of organic anions by roots of cabbage, carrot, and potato as influenced by environmental factors and plant age. J Plant Nutr Soil Sci 167: 623–629

Dechassa N, Schenk MK, Claassen N, Steingrobe B (2003) Phosphorus efficiency of cabbage (*Brassica oleraceae* L. var. *capitata*), carrot (*Daucus carota* L.), and potato (*Solanum tuberosum* L.). Plant Soil 250: 215–224

Defra (2000) Fertiliser Recommendations for Agricultural and Horticultural Crops (RB209) (7th edition). HMSO, Norwich

Delhaize E, Randall PJ (1995) Characterization of a phosphate-accumulator mutant of *Arabidopsis thaliana*. Plant Physiol 107: 207–213

Delhaize E, Hebb DM, Ryan PR (2001) Expression of a *Pseudomonas aeruginosa* citrate synthase gene is not associated with either enhanced citrate accumulation or efflux. Plant Physiol 125: 2059–2067

Delhaize E, Gruber BD, Ryan PR (2007) The roles of organic anion permeases in aluminium resistance and mineral nutrition. FEBS Lett 581: 2255–2262

Dennis DT, Blakeley SD (2000) Carbohydrate metabolism. In: Buchanan BB, Gruissem W, Jones RL (eds), Biochemistry & Molecular Biology of Plants. ASPP, Rockville, MD, pp 630–675

Dinkelaker B, Hengeler C, Marschner H (1995) Distribution and function of proteoid roots and other root clusters. Bot Acta 108: 183–200

Dörmann P, Benning C (2002) Galactolipids rule in seed plants. Trends Plant Sci 7: 112–118

Drew MC (1975) Comparison of the effects of a localized supply of phosphate, nitrate, ammonium and potassium on the growth of the seminal root system, and the shoot, in barley. New Phytol 75: 479–490

Elser JJ, Fagan WF, Denno RF, Dobberfuhl DR, Folarin A, Huberty A, Interlandi S, Kilham SS, McCauley E, Schulz KL, Siemann EH, Sterner RW (2000a) Nutritional constraints in terrestrial and freshwater food webs. Nature 408: 578–580

Elser JJ, Sterner RW, Gorokhova E, Fagan WF, Markow TA, Cotner JB, Harrison JF, Hobbie SE, Odell GM, Weider LJ (2000b) Biological stoichiometry from genes to ecosystems. Ecol Lett 3: 540–550

Epstein E (1972) Mineral Nutrition of Plants: Principles and Perspectives. Wiley, New York

Essigmann B, Güler S, Narang RA, Linke D, Benning C (1998) Phosphate availability affects the thylakoid lipid composition and the expression of *SQD1*, a gene required for sulfolipid biosynthesis in *Arabidopsis thaliana*. Proc Natl Acad Sci USA 95: 1950–1955

Fageria NK, Baligar VC (1997) Phosphorus use efficiency by corn genotypes. J Plant Nutr 20: 1267–1277

Fageria NK, Baligar VC (1999) Phosphorus-use efficiency in wheat genotypes. J Plant Nutr 22: 331–340

Fageria NK, Wright RJ, Baligar VC (1988) Rice cultivar evaluation for phosphorus use efficiency. Plant Soil 111: 105–109

Fenner M (1986) The allocation of minerals to seeds in *Senecio vulgaris* plants subjected to nutrient shortage. J Ecol 74: 385–392

Fixen PE (2005) Proceedings of the International Fertiliser Society 569. Decision Support Systems in Integrated Crop Nutrient Management. IFS, York

Flügge U-I, Häusler RE, Ludewig F, Fischer K (2003) Functional genomics of phosphate antiport systems of plastids. Physiol Plant 118: 475–482

Föhse D, Claassen N, Jungk A (1988) Phosphorus efficiency of plants. I. External and internal P requirement and P uptake efficiency of different plant species. Plant Soil 110: 101–109

Föhse D, Claassen N, Jungk A (1991) Phosphorus efficiency of plants. II. Significance of root radius, root hairs and cation-anion balance for phosphorus influx in seven plant species. Plant Soil 132: 261–272

Forde B, Lorenzo H (2001) The nutritional control of root development. Plant Soil 232: 51–68

Franco-Zorrilla JM, Martín AC, Solano R, Rubio V, Leyva A, Paz-Ares J (2002) Mutations at *CRE1* impair cytokinin-induced repression of phosphate starvation responses in *Arabidopsis*. Plant J 32: 353–360

Franco-Zorrilla JM, González E, Bustos R, Linhares F, Leyva A, Paz-Ares J (2004) The transcriptional control of plant responses to phosphate limitation. J Exp Bot 55: 285–293

Franco-Zorrilla JM, Martín AC, Leyva A, Paz-Ares J (2005) Interaction between phosphate-starvation, sugar, and cytokinin signaling in Arabidopsis and the roles of cytokinin receptors CRE1/AHK4 and AHK3. Plant Physiol 138: 847–857

Franco-Zorrilla JM, Valli A, Todesco M, Mateos I, Puga MI, Rubio-Somoza I, Leyva A, Weigel D, García JA, Paz-Ares J (2007) Target mimicry provides a new mechanism for regulation of microRNA activity. Nature Genet 39: 1033–1037

Frentzen M (2004) Phosphatidylglycerol and sulfoquinovosyldiacylglycerol: anionic membrane lipids and phosphate regulation. Curr Opin Plant Biol 7: 270–276

Fujii H, Chiou TJ, Lin SI, Aung K, Zhu JK (2005) A miRNA involved in phosphate-starvation response in *Arabidopsis*. Current Biol 15: 2038–2043

Gahoonia TS, Nielsen NE (2004a) Root traits as tools for creating phosphorus efficient crop varieties. Plant Soil 260: 47–57

Gahoonia TS, Nielsen NE (2004b) Barley genotypes with long root hairs sustain high grain yields in low-P field. Plant Soil 262: 55–62

Garten CT (1976) Correlations between concentrations of elements in plants. Nature 261: 686–688

Gaude N, Tippmann H, Flemetakis E, Katinakis P, Udvardi M, Dormann P (2004) The galactolipid digalactosyldiacylglycerol accumulates in the peribacteroid membrane of nitrogen-fixing nodules of soybean and *Lotus*. J Biol Chem 279: 34624–34630

Gaume A, Mächler F, De León C, Narro L, Frossard E (2001) Low-P tolerance by maize (*Zea mays* L.) genotypes: significance of root growth, and organic acids and acid phosphatase root exudation. Plant Soil 228: 253–264

George TS, Richardson AE (2008) Potential and limitations to improving crops for enhanced phosphorus utilization. In: White PJ, Hammond JP (eds), The Ecophysiology of Plant-Phosphorus Interactions. Springer, Dordrecht, The Netherlands, pp 247–270

George TS, Richardson AE, Hadobas PA, Simpson RJ (2004) Characterisation of transgenic *Trifolium subterraneum* L. which expresses *phyA* and releases extracellular phytase: growth and P nutrition in laboratory media and soil. Plant Cell Environ 27: 1351–1361

George TS, Richardson AE, Smith JB, Hadobas PA, Simpson RJ (2005a) Limitations to the potential of transgenic *Trifolium subterraneum* L. plants that exude phytase when grown in soils with a range of organic P content. Plant Soil 278: 263–274

George TS, Simpson RJ, Hadobas PA, Richardson AE (2005b) Expression of a fungal phytase gene in *Nicotiana tabacum* improves phosphorus nutrition in plants grown in amended soil. Plant Biotech J 3: 129–140

George TS, Gregory PJ, Hocking PJ, Richardson AE (2008) Variation in root-associated phosphatase activities in wheat contributes to the utilisation of organic P substrates in vitro, but does not explain differences in the P nutrition of plants when grown in soil. Environ Exp Bot (in press)

Górny AG, Sodkiewicz T (2001) Genetic analysis of the nitrogen and phosphorus utilization efficiencies in mature spring barley plants. Plant Breeding 120: 129–132

Gourley CJP, Allan DL, Russelle MP (1994) Plant nutrient efficiency: a comparison of definitions and suggested improvement. Plant Soil 158: 29–37

Graham JH (2000) Assessing the costs of arbuscular mycorrhizal symbiosis in agroecosystems. In: Podilha GK, Douds DD (eds), Current Advances in Mycorrhizae Research. APS Press, St Paul, MN, pp 127–142

Green DG, Ferguson WS, Warder FG (1973) Accumulation of toxic levels of phosphorus in the leaves of phosphorus-deficient barley. Can J Plant Sci 53: 241–246

Greenwood DJ, Cleaver TJ, Turner MK, Hunt J, Niendorf KB, Loquens SMH (1980) Comparison of the effects of phosphate fertilizer on the yield, phosphate content and quality of 22 different vegetable and agricultural crops. J Agric Sci 95: 457–469

Greenwood DJ, Stellacci AM, Meacham MC, Broadley MR, White PJ (2005) Phosphorus response components of different *Brassica oleracea* genotypes are reproducible in different environments. Crop Sci 45: 1728–1735

Greenwood DJ, Stellacci AM, Meacham MC, Mead A, Broadley MR, White PJ (2006) Relative values of physiological parameters of P response of different genotypes can be measured in experiments with only two P treatments. Plant Soil 281: 159–172

Gregory PJ, George TS (2005) Soil Management for Nutrient Use Efficiency - An Overview. Proceedings 564, International Fertiliser Society, York.

Grime JP (2001) Plant Strategies, Vegetation Processes, and Ecosystem Properties (2nd edition). Wiley, Chichester

Grime JP, Thompson K, Hunt R, Hodgson JG, Cornelissen JHC, Rorison IH, Hendry GAF, Ashenden TW, Askew AP, Band SR, Booth RE, Bossard CC, Campbell BD, Cooper JEL, Davison AW, Gupta PL, Hall W, Hand DW, Hannah MA, Hillier SH, Hodkinson DJ, Jalili A, Liu Z, Mackey JML, Matthews N, Mowforth MA, Neal AM, Reader RJ, Reiling K, Ross-Fraser W, Spencer RE, Sutton F, Tasker DE, Thorpe PC, Whitehouse J (1997) Integrated screening validates primary axes of specialisation in plants. Oikos 79: 259–281

Gunawardena SFBN, Danso SKA, Zapata F (1993) Phosphorus requirement and sources of nitrogen in three soybean (*Glycine max*) genotypes, Bragg, nts 382 and Chippewa. Plant Soil 151: 1–9

Güsewell S (2004) N:P ratios in terrestrial plants: variation and functional significance. New Phytol 164: 243–266

Güsewell S (2005) Nutrient resorption of wetland graminoids is related to the type of nutrient limitation. Funct Ecol 19: 344–354

Güsewell S, Koerselman W (2002) Variation in nitrogen and phosphorus concentrations of wetland plants. Perspect Plant Ecol Evol Syst 5: 37–61

Güsewell S, Bollens U, Ryser P, Klötzli F (2003) Contrasting effects of nitrogen, phosphorus and water regime on first- and second-year growth of 16 wetland plant species. Funct Ecol 11: 754–765

Güsewell S, Bailey KM, Roem WJ, Bedford BL (2005) Nutrient limitation and botanical diversity in wetlands: can fertilisation raise species richness? Oikos 109: 71–80

Hammond JP, White PJ (2008a) Sucrose transport in the phloem: integrating root responses to phosphorus starvation. J Exp Bot 59: 93–109

Hammond JP, White PJ (2008b) Diagnosing phosphorus deficiency in crops. In: White PJ, Hammond JP (eds), The Ecophysiology of Plant-Phosphorus Interactions. Springer, Dordrecht, The Netherlands, pp 225–246

Hammond JP, Bennett MJ, Bowen HC, Broadley MR, Eastwood DC, May ST, Rahn C, Swarup R, Woolaway KE, White PJ (2003) Changes in gene expression in Arabidopsis shoots during phosphate starvation and the potential for developing smart plants. Plant Physiol 132: 578–596

Hammond JP, Broadley MR, White PJ (2004) Genetic responses to phosphorus deficiency. Ann Bot 94: 323–332

Hammond JP, Broadley MR, Craigon DJ, Higgins J, Emmerson Z, Townsend H, White PJ, May ST (2005) Using genomic DNA-based probe-selection to improve the sensitivity of high-density oligonucleotide arrays when applied to heterologous species. Plant Methods 1: 10

Han W, Fang J, Guo D, Zhang Y (2005) Leaf nitrogen and phosphorus stoichiometry across 753 terrestrial plant species in China. New Phytol 168: 377–385

Haran S, Logendra S, Seskar M, Bratanova M, Raskin I (2000) Characterization of Arabidopsis acid phosphatase promoter and regulation of acid phosphatase expression. Plant Physiol 124: 615–626

Harrison MJ (1999) Molecular and cellular aspects of the arbuscular mycorrhizal symbiosis. Annu Rev Plant Physiol Plant Mol Biol 50: 361–389

Härtel H, Dörmann P, Benning C (2000) DGD1-independent biosynthesis of extraplastidic galactolipids after phosphate deprivation in *Arabidopsis*. Proc Natl Acad Sci USA 97: 10649–10654

He Z, Ma Z, Brown KM, Lynch JP (2005) Assessment of inequality of root hair density in *Arabidopsis thaliana* using the Gini coefficient: a close look at the effect of phosphorus and its interaction with ethylene. Ann Bot 95: 287–293

Heathwaite L, Sharpley A, Bechmann M (2003) The conceptual basis for a decision support framework to assess the risk of phosphorus loss at the field scale across Europe. J Plant Nutr Soil Sci 166: 447–458

Hedley MJ, Mortvedt JJ, Bolan NS, Syers JK (1995) Phosphorus fertility management in agroecosystems. In: Tiessen H (ed), Phosphorus in the Global Environment. Transfers Cycles and Management, Wiley, Chichester, pp 59–92

Hermans C, Hammond JP, White PJ, Verbruggen N (2006) How do plants respond to nutrient shortage by biomass allocation? Trends Plant Sci 11: 610–617

Hernández G, Ramírez M, Valdés-López O, Tesfaye M, Graham MA, Czechowski T, Schlereth A, Wandrey M, Erban A, Cheung F, Wu HC, Lara M, Town CD, Kopka J, Udvardi MK, Vance CP (2007) Phosphorus stress in common bean: root transcript and metabolic responses. Plant Physiol 144: 752–767

Hewitt MM, Carr JM, Williamson CL, Slocum RD (2005) Effects of phosphate limitation on expression of genes involved in pyrimidine synthesis and salvaging in *Arabidopsis*. Plant Physiol Biochem 43: 91–99

Hinsinger P (2001) Bioavailability of soil inorganic P in the rhizosphere as affected by root-induced chemical changes: a review. Plant Soil 237: 173–195

Hipps NA, Davies MJ, Dodds P, Buckley GP (2005) The effects of phosphorus nutrition and soil pH on the growth of some ancient woodland indicator plants and their interaction with competitor species. Plant Soil 271: 131–141

Ho MD, Rosas JC, Brown KM, Lynch JP (2005) Root architectural tradeoffs for water and phosphorus acquisition. Funct Plant Biol 32: 737–748

Hoch WA, Zeldin EL, McCown BH (2001) Physiological significance of anthocyanins during autumnal leaf senescence. Tree Physiol 21: 1–8

Hocking P (2001) Organic acids exuded from roots in phosphorus uptake and aluminium tolerance of plants in acid soils. Adv Agron 74: 63–97

Hodge A (2004) The plastic plant: root responses to heterogeneous supplies of nutrients. New Phytol 162: 9–24

Hoffland E (1992) Quantitative evaluation of the role of organic acid exudation in the mobilization of rock phosphate by rape. Plant Soil 140: 279–289
Horst WJ, Kamh M, Jibrin JM, Chude VO (2001) Agronomic measures for increasing P availability to crops. Plant Soil 237: 211–223
Hou XL, Wu P, Jiao FC, Jia QJ, Chen HM, Yu J, Song XW, Yi KK (2005) Regulation of the expression of *OsIPS1* and *OsIPS2* in rice via systemic and local Pi signalling and hormones. Plant Cell Environ 28: 353–364
Hutchings MJ, John EA (2004) The effects of environmental heterogeneity on root growth and root/shoot partitioning. Ann Bot 94: 1–8
Ishikawa S, Adu-Gyamfi JJ, Nakamura T, Yoshihara T, Watanabe T, Wagatsuma T (2002) Genotypic variability in phosphorus solubilising activity of root exudates by pigeon pea grown in low-nutrient environments. Plant Soil 245: 71–81
Jain A, Poling MD, Karthikeyan AS, Blakeslee JJ, Peer WA, Titapiwatanakun B, Murphy AS, Raghothama KG (2007a) Differential effects of sucrose and auxin on localized phosphate deficiency-induced modulation of different traits of root system architecture in Arabidopsis. Plant Physiol 144: 232–247
Jain A, Vasconcelos MJ, Raghothama KG, Sahi SV (2007b) Molecular mechanisms of plant adaptation to phosphate deficiency. Plant Breeding Rev 29: 359–419
Janssens F, Peeters A, Tallowin JRB, Bakker JP, Bekker RM, Fillat F, Oomes MJM (1998) Relationship between soil chemical factors and grassland diversity. Plant Soil 202: 69–78
Jeschke WD, Kirkby EA, Peuke AD, Pate JS, Hartung W (1997) Effects of P deficiency on assimilation and transport of nitrate and phosphate in intact plants of castor bean (*Ricinus communis* L.). J Exp Bot 48: 75–91
Johnson NC, Graham JH, Smith FA (1997) Functioning of mycorrhizal associations along the mutualism-parasitism continuum. New Phytol 135: 575–586
Johnston AE, Lane PW, Mattingly GEG, Poulton PR, Hewitt MV (1986) Effects of soil and fertilizer P on yields of potatoes, sugar beet, barley and winter wheat on a sandy clay loam soil at Saxmundham, Suffolk. J Agric Sci 106: 155–167
Jones DL (1998) Organic acids in the rhizosphere - a critical review. Plant Soil 205: 25–44
Jones DL, Dennis PG, Owen AG, van Hees PAW (2003) Organic acid behavior in soils - misconceptions and knowledge gaps. Plant Soil 248: 31–41
Jouhet J, Maréchal E, Baldan B, Bligny R, Joyard J, Block MA (2004) Phosphate deprivation induces transfer of DGDG galactolipid from chloroplast to mitochondria. J Cell Biol 167: 863–874
Jungk A (2001) Root hairs and the acquisition of plant nutrients from soil. J Plant Nutr Soil Sci 164: 121–129
Karandashov V, Bucher M (2005) Symbiotic phosphate transport in arbuscular mycorrhizas. Trends Plant Sci 10: 22–29
Karthikeyan AS, Varadarajan DK, Jain A, Held MA, Carpita NC, Raghothama KG (2007) Phosphate starvation responses are mediated by sugar signaling in *Arabidopsis*. Planta 225: 907–918
Kerkhoff AJ, Fagan WF, Elser JJ, Enquist BJ (2006) Phylogenetic and growth form variation in the scaling of nitrogen and phosphorus in the seed plants. Am Nat 168: E103–E122
Kirkby EA, Johnston AE (2008) Soil and fertilizer phosphorus in relation to crop nutrition. In: White PJ, Hammond JP (eds), The Ecophysiology of Plant-Phosphorus Interactions. Springer, Dordrecht, The Netherlands, pp 177–223
Klepper B (1992) Development and growth of crop root systems. Adv Soil Sci 19: 1–25
Kobayashi K, Masuda T, Takamiya K-I, Ohta H (2006) Membrane lipid alteration during phosphate starvation is regulated by phosphate signaling and auxin/cytokinin cross-talk. Plant J 47: 238–248
Koyama H, Kawamura A, Kihara T, Hara T, Takita E, Shibata D (2000) Overexpression of mitochondrial citrate synthase in *Arabidopsis thaliana* improved growth on a phosphorus-limited soil. Plant Cell Physiol 41: 1030–1037
Laegreid M, Bøckman OC, Kaarstad O (1999) Agriculture, Fertilizers and the Environment. CABI, Wallingford

Lambers H, Shane MW, Cramer MD, Pearse SJ, Veneklaas EJ (2006) Root structure and functioning for efficient acquisition of phosphorus: matching morphological and physiological traits. Ann Bot 98: 693–713

Lamont BB (2003) Structure, ecology and physiology of root clusters – a review. Plant Soil 248: 1–19

Lee RB, Ratcliffe RG, Southon TE (1990) ^{31}P NMR measurements of the cytoplasmic and vacuolar Pi content of mature maize roots: relationships with phosphorus status and phosphate fluxes. J Exp Bot 41: 1063–1078

Li M, Osaki M, Rao IM, Tadano T (1997) Secretion of phytase from the roots of several plant species under phosphorus-deficient conditions. Plant Soil 195: 161–169

Li M, Welti R, Wang X (2006) Quantitive profiling of Arabidopsis polar glycerolipids in response to phosphorus starvation. Roles of phospholipases Dζ1 and Dζ2 in phosphatidylcholine hydrolysis and digalactosyldiacylglycerol accumulation in phosphorus-starved plants. Plant Physiol 142: 750–761

Liao H, Yan X, Rubio G, Beebe SE, Blair MW, Lynch JP (2004) Genetic mapping of basal root gravitropism and phosphorus acquisition efficiency in common bean. Funct Plant Biol 31: 959–970

Liao H, Wan H, Shaff J, Wang X, Yan X, Kochian LV (2006) Phosphorus and aluminum interactions in soybean in relation to aluminum tolerance. Exudation of specific organic acids from different regions of the intact root system. Plant Physiol 141: 674–684

Liu C, Muchhal US, Raghothama KG (1997) Differential expression of TPSI1, a phosphate starvation-induced gene in tomato. Plant Mol Biol 33: 867–874

Liu J, Li Y, Tong Y, Gao J, Li B, Li J, Li Z (2001) Chromosomal location of genes conferring the tolerance to Pi starvation stress and acid phosphatase (APase) secretion in the genome of rye (*Secale cereale* L.) Plant Soil 237: 267–274

Liu J, Samac DA, Bucciarelli B, Allan DL, Vance CP (2005) Signaling of phosphorus deficiency-induced gene expression in white lupin requires sugar and phloem transport. Plant J 41: 257–268

Lloyd JC, Zakhleniuk OV (2004) Responses of primary and secondary metabolism to sugar accumulation revealed by microarray expression analysis of the *Arabidopsis* mutant, *pho3*. J Exp Bot 55: 1221–1230

Loneragan JF, Asher CH (1967) Responses of plants to phosphate concentration in solution culture. II. Role of phosphate absorption and its relation to growth. Soil Sci 103: 311–318

Loneragan JF, Grunes DL, Welch RM, Aduayi EA, Tengah A, Lazar VA, Cary EE (1982) Phosphorus accumulation and toxicity in leaves in relation to zinc supply. Soil Sci Soc Am J 46: 345–352

López-Bucio J, de la Vega OM, Guevara-García A, Herrera-Estrella L (2000a) Enhanced phosphorus uptake in transgenic tobacco plants that overproduce citrate. Nature Biotech 18: 450–453

López-Bucio J, Nieto-Jacobo MF, Ramírez-Rodríguez V, Herrera-Estrella L (2000b) Organic acid metabolism in plants: from adaptive physiology to transgenic varieties for cultivation in extreme soils. Plant Sci 160: 1–13

López-Bucio J, Hernández-Abreu E, Sánchez-Calderón L, Nieto-Jacobo MF, Simpson J, Herrera-Estrella L (2002) Phosphate availability alters architecture and causes changes in hormone sensitivity in the Arabidopsis root system. Plant Physiol 129: 244–256

López-Bucio J, Cruz-Ramírez A, Herrera-Estrella L (2003) The role of nutrient availability in regulating root architecture. Curr Opin Plant Biol 6: 280–287

López-Bucio J, Hernández-Abreu E, Sánchez-Calderón L, Pérez-Torres A, Rampey RA, Bartel B, Herrera-Estrella L (2005) An auxin transport independent pathway is involved in phosphate stress-induced root architectural alterations in Arabidopsis. Identification of *BIG* as a mediator of auxin pericycle cell activation. Plant Physiol 137: 681–691

Lott JNA, Ockenden I, Raboy V, Batten GD (2000) Phytic acid and phosphorus in crop seeds and fruits: a global estimate. Seed Sci Res 10: 11–33

Lynch JP, Brown KM (2001) Topsoil foraging - an architectural adaptation of plants to low phosphorus availability. Plant Soil 237: 225–237

Ma Z, Baskin TI, Brown KM, Lynch JP (2003) Regulation of root elongation under phosphorus stress involves changes in ethylene responsiveness. Plant Physiol 131: 1381–1390

Malamy JE (2005) Intrinsic and environmental response pathways that regulate root system architecture. Plant Cell Environ 28: 67–77

Malkin R, Niyogi K (2000) Photosynthesis. In: Buchanan BB, Gruissem W, Jones RL (eds), Biochemistry & Molecular Biology of Plants. ASPP, Rockville, MD, pp 568–628

Mamolos AP, Veresoglou DS, Barbayiannis N (1995) Plant species abundance and tissue concentrations of limiting nutrients in low-nutrient grasslands: a test of competition theory. J Ecol 83: 485–495

Manske GGB, Ortiz-Monasterio JI, Van Grinkel M, González R, Rajaram S, Molina E, Vlek PLG (2000) Traits associated with improved P-uptake efficiency in CIMMYT's semidwarf spring bread wheat grown on an acid Andisol in Mexico. Plant Soil 221: 189–204

Marschner H (1995) Mineral Nutrition of Higher Plants (2nd edition). Academic, London

Marschner P (2008) The effect of rhizosphere microorganisms on P uptake by plants. In: White PJ, Hammond JP (eds), The Ecophysiology of Plant-Phosphorus Interactions. Springer, Dordrecht, The Netherlands, pp 165–176

Marschner P, Solaiman Z, Rengel Z (2007) Brassica genotypes differ in growth, phosphorus uptake and rhizosphere properties under P-limiting conditions. Soil Biol Biochem 39: 87–98

Martín AC, del Pozo JC, Iglesias J, Rubio V, Solano R, de la Peña A, Leyva A, Paz-Ares J (2000) Influence of cytokinins on the expression of phosphate starvation responsive genes in *Arabidopsis*. Plant J 24: 559–567

McCrea AR, Trueman IC, Fullen MA (2004) Factors relating to soil fertility and species diversity in both semi-natural and created meadows in the West Midlands of England. Eur J Soil Sci 55: 335–348

McGroddy ME, Daufresne T, Hedin LO (2004) Scaling of C:N:P stoichiometry in forests worldwide: implications of terrestrial Redfield-type ratios. Ecology 85: 2390–2401

Mengel K (1997) Agronomic measures for better utilization of soil and fertilizer phosphates. Eur J Agron 7: 221–233

Mengel K, Kirkby EA (2001) Principles of Plant Nutrition (5th edition). Kluwer, Dordrecht, The Netherlands

Miller SS, Liu J, Allan DL, Menzhuber CJ, Fedorova M, Vance CP (2001) Molecular control of acid phosphatase secretion into the rhizosphere of proteoid roots from phosphorus-stressed white lupin. Plant Physiol 127: 594–606

Mimura T (1999) Regulation of phosphate transport and homeostasis in plant cells. Int Rev Cytol 191: 149–200

Misson J, Raghothama KG, Jain A, Jouhet J, Block MA, Bligny R, Ortet P, Creff A, Somerville S, Rolland N, Doumas P, Nacry P, Herrerra-Estrella L, Nussaume L, Thibaud M-C (2005) A genome-wide transcriptional analysis using *Arabidopsis thaliana* Affymetrix gene chips determined plant responses to phosphate deprivation. Proc Natl Acad Sci USA 102: 11934–11939

Miura K, Rus A, Sharkhuu A, Yokoi S, Karthikeyan AS, Raghothama KG, Baek D, Koo YD, Jin JB, Bressan RA, Yun D-J, Hasegawa PM (2005) The *Arabidopsis* SUMO E3 ligase SIZ1 controls phosphate deficiency responses. Proc Natl Acad Sci USA 102: 7760–7765

Morcuende R, Bari R, Gibon Y, Zheng WM, Pant BD, Bläsing O, Usadel B, Czechowski T, Udvardi MK, Stitt M, Scheible WR (2007) Genome-wide reprogramming of metabolism and regulatory networks of *Arabidopsis* in response to phosphorus. Plant Cell Environ 30: 85–112

Morgan JAW, Bending GD, White PJ (2005) Biological costs and benefits to plant-microbe interactions in the rhizosphere. J Exp Bot 56: 1729–1739

Mudge SR, Smith FW, Richardson AE (2003) Root-specific and phosphate-regulated expression of phytase under the control of a phosphate transporter promoter enables *Arabidopsis* to grow on phytate as a sole phosphorus source. Plant Sci 165: 871–878

Müller R, Nilsson L, Krintel C, Nielsen TH (2004) Gene expression during recovery from phosphate starvation in roots and shoots of *Arabidopsis thaliana*. Physiol Plant 122: 233–243

Müller R, Nilsson L, Nielsen LK, Nielsen TH (2005) Interaction between phosphate starvation signalling and hexokinase-independent sugar sensing in *Arabidopsis* leaves. Physiol Plant 124: 81–90

Müller R, Morant M, Jarmer H, Nilsson L, Nielsen TH (2007) Genome-wide analysis of the Arabidopsis leaf transcriptome reveals interaction of phosphate and sugar metabolism. Plant Physiol 143: 156–171

Nacry P, Canivenc G, Muller B, Azmi A, Van Onckelen H, Rossignol M, Doumas P (2005) A role for auxin redistribution in the responses of the root system architecture to phosphate starvation in Arabidopsis. Plant Physiol 138: 2061–2074

Narang RA, Bruene A, Altmann T (2000) Analysis of phosphate acquisition efficiency in different Arabidopsis accessions. Plant Physiol 124: 1786–1799

Neumann G, Römheld V (1999) Root excretion of carboxylic acids and protons in phosphorus-deficient plants. Plant Soil 211: 121–130

Niklas KJ (2008) Carbon/nitrogen/phosphorus allometric relations across species. In: White PJ, Hammond JP (eds), The Ecophysiology of Plant-Phosphorus Interactions. Springer, Dordrecht, The Netherlands, pp 9–30

Niklas KJ, Owens T, Reich PB, Cobb ED (2005) Nitrogen/phosphorus leaf stoichiometry and the scaling of plant growth. Ecol Lett 8: 636–642

Oberson A, Joner EJ (2005) Microbial turnover of phosphorus in soil. In: Turner BL, Frossard E, Baldwin DS (eds), Organic Phosphorus in the Environment. CABI, Wallingford, pp 133–164

Ohwaki Y, Hirata H (1992) Differences in carboxylic acid exudation among P-starved leguminous crops in relation to carboxylic acid contents in plant tissues and phospholipid level in roots. Soil Sci Plant Nutr 38: 235–243

Oracka T, Łapiñski B (2006) Nitrogen and phosphorus uptake and utilization efficiency in D(R) substitution lines of hexaploid triticale. Plant Breeding 125: 221–224

Osborne LD, Rengel Z (2002) Screening cereals for genotypic variation in efficiency of phosphorus uptake and utilisation. Aust J Agric Res 53: 295–303

O'Toole JC, Bland WL (1987) Genotypic variation in crop plant root systems. Adv Agron 41: 91–145

Oyanagi A (1994) Gravitropic response growth angle and vertical distribution of roots of wheat (*Triticum aestivum* L.). Plant Soil 165: 323–326

Ozanne PG, Keay J, Biddiscombe EF (1969) The comparative applied phosphate requirement of eight annual pasture species. Aust J Biol Sci 20: 809–818

Ozturk L, Eker S, Torun B, Cakmak I (2005) Variation in phosphorus efficiency among 73 bread and durum wheat genotypes grown in a phosphorus-deficient calcareous soil. Plant Soil 269: 69–80

Paterson E, Sim A, Standing D, Dorward M, McDonald AJS (2006) Root exudation from *Hordeum vulgare* in response to localized nitrate supply. J Exp Bot 57: 2413–2420

Paul MJ, Pellny TK (2003) Carbon metabolite feedback regulation of leaf photosynthesis and development. J Exp Bot 54: 539–547

Pearse SJ, Venaklaas EJ, Cawthray G, Bolland MDA, Lambers H (2007) Carboxylate composition of root exudates does not relate consistently to a crop species' ability to use phosphorus from aluminium, iron or calcium phosphate sources. New Phytol 173: 181–190

Pearse SJ, Veneklaas EJ, Cawthray G, Bolland MDA, Lambers H (2008) Rhizosphere processes do not explain variation in P acquisition from sparingly soluble forms of P among *Lupinus albus* accessions. Aust J Agric Res 59: (in press)

Perrott KW, Kear MJ (2000) Laboratory comparison of nutrient release rates from fertilizers. Commun Soil Sci Plant Anal 31: 2007–2017

Petters J, Göbel C, Scheel D, Rosahl S (2002) A pathogen-responsive cDNA from potato encodes a protein with homology to a phosphate starvation-induced phosphatase. Plant Cell Physiol 43: 1049–1053

Plaxton WC, Carswell MC (1999) Metabolic aspects of the phosphate starvation response in plants. In: Lerner HR (ed), Plant Responses to Environmental Stresses: From Phytohormones to Genome Reorganisation. Dekker, New York, pp 349–372

Raghothama KG, Karthikeyan AS (2005) Phosphate acquisition. Plant Soil 274: 37–49

Rao IM, Fredeen AL, Terry N (1990) Leaf phosphate status, photosynthesis, and carbon partitioning in sugar beet. Plant Physiol 92: 29–36

Raven JA (2008) Phosphorus and the future. In: White PJ, Hammond JP (eds), The Ecophysiology of Plant-Phosphorus Interactions. Springer, Dordrecht, The Netherlands, pp 271–283

Rengel Z, Marschner P (2005) Nutrient availability and management in the rhizosphere: exploiting genotypic differences. New Phytol 168: 305–312

Reymond M, Svistoonoff S, Loudet O, Nussaume L, Desnos T (2006) Identification of QTL controlling root growth response to phosphate starvation in *Arabidopsis thaliana*. Plant Cell Environ 29: 115–125

Richardson AE, Hadobas PA, Hayes JE (2001) Extracellular secretion of *Aspergillus* phytase from *Arabidopsis* roots enables plants to obtain phosphorus from phytate. Plant J 25: 641–649

Robinson D (1994) The responses of plants to non-uniform supplies of nutrients. New Phytol 127: 635–674

Rook F, Hadingham SA, Li Y, Bevan MW (2006) Sugar and ABA response pathways and the control of gene expression. Plant Cell Environ 29: 426–434

Rorison IM (1968) The response to phosphorus of some ecologically distinct plant species. I. Growth rates and phosphorus absorption. New Phytol 67: 913–923

Rubio G, Liao H, Yan XL, Lynch JP (2003) Topsoil foraging and its role in plant competitiveness for phosphorus in common bean. Crop Sci 43: 598–607

Rubio G, Sorgonà A, Lynch JP (2004) Spatial mapping of phosphorus influx in bean root systems using digital autoradiography. J Exp Bot 55: 2269–2280

Rubio V, Linhares F, Solano R, Martín AC, Iglesias J, Leyva A, Paz-Ares J (2001) A conserved MYB transcription factor involved in phosphate starvation signaling both in vascular plants and in unicellular algae. Genes Dev 15: 2122–2133

Ryan PR, Delhaize E, Jones DL (2001) Function and mechanism of organic anion exudation from plant roots. Annu Rev Plant Physiol Plant Mol Biol 52: 527–560

Sánchez-Calderón L, López-Bucio J, Chacón-López A, Cruz-Ramírez A, Nieto-Jacobo F, Dubrovsky JG, Herrera-Estrella L (2005) Phosphate starvation induces a determinate developmental program in the roots of *Arabidopsis thaliana*. Plant Cell Physiol 46: 174–184

Sánchez-Calderón L, López-Bucio J, Chacón-López A, Gutiérrez-Ortega A, Hernández-Abreu E, Herrera-Estrella L (2006) Characterization of *low phosphorus insensitive* mutants reveals a crosstalk between low P-induced determinate root development and the activation of genes involved in the adaptation of Arabidopsis to P deficiency. Plant Physiol 140: 879–889

Sanginga N, Lyasse O, Singh BB (2000) Phosphorus use efficiency and nitrogen balance of cowpea breeding lines in a low P soil derived from savanna zone in West Africa. Plant Soil 220: 119–128

Sattelmacher B, Kuene R, Malagamba P, Moreno U (1990) Evaluation of tuber bearing *Solanum* species belonging to different ploidy levels for its yielding potential at low soil fertility. Plant Soil 129: 227–233

Schachtman DP, Reid RJ, Ayling SM (1998) Phosphorus uptake by plants: from soil to cell. Plant Physiol 116: 447–453

Schulze J, Temple G, Temple SJ, Beschow H, Vance CP (2006) Nitrogen fixation by white lupin under phosphorus deficiency. Ann Bot 98: 731–740

Schünmann PHD, Richardson AE, Vickers CE, Delhaize E (2004) Promoter analysis of the barley *Pht1;1* phosphate transporter gene identifiesregions controlling root expression and responsiveness to phosphate deprivation. Plant Physiol 136: 4205–4214

Shane MW, de Vos M, De Roock S, Lambers H (2003) Shoot P status regulates cluster-root growth and citrate exudation in *Lupinus albus* grown with a divided root system. Plant Cell Environ 26: 265–273

Shane MW, McCully ME, Lambers H (2004) Tissue and cellular phosphorus storage during development of phosphorus toxicity in *Hakea prostrata* (Proteaceae). J Exp Bot 55: 1033–1044
Shen J, Li H, Neumann G, Zhang F (2005) Nutrient uptake, cluster root formation and exudation of protons and citrate in *Lupinus albus* as affected by localized supply of phosphorus in a split-root system. Plant Sci 168: 837–845
Shin H, Shin HS, Chen R, Harrison MJ (2006) Loss of *At4* function impacts phosphate distribution between the roots and the shoots during phosphate starvation. Plant J 45: 712–726
Siedow JN, Day DA (2000) Respiration and photorespiration. In: Buchanan BB, Gruissem W, Jones RL (eds), Biochemistry & Molecular Biology of Plants. ASPP, Rockville, MD, pp 676–728
Smith FW, Mudge SR, Rae AL, Glassop D (2003) Phosphate transport in plants. Plant Soil 248: 71–83
Smith SE, Read DJ (2007) Mycorrhizal Symbiosis (3rd edition). Elsevier, London
Smith SE, Smith FA, Jakobsen I (2003) Mycorrhizal fungi can dominate phosphate supply to plants irrespective of growth responses. Plant Physiol 133: 16–20
Smith VH (1992) Effects of nitrogen: phosphorus supply ratios on nitrogen fixation in agricultural and pastoral ecosystems. Biogeochemistry 18: 19–35
Solfanelli C, Poggi A, Loreti E, Alpi A, Perata P (2006) Sucrose-specific induction of the anthocyanin biosynthetic pathway in Arabidopsis. Plant Physiol 140: 637–646
Somerville C, Browse J, Jaworski JG, Ohlrogge JB (2000) Lipids. In: Buchanan BB, Gruissem W, Jones RL (eds), Biochemistry & Molecular Biology of Plants. ASPP, Rockville, MD, pp 456–527
Stalham MA, Allen EJ (2001) Effect of variety, irrigation regime and planting date on depth, rate, duration and density of root growth in the potato (*Solanum tuberosum*) crop. J Agric Sci 137: 251–270
Steen I (1998) Phosphorus availability in the 21st century: management of a non-renewable resource. Phosphorus Potassium 217: 25–31
Stöcklin J, Schweizer K, Körner C (1998) Effects of elevated CO_2 and phosphorus addition on productivity and community composition of intact monoliths from calcareous grassland. Oecologia 116: 50–56
Ström L (1997) Root exudation of organic acids: importance to nutrient availability and the calcifuge and calcicole behaviour of plants. Oikos 80: 459–466
Su J, Xiao Y, Li M, Liu Q, Li B, Tong Y, Jia J, Li Z (2006) Mapping QTLs for phosphorus-deficiency tolerance at wheat seedling stage. Plant Soil 281: 25–36
Subbarao GV, Ae N, Otani T (1997) Genotypic variation in the iron- and aluminium-phosphate solubilising activity of pigeon pea root exudates under P deficient conditions. Soil Sci Plant Nutr 43: 295–305
Sunkar R, Zhu J-K (2004) Novel and stress-regulated microRNAs and other small RNAs from Arabidopsis. Plant Cell 16: 2001–2019
Svistoonoff S, Creff A, Reymond M, Sigoillot-Claude C, Ricaud L, Blanchet A, Nussaume L, Desnos T (2007) Root tip contact with low-phosphate media reprograms plant root architecture. Nat Genet 39: 792–796
Tadano T, Sasaki H (1991) Secretion of acid phosphatase by the roots of several crop species under phosphorus-deficient conditions. Soil Sci Plant Nutr 37: 129–140
Teng S, Keurentjes J, Bentsink L, Koornneef M, Smeekens S (2005) Sucrose-specific induction of anthocyanin biosynthesis in Arabidopsis requires the MYB75/PAP1 gene. Plant Physiol 139: 1840–1852
Tesfaye M, Liu J, Allan DL, Vance CP (2007) Genomic and genetic control of phosphate stress in legumes. Plant Physiol 144: 594–603
Tessier JT, Raynal DJ (2003) Use of nitrogen to phosphorus ratios in plant tissue as an indicator of nutrient limitation and nitrogen saturation. J Appl Ecol 40: 523–534
Thomas C, Sun Y, Naus K, Lloyd A, Roux S (1999) Apyrase functions in plant phosphate nutrition and mobilizes phosphate from extracellular ATP. Plant Physiol 119: 543–551
Thompson K, Parkinson JA, Band SR, Spencer RE (1997) A comparative study of leaf nutrient concentrations in a regional herbaceous flora. New Phytol 136: 679–689

Thornton B, Paterson E, Midwood AJ, Sim A, Pratt SM (2004) Contribution of current carbon assimilation in supplying root exudates of *Lolium perenne* measured using steady-state ^{13}C labeling. Physiol Plant 120: 434–441

Tian G, Carsky RJ, Kang BT (1998) Differential phosphorus responses of leguminous cover crops on soil with variable history. J Plant Nutr 21: 1641–1653

Ticconi CA, Abel S (2004) Short on phosphate: plant surveillance and countermeasures. Trends Plant Sci 9: 548–555

Ticconi CA, Delatorre CA, Lahner B, Salt DE, Abel S (2004) Arabidopsis *pdr2* reveals a phosphate-sensitive checkpoint in root development. Plant J 37: 801–814

Tomscha JL, Trull MC, Deikman J, Lynch JP, Guiltinan MJ (2004) Phosphatase under-producer mutants have altered phosphorus relations. Plant Physiol 135: 334–345

Trehan SP, Sharma RC (2003) External phosphorus requirement of different potato (*Solanum tuberosum*) cultivars resulting from different internal requirements and uptake efficiencies. Indian J Agric Sci 73: 54–56

Turner BL (2007) Inositol phosphates in soil: amounts, forms and significance of the phosphorylated inositol stereoisomers. In: Turner BL, Richardson AE, Mullaney EJ (eds), Inositol Phosphates: Linking Agriculture and the Environment. CABI, Wallingford, pp 186–206

Uhde-Stone C, Zinn KE, Ramirez-Yáñez M, Li A, Vance CP, Allan DL (2003) Nylon filter arrays reveal differential gene expression in proteoid roots of white lupin in response to phosphorous deficiency. Plant Physiol 131: 1064–1079

Vádez V, Lasso JH, Beck DP, Drevon JJ (1999) Variability of N_2-fixation in common bean (*Phaseolus vulgaris* L.) under P deficiency is related to P use efficiency. Euphytica 106: 231–242

Vance CP, Uhde-Stone C, Allan DL (2003) Phosphorus acquisition and use: critical adaptations by plants for securing a nonrenewable resource. New Phytol 157: 423–447

Van ura V, Hovadík A (1965) Root exudates of plants. II. Composition of root exudates of some vegetables. Plant Soil 22: 21–32

Wasaki J, Yonetani R, Kuroda S, Shinano T, Yazaki J, Fujii F, Shimbo K, Yamamoto K, Sakata K, Sasaki T, Kishimoto N, Kikuchi S, Yamagishi M, Osaki M (2003) Transcriptomic analysis of metabolic changes by phosphorus stress in rice plant roots. Plant Cell Environ 26: 1515–1523

Wasaki J, Shinano T, Onishi K, Yonetani R, Yazaki J, Fujii F, Shimbo K, Ishikawa M, Shimatani Z, Nagata Y, Hashimoto A, Ohta T, Sato Y, Miyamoto C, Honda S, Kojima K, Sasaki T, Kishimoto N, Kikuchi S, Osaki M (2006) Transcriptomic analysis indicates putative metabolic changes caused by manipulation of phosphorus availability in rice leaves. J Exp Bot 57: 2049–2059

Wassen MJ, Olde Venterink H, Lapshina ED, Tanneberger F (2005) Endangered plants persist under phosphorus limitation. Nature 437: 547–550

White PJ (2003) Ion transport. In: Thomas B, Murphy DJ, Murray BG (eds), Encyclopaedia of Applied Plant Sciences. Academic, London, pp 625–634

White PJ, Broadley MR, Greenwood DJ, Hammond JP (2005a) Genetic Modifications to Improve Phosphorus Acquisition by Roots. Proceedings 568, International Fertiliser Society, York

White PJ, Broadley MR, Hammond JP, Thompson AJ (2005b) Optimising the potato root system for phosphorus and water acquisition in low-input growing systems. Aspects Appl Biol 73: 111–118

Wissuwa M (2003) How do plants achieve tolerance to phosphorus deficiency? Small causes with big effects. Plant Physiol 133: 1947–1958

Wissuwa M (2005) Combining a modeling with a genetic approach in establishing associations between genetic and physiological effects in relation to phosphorus uptake. Plant Soil 269: 57–68

Wissuwa M, Gamat G, Ismail AM (2005) Is root growth under phosphorus deficiency affected by source or sink limitations? J Exp Bot 56: 1943–1950

Wright IJ, Reich PB, Cornelissen JHC, Falster DS, Garnier E, Hikosaka K, Lamont BB, Lee W, Oleksyn J, Osada N, Poorter H, Villar R, Warton DI, Westoby M (2005) Assessing the generality of global leaf trait relationships. New Phytol 166: 485–496

Wu P, Ma L, Hou X, Wang M, Wu Y, Liu F, Deng XW (2003) Phosphate starvation triggers distinct alterations of genome expression in Arabidopsis roots and leaves. Plant Physiol 132: 1260–1271

Xiao K, Harrison MJ, Wang Z-Y (2005) Transgenic expression of a novel *M. truncatula* phytase gene results in improved acquisition of organic phosphorus by *Arabidopsis*. Planta 222: 27–36

Xiao K, Katagi H, Harrison M, Wang ZY (2006) Improved phosphorus acquisition and biomass production in Arabidopsis by transgenic expression of a purple acid phosphatase gene from *M. truncatula*. Plant Sci 170: 191–202

Yan X, Beebe SE, Lynch JP (1995a) Genetic variation for phosphorus efficiency of common bean in contrasting soil types. II. Yield response. Crop Sci 35: 1094–1099

Yan X, Lynch JP, Beebe SE (1995b) Genetic variation for phosphorus efficiency of common bean in contrasting soil types. I. Vegetative response. Crop Sci 35: 1086–1093

Yan X, Liao H, Beebe SE, Blair MW, Lynch JP (2004) QTL mapping of root hair and acid exudation traits and their relationship to phosphorus uptake in common bean. Plant Soil 265: 17–29

Yan Y, Wu P, Ling H, Xu G, Xu F, Zhang Q (2006) Plant nutriomics in China: an overview. Ann Bot 98: 473–482

Yu B, Xu C, Benning C (2002) Arabidopsis disrupted in SQD2 encoding sulfolipid synthase is impaired in phosphate-limited growth. Proc Natl Acad Sci USA 99: 5732–5737

Zhang K, Greenwood DJ, White PJ, Burns IG (2007) A dynamic model for the combined effects of N, P and K fertilizers on yield and mineral composition; description and experimental test. Plant Soil 298: 81–98

Zhang YJ, Lynch JP, Brown KM (2003) Ethylene and phosphorus availability have interacting yet distinct effects on root hair development. J Exp Bot 54: 2351–2361

Zhao J, Fu J, Liao H, He Y, Nian H, Hu Y, Qiu L, Dong Y, Yan X (2004) Characterisation of root architecture in an applied core collection for phosphorus efficiency of soybean germplasm. Chinese Sci Bull 49: 1611–1620

Zhu J, Lynch JP (2004) The contribution of lateral rooting to phosphorus acquisition efficiency in maize (*Zea mays*) seedlings. Funct Plant Biol 31: 949–958

Zhu J, Kaeppler SM, Lynch JP (2005) Topsoil foraging and phosphorus acquisition efficiency in maize (*Zea mays*). Funct Plant Biol 32: 749–762

Zhu J, Mickelson SM, Kaeppler SM, Lynch JP (2006) Detection of quantitative trait loci for seminal root traits in maize (*Zea mays* L.) seedlings grown under differential phosphorus levels. Theor Appl Genet 113: 1–10

Zimmermann P, Zardi G, Lehmann M, Zeder C, Amrhein N, Frossard E, Bucher M (2003) Engineering the root-soil interface via targeted expression of a synthetic phytase gene in trichoblasts. Plant Biotech J 1: 353–360

Zimmermann P, Regierer B, Kossmann J, Frossard E, Amrhein N, Bucher M (2004) Differential expression of three purple acid phosphatases from potato. Plant Biol 6: 519–528

Chapter 5
ROOT STRATEGIES FOR PHOSPHORUS ACQUISITION

Jonathan P. Lynch and Kathleen M. Brown

LOW SOIL P AVAILABILITY IS A PRIMARY CONSTRAINT TO PLANT PRODUCTIVITY

Soil infertility is a primary constraint to plant productivity over the majority of the earth's land surface. Nitrogen is often limiting in young soils of the temperate zone, while phosphorus (P) is a primary limitation in most forests, weathered soils and the humid tropics, which support the majority of terrestrial plant biomass (Walker 1965; Lynch and Deikman 1998; Figure 5.1). Low soil P availability is caused by several factors, including the reactivity of orthophosphate (Pi) with common soil constituents such as Fe and Al oxides, resulting in compounds of limited bioavailability, especially as soil weathering progresses, and the open-ended P cycle that tends towards depletion. Human activity in many managed ecosystems has reduced P bioavailability further through topsoil erosion, acidification, and nutrient mining, especially in developing countries (Hartemink 2003). Approximately 50% of the agricultural soils in the world have been degraded significantly by human activity, including 75% of the agricultural soils of Africa (Oldeman *et al.* 1991; Wood *et al.* 2000). Replenishment of soil P reserves through fertilization is common in developed countries, but the economic sustainability of this practice is in question, as economically recoverable P reserves are estimated to be 50% depleted by the middle of this century (Steen 1998; Abelson 1999). In many developing countries, especially in Africa, fertilizer use is negligible (World Bank 2004), and the productivity of many of these agroecosystems is P-limited. The development of crops and cropping systems with greater productivity on soils of low P bioavailability would substantially improve global food security (Lynch 2007). The response of terrestrial ecosystems to global climate change will depend on interactions of climate change variables with edaphic limitations to plant productivity, including P (Lynch and St. Clair 2004). The adaptation of plants to low P availability is therefore of considerable interest in both basic and applied plant biology.

P.J. White, J.P. Hammond (eds.) *The Ecophysiology of Plant-Phosphorus Interactions*,

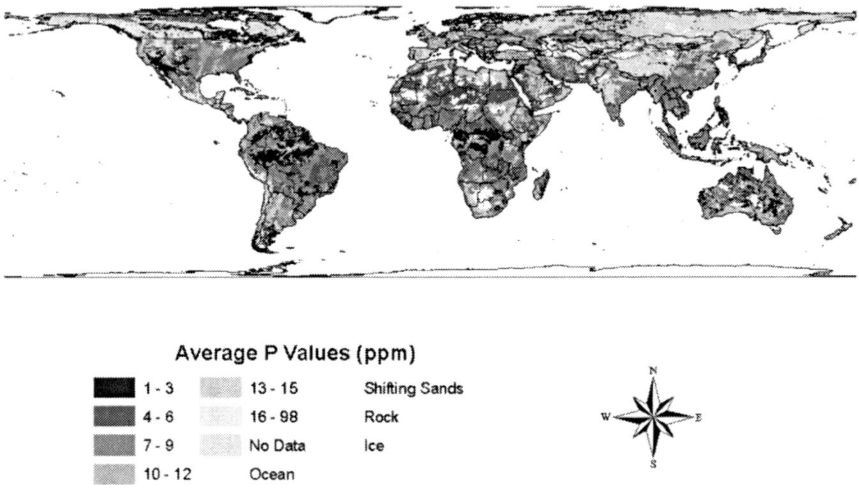

Fig. 5.1 Map of soil phosphorus availability. (Jaramillo and Lynch, unpublished, 2008.)

PHOSPHORUS IS AN IMMOBILE SOIL NUTRIENT

The large majority of P taken up by plant roots is in the form of orthophosphate (Pi; Smith *et al.* 2003). Because of its reactivity with many chemical and biological components of the soil, only a small part of the total P content of the soil occurs as Pi (Comerford 1998), and movement of Pi in the soil solution is slow. Transport of Pi through mass flow of water to the root surface is negligible, and diffusion of Pi in soils is typically in the order of fractions of a millimeter per day (Barber 1962, 1995). Phosphorus uptake creates 'depletion zones' around the root with little bioavailable P, that will only slowly recharge through diffusion and mineralization (Hinsinger *et al.* 2005). This makes P the most immobile, and often the least bioavailable, of the macronutrients. The immobility of P in soil typically results in large spatial variation in P bioavailability. Mature soils display large vertical gradients in P content and bioavailability, caused by continual deposition of P on the soil surface in shoot residues, and greater organic matter content, biological activity, and P turnover in the topsoil (Lynch and Brown 2001). Spatial variation in P bioavailability may also be created by soil fauna, especially ants and termites. A fundamental challenge in plant acquisition of P is the need to explore a heterogeneous substrate and the need to place roots or root symbionts within millimeters of fresh P sources that have not already been depleted.

ROOT TRAITS ARE KEY ADAPTATIONS TO LOW P AVAILABILITY

We focus here on root strategies for P-acquisition. By 'strategy' we mean traits or sets of traits that have adaptive value in acquiring P. By 'trait' we mean a 'phene' as a distinct element of an organism's phenotype. Just as a genotype is comprised of many distinct genes, a phenotype is comprised of many distinct phenes. Phenes are generally more abstract and indistinct than genes, and have received less research attention, but phenes rather than genes determine fitness. Phenes rather than genes are subject to selection in plant domestication, and even today, are the basis of the vast majority of crop breeding. As an example of a phene that is important for P-acquisition, root hair length is a phene controlled by multiple genes. Intraspecific variation for root hair length is an important determinant of P-acquisition (see below). The adaptive value of phenes may be affected by interactions with other phenes. In the case of root hairs, root hair length has strong synergism with root hair density (i.e. number of root hairs per unit root epidermal surface area) for P acquisition (Ma *et al.* 2001b). This type of synergism is useful to consider in the context of an 'integrated phenotype', i.e. a set of phenes whose interactions determine fitness in a particular environment. We consider a phene or integrated phenotype that enhances P-acquisition a 'strategy'. The identification of adaptive strategies, and an understanding of the physiological and ecological tradeoffs associated with them, are essential in understanding plant adaptation to P-limited ecosystems. This is especially important for breeding more P-efficient crops, an enterprise of great importance for global food security (Lynch 2007).

Given the low bioavailability and mobility of P in most soils, the ability of root systems to explore effectively the soil and exploit the rhizosphere, at minimal metabolic cost, is essential to plant fitness. Roots display a variety of adaptations to low P availability (Lynch and Brown 2006), including mycorrhizal symbioses (Smith and Read 1997), root hair elongation and proliferation (Bates and Lynch 1996; Ma *et al.* 2001a,b), rhizosphere modification through secretion of organic acids (Jones 1998; Ryan *et al.* 2001), protons (Hinsinger 2001), and phosphatases (Hayes *et al.* 1999), and modification of root architecture to maximize P acquisition efficiency (Figure 5.2; Lynch and Brown 2001; Lynch 2005; George and Richardson 2008; Kirkby and Johnston 2008; Vance 2008; White and Hammond 2008). Phosphorus-deficient plants typically have higher root to shoot ratios than high-P plants, either because of allometric relationships (Niklas 1994) or because of increased biomass allocation to roots (Gutschick 1993; Nielsen *et al.* 2001). Increased root growth is obviously beneficial for P-acquisition, but can slow overall plant growth because of the increased respiratory burden of root tissue (Van der Werf *et al.* 1992; Hansen *et al.* 1998; Nielsen *et al.* 1998, 2001; Lynch and St. Clair 2004).

These topics encompass several active fields of research with large literatures that cannot be adequately reviewed here. Several recent reviews summarize plant responses to low P availability, emphasizing cellular and biochemical or molecular

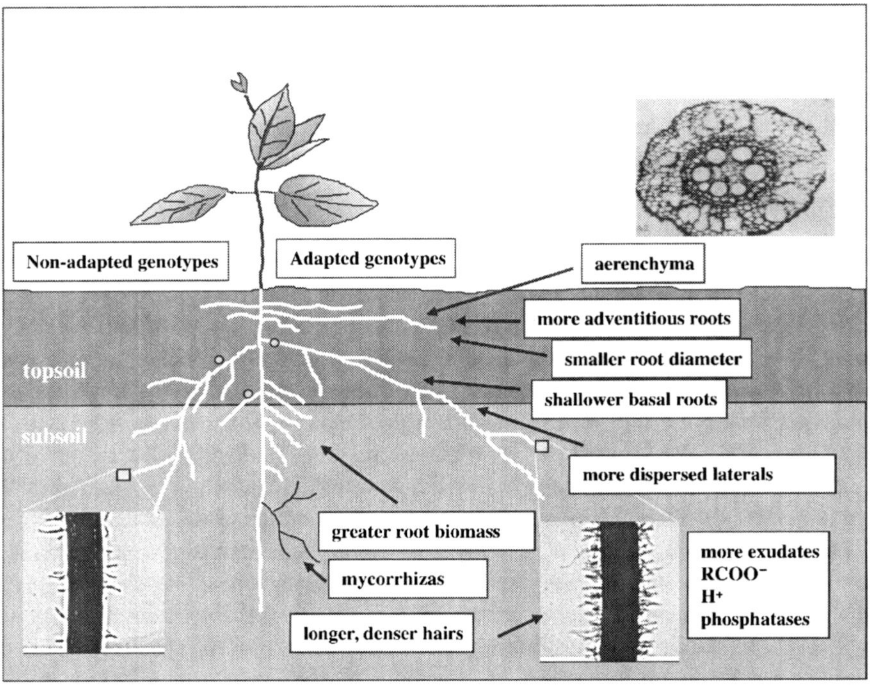

Fig. 5.2 Changes in root architecture, morphology, and anatomy associated with adaptation to low phosphorus in common bean

processes (Abel *et al.* 2002; Vance *et al.* 2003; Ticconi and Abel 2004; Tesfaye *et al.* 2007), and specific adaptations to low P availability, such as cluster roots (Diem *et al.* 2000; Lamont 2003; Neumann and Martinoia 2002; Shane and Lambers 2005; Lambers *et al.* 2006), mycorrhizas (Harrison 1999; Smith *et al.* 2003; Oldroyd *et al.* 2005) and rhizosphere modification (Hinsinger *et al.* 2003; Kochian *et al.* 2005). Our focus here is on strategies of the roots themselves, at the scale of the organ and organism, with particular attention to the functional importance of phenes involved with root growth and architecture for P-acquisition. While this subject has received less research attention than mycorrhizas, rhizosphere modification, cluster roots, and genetic responses to P stress, root growth and architecture have overarching importance at the organismal scale by determining the extent and localization of soil exploration, and by locating the expression of other root traits in specific soil domains.

BIOMASS ALLOCATION TO ROOTS

A common response of plants to P-deficiency is to increase their root to shoot dry-weight (DW) ratio, resulting from a greater inhibition of shoot growth than root growth (Whiteaker *et al.* 1976; Lynch *et al.* 1991; Mollier and Pellerin 1999).

A portion of this apparent change in root to shoot DW ratio is allometric, i.e. root: shoot ratios normally decline with growth, and since plants supplied with low P grow more slowly, their root to shoot ratios are higher at a given plant age. However, when this factor is eliminated by comparison of allometric coefficients among plants grown at different P levels, some genotypes show a greater allometric coefficient (larger increase in root DW relative to increases in shoot DW) with low P, while others do not. In the study of Nielsen *et al.* (2001), genotypes of common bean that were P-efficient (less yield depression under low P) maintained a higher root to shoot ratio (higher allometric coefficient) with continued growth, supporting the idea that root growth is valuable for P acquisition. Low P availability reduces leaf appearance, leaf expansion, and shoot branching (Radin and Eidenbock 1984; Lynch *et al.* 1991). Among annual plants, P-stress decreases shoot growth in dicots more than in monocots, possibly because of differences in leaf morphology (Halsted and Lynch 1996).

Root growth is a key trait for optimizing the efficiency of P acquisition and use in plants (Lynch 1995; Manske *et al.* 2000). Low P availability changes the distribution of growth among various root types. In common bean, growth of main root (primary and basal root) axes was maintained under low P, while initiation of lateral roots is reduced, so that lateral root density declines (Borch *et al.* 1999). Mean lateral root length was unaffected. In experiments with maize subjected to P-starvation, axile (seminal and nodal) root elongation and lateral root density were unaffected, but lateral root elongation was first promoted slightly, then severely retarded, as P-starvation proceeded (Mollier and Pellerin 1999). Initiation of new axile roots also ceased after six days of P-starvation. The maintenance of main root elongation in maize and bean could be interpreted as exploratory behavior, allowing these roots to grow maximally until they encounter localized patches of higher P availability. When the main root of a P-deficient plant encounters a patch of higher P availability, lateral roots may proliferate within the patch (see below; Robinson 2005). In Arabidopsis seedlings deprived of added P, lateral root number was also reduced, but in this case the remaining lateral roots elongated at the expense of the primary root (Williamson *et al.* 2001; Lopez-Bucio *et al.* 2002; Al-Ghazi *et al.* 2003). One reason for the discrepancy may be that Arabidopsis plants lack main roots other than the primary root (i.e. they have no root type analogous to the basal roots of legumes or the seminal roots of grasses), so that a subset of lateral roots must take over the functions of the other main roots in species with more complex root systems. We have observed genetic variation for the effect of buffered low P on lateral root length and number in maize, with some genotypes showing an increase and others showing a decrease in these variables (Zhu and Lynch 2004; Zhu *et al.* 2005a). Genotypes with increased or sustained lateral root development under P-deficiency had superior ability to acquire P and maintain growth.

In common bean, some genotypes preferentially increase growth of adventitious roots, which have the advantages of low construction cost and location in topsoil (see below; Miller *et al.* 2003; Ochoa *et al.* 2006). Adventitious rooting has long been associated with adaptation to waterlogging (Vartapetian and Jackson 1997) and has recently been associated with root rot resistance (Snapp *et al.* 2003; Roman-Aviles *et al.* 2004) and responses to root herbivory (Riedell and Reese

1999). In some crops, such as maize, a high proportion of the mature root system consists of adventitious roots, so prevention of adventitious rooting reduces water uptake even in well-watered plants (Jeschke *et al.* 1997). Under low P conditions, adventitious root development may be delayed or reduced, primarily as a result of overall growth inhibition (Pellerin *et al.* 2000; Miller *et al.* 2003; Ochoa *et al.* 2006). In some genotypes of bean, the maintenance of adventitious root formation, when overall growth is inhibited by P-deficiency, results in an increased proportion of root length in the adventitious root system (Miller *et al.* 2003). This characteristic was associated with plant P-efficiency traits in soils with poor P availability.

ROOT TRAITS AFFECTING SOIL EXPLORATION

Phosphorus distribution is highly heterogeneous in most soils, generally being greatest in surface horizons and decreasing with depth (Chu and Chang 1966; Anderson 1980; Pothuluri *et al.* 1986). The availability of soil P is also highly heterogeneous because of spatial heterogeneity of pH, eH, microbial activity, temperature, etc. (Barber 1995). Phosphorus mobilization and uptake by the root creates zones of P depletion that vary sharply on the scale of millimeters (Joner *et al.* 1995; Hinsinger *et al.* 2005). As a result of the development of P depletion zones around existing roots, P acquisition is highly dependent on continued root growth and exploration of new soil domains that have not yet been depleted of P (Barber 1995).

Topsoil exploration

Since the topsoil is generally the soil stratum with greatest P bioavailability, adaptation to low soil P availability is associated with the extent of topsoil foraging among genotypes of maize and bean (Bonser *et al.* 1996; Ge *et al.* 2000; Liao *et al.* 2001; Ho *et al.* 2005; Zhu *et al.* 2005b). We recently reviewed the importance of topsoil exploration for plant P-efficiency traits (Lynch and Brown 2001).

Architectural traits associated with enhanced topsoil foraging in common bean include shallower growth of basal roots, enhanced adventitious rooting, and greater dispersion of lateral roots (Figure 5.2). There are several lines of evidence that shallower basal root growth enhances topsoil foraging and thereby P acquisition efficiency. The geometric simulation model *SimRoot* was used to model the effect of changing basal root gravitropism on inter-root competition for P (Ge *et al.* 2000). This study showed that in soils with uniform P distribution, shallower root systems explored more soil per unit of root biomass than deeper systems, because shallower systems have more dispersed basal roots and therefore less inter-root competition, which occurs when neighboring roots have overlapping P depletion zones (Ge *et al.* 2000). In stratified soils with more P in the topsoil, the simulations showed that shallower root systems acquired more P than deep ones, by concentrating root

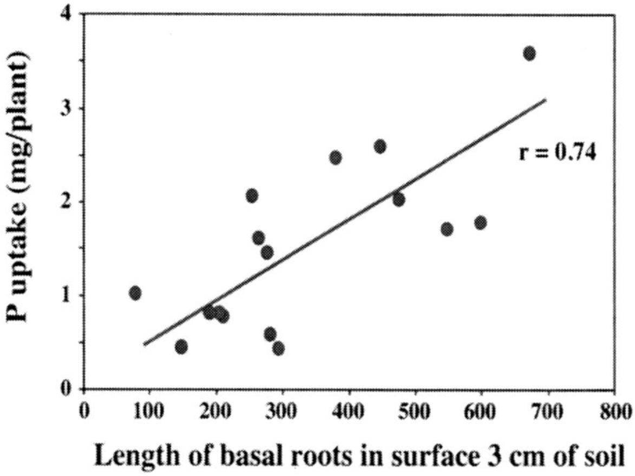

Fig. 5.3 Relationship between root shallowness and phosphorus uptake in recombinant inbred lines of common bean (*Phaseolus vulgaris* L.) growing in low-P soil. (Data extracted from Liao *et al.* (2001). With permission.)

foraging in the topsoil (Ge *et al.* 2000). These modeling results are supported by the significant correlation of basal root growth angle of young bean genotypes grown in growth pouches, with their yield in field trials in low P tropical soils (Bonser *et al.* 1996). In comparisons of individual plants grown in pots of soil, genotypes with shallower basal roots had greater Pi uptake than those with deeper root systems (Figure 5.3; Liao *et al.* 2001). Bean genotypes with shallower basal roots had superior growth in a low P field trial in Honduras (Ho *et al.* 2005). Genetic analysis of bean lines segregating for basal root shallowness showed co-segregation of QTL for root shallowness and Pi uptake in the field in Colombia (Liao *et al.* 2004). In maize, genotypes with shallower seminal roots (analogous to basal roots in dicots) had superior growth in low P soils in the field and glasshouse (Zhu *et al.* 2005b). Similar results have been observed with soybean (Zhao et al. 2004). It therefore appears that basal root shallowness is an important trait for topsoil foraging and P acquisition efficiency in annual crops.

In crops such as bean, adventitious roots emerge from the subterranean portion of the hypocotyl and grow horizontally through the topsoil. Adventitious rooting is therefore an important element of topsoil exploration by the root system. Bean genotypes differ in their extent of adventitious rooting and in the regulation of this trait by P (Miller *et al.* 2003; Ochoa *et al.* 2006). As with basal root gravitropism, genotypic and P-induced adventitious rooting vary widely, from virtually no adventitious rooting in some conditions to dozens of adventitious roots in others (Miller *et al.* 2003; Ochoa *et al.* 2006). A field study in a tropical low P soil showed that bean genotypes with greater growth and Pi uptake had more adventitious rooting relative to basal root growth than did P-inefficient genotypes (Miller *et al.* 2003). Adventitious roots may have several benefits for topsoil exploration. Obviously, their horizontal growth concentrates their foraging activity in the topsoil. Other advantages may relate to the anatomical and morphological differences between adventitious roots and basal roots. In bean, adventitious roots have greater specific

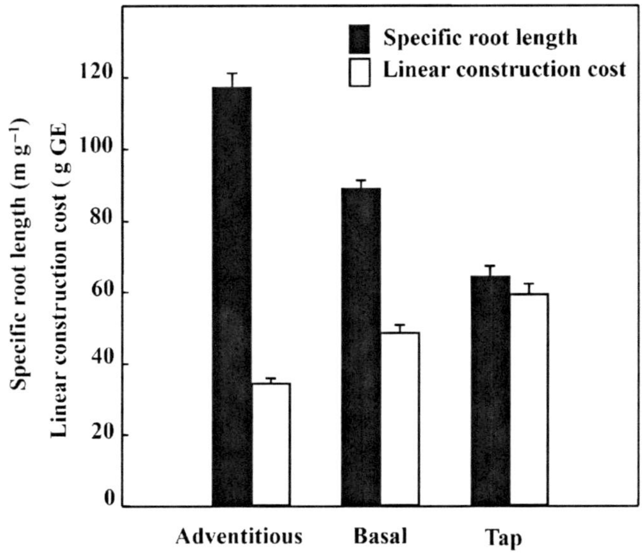

Fig. 5.4 Specific root length and linear construction cost (in glucose equivalents per cm root length) of root classes of common bean (*Phaseolus vulgaris* L.). Each bar is the mean of four replicates, error bars = SEM. (From Lynch and Ho (2005). With permission.)

root length (root length per unit root mass) than other root types (Figure 5.4). This is advantageous for topsoil exploration because it enables the plant to explore a larger volume of soil per unit of metabolic investment in root tissue (Lynch and St. Clair 2004). Also, adventitious roots may have a greater abundance of aerenchyma than other root types (Vartapetian and Jackson 1997), which may be a mechanism of reducing the metabolic costs of soil exploration (see below). Finally, adventitious roots also have less lateral branching than basal roots, which would again serve to disperse root foraging over larger soil volumes for a given metabolic investment (Miller *et al.* 2003).

Reducing root metabolic costs of soil exploration

A number of studies have shown that the metabolic costs of soil exploration by root systems (which generally include mycorrhizal symbioses) are quite substantial, and can exceed 50% of daily photosynthesis (Lambers *et al.* 2002). Following the economic paradigm of plant resource allocation (Bloom *et al.* 1985), we use the term "cost" to denote metabolic investment, including the production and maintenance of tissues, which is measurable in units of carbon (C is a convenient 'currency' for our analysis- other 'currencies', including P itself, may also be useful in some contexts; Koide and Elliott 1989; Snapp *et al.* 1995; Koide *et al.* 2000).

Plant resource allocation to root growth typically increases under nutrient stress, and therefore the metabolic costs of root growth can be a significant component of plant fitness and adaptation under nutrient stress. All else being equal, a plant that is able to acquire a limiting soil resource at reduced metabolic cost will have superior fitness, because it will have more metabolic resources available for defense, growth, and reproduction.

The importance of root C costs in plant adaptation to low P is illustrated by our work with common bean. In bean, low P availability increases the fraction of daily photosynthate respired by roots by 75% in both P-efficient and P-inefficient genotypes (Nielsen *et al.* 1998, 2001). However, P-efficient genotypes had greater root growth per unit root respiration than did P-inefficient genotypes (Figure 5.5; Nielsen *et al.* 2001), which enabled P-efficient genotypes to develop more than twice as much root biomass at low P than the P-inefficient genotypes. Phosphorus-stress slightly increased the specific respiration rate (i.e. respiration per unit biomass) of roots of the P-inefficient genotype, but halved the respiration rate of roots of the P-efficient genotype (Lynch and St. Clair 2004). Thus, adaptation to low P availability in this species is associated with the ability to explore the soil at minimal metabolic cost. We refer to the metabolic cost of P acquisition as 'P acquisition efficiency', or PAE.

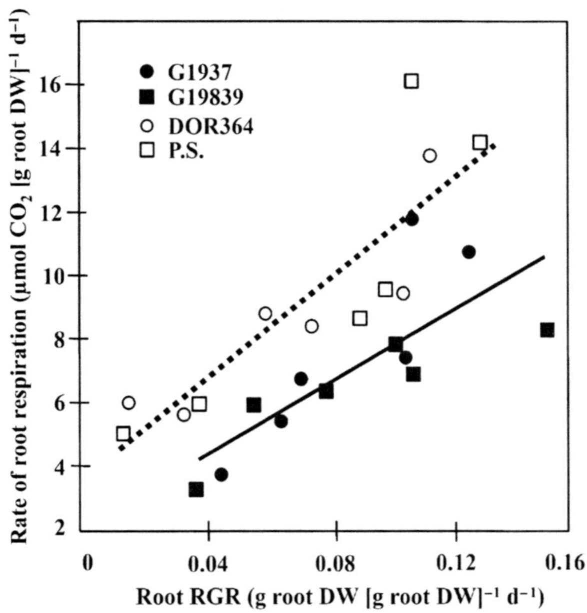

Fig. 5.5 Relationship of root respiration and root relative growth rate in four genotypes of common bean (*Phaseolus vulgaris* L.). Open symbols represent genotypes that are P-inefficient (i.e. have poor growth in low-P media), closed symbols represent genotypes that are P-efficient. (From Nielsen *et al.* (2001). With permission)

Table 5.1 Maintenance respiration dominates root respiration under low P in common bean. R_g = growth respiration, R_{iu} = ion uptake respiration, R_m = maintenance respiration. (Lynch and Ho 2005.)

		% of total root respiration		
Days after planting	Phosphorus level	R_g	R_{iu}	R_m
14	High	29	14	57
	Low	19	9	72
28	High	25	11	64
	Low	6	4	89

Several types of root traits could alter the relationship of root growth and root C cost. Geometric modeling suggests that root architecture can alter the C cost of soil exploration by regulating the extent of root competition within and among root systems (Ge *et al.* 2000; Rubio *et al.* 2001). The importance of root architecture for interplant competition for P was confirmed in field studies (Rubio *et al.* 2003). Morphological traits such as root hairs could enhance P acquisition at minimal root C cost (Bates and Lynch 2000a,b; Ma *et al.* 2001b). One mechanism of reducing root costs is to allocate more biomass to root classes that are less metabolically demanding per unit of P acquisition. We have shown that adventitious roots acquire P at less metabolic cost than basal and tap roots, and that P-stress increases relative biomass allocation to adventitious roots, especially in P-efficient genotypes (Miller *et al.* 2003).

Root respiration can be divided into three components: growth of new tissue, maintenance of existing tissue, and ion uptake and assimilation (Bouma *et al.* 1996; Amthor 2000; Lambers *et al.* 2002). As root systems mature and the proportion of non-growing tissue increases, maintenance respiration becomes an increasing fraction of total respiration. Even in young bean plants, maintenance respiration comprises 90% of total root respiration (Table 5.1). In this context it is noteworthy that a under P-stress, a P-efficient bean genotype had 50% lower maintenance respiration than a P-inefficient genotype (Ho et al. 2003; unpublished). Reduced maintenance respiration of root tissue under P-stress is an important adaptation to low P availability, by making more fixed C available for continued root growth.

Aerenchyma reduces the metabolic costs of soil exploration

Aerenchyma is a series of air spaces formed in the root cortex, and the lysigenous type, found in crop plants such as maize, is formed by a regular pattern of collapsed cortical cells (Esau 1977; Jackson and Armstrong 1999). Root aerenchyma is an adaptation to hypoxia (reviewed in Jackson and Armstrong 1999). In C3 plants, aerenchyma may also provide a photosynthetic benefit by concentrating CO_2 from root respiration and channeling it to leaf intercellular spaces (Constable and Longstreth 1994; Constable *et al.* 1992). Although the overwhelming majority of research on root aerenchyma has focused on its importance in hypoxia, root aerenchyma can also be induced by suboptimal nutrient availability. In aerated solution, aerenchyma was observed in maize roots when N, P or S were omitted from the medium (Konings and Verschuren 1980; Drew *et al.* 1989; Bouranis *et al.* 2003; Fan *et al.* 2003). The response to low P was also observed in common bean (Eshel *et al.* 1995; Fan *et al.* 2003) and rice (Lu *et al.* 1999). In maize the induction of aerenchyma by low P may be related to increased ethylene sensitivity of P-stressed roots (He *et al.* 1992).

Aerenchyma could be an important trait for plants experiencing edaphic stress, since living cortical cells are lost. This reduces root C costs by dramatically reducing maintenance respiration (Lynch and Brown 1998; Fan *et al.* 2003). Besides reducing the ongoing C cost of root maintenance, lysis of cortical cells may contribute prefixed C to root apices. An additional benefit from aerenchyma formation would be the reduced P requirement of root growth, which in conditions of P limitation can be as significant as C costs for metabolic efficiency (Snapp *et al.* 1995; Koide *et al.* 2000). Phosphorus released from cortical tissue by aerenchyma formation would be useful in meeting the P demands of new root elongation. A similar concept has been proposed for cortical senescence in grasses (Gillespie and Deacon 1988; although see Lascaris and Deacon 1991).

Results from our laboratory support the hypothesis that aerenchyma formation is a useful adaptation to low P. In bean and maize, we observed substantial genotypic variation in the induction of cortical aerenchyma by P-stress (Fan *et al.* 2003). Differences in aerenchyma formation induced by ethylene treatments and genotypic variation were correlated with proportionate reductions in root P concentration in low-P roots. Reduced P requirement for soil exploration would be advantageous in conditions of low P availability. Phosphorus liberated by senescing cortical cells could be used for continued apical growth. In low P conditions, most of the Pi taken up by roots is utilized to meet local tissue demand (Snapp and Lynch 1996). Variation in aerenchyma formation was disproportionately correlated with root respiration (Figure 5.6; Fan *et al.* 2003). Root segments with 20% cross sectional area as aerenchyma had half the respiration of roots without aerenchyma. The disproportionate effect of aerenchyma on respiration may reflect the fact that the cortical cells lost during the formation of aerenchyma are metabolically active, while inactive tissues such as sclerenchyma and xylem vessels do not contribute to maintenance respiration. Results with isolated root segments were confirmed in intact plants; whole root systems of a maize genotype with abundant aerenchyma has less root respiration per unit of root length than did a genotype with less aerenchyma (Fan *et al.* 2003). In glasshouse and field studies, the high-aerenchyma maize genotype Oh43 had greater root growth in low P soil than the low-aerenchyma genotype, W64a (Zhu *et al.* 2004; unpublished). In maize, root porosity was highly correlated with sustained root growth under low P (Figure 5.7).

Genetic variation for aerenchyma induction in response to waterlogging has been observed in many species, including banana (Aguilar *et al.* 1999), wheat (Huang *et al.* 1994), barley (Garthwaite *et al.* 2003) and maize (Lizaso *et al.* 2001; Mano *et al.* 2006). Related species may also vary in constitutive (non-stressed) aerenchyma formation (Ray *et al.* 1998; Visser *et al.* 2000; Mano *et al.* 2007). We

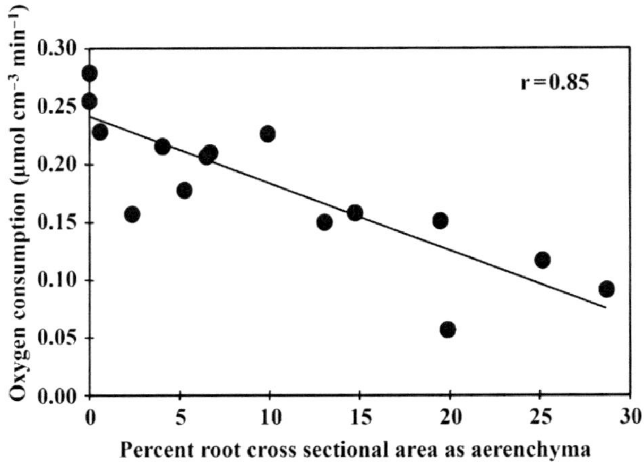

Fig. 5.6 Increasing abundance of aerenchyma is associated with reduced respiration in maize (*Zea mays* L.) roots. Each data point is the mean of 6 measurements of respiration and 10–12 measurements of aerenchyma on comparable root segments. (From Fan *et al.* (2003). with permission.)

Fig. 5.7 Maintenance of root growth in a low-P field as related to cortical aerenchyma formation in unrelated maize (*Zea mays* L.) genotypes. Root weights are expressed as the proportion of corresponding high-P roots. Each point is the mean of four replicates. (Zhu, Kaeppler, and Lynch, unpublished, 2004.)

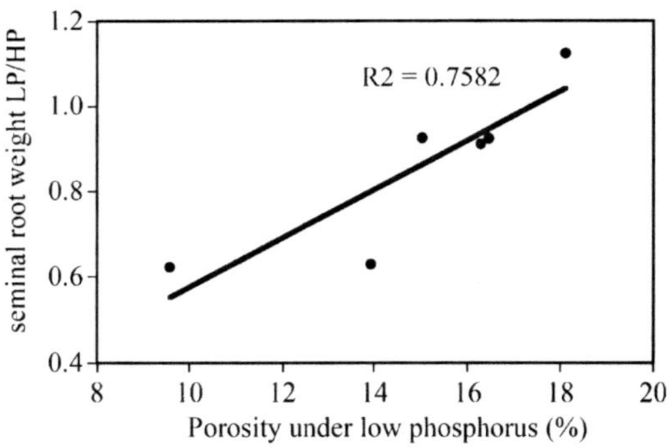

observed large genotypic variation (200–300%) in aerenchyma formation in response to P-stress in both maize and bean (Fan *et al.* 2003). Such variation raises interesting questions regarding the adaptive importance and functional tradeoffs for aerenchyma in diverse environments. Possible tradeoffs to aerenchyma formation include reduced physical resistance to crushing (Striker *et al.* 2006, 2007), reduced radial water transport (Fan *et al.* 2007), reduced mycorrhizal habitat, and increased axial spread of root pathogens. The large intraspecific variation in important crop species (Fan *et al.* 2003; Mano *et al.* 2006) also makes aerenchyma amenable to plant breeding, as is currently underway to enhance flooding tolerance in maize and other grains (Ray *et al.* 1999; Setter and Waters 2003).

Root etiolation

Shoots respond to low light intensity by etiolation, enhanced elongation at the expense of radial thickening and lateral shoot growth. This response is adaptive by increasing the likelihood of the shoot growing into better illumination and by increasing light capture in competitive situations. We hypothesize that an analogous process occurs in roots sensing low P availability.

There are many reports of increased fineness of roots under nutrient deficiency, usually described as increased specific root length (SRL, root length per unit weight). However, increased SRL could result from the increased proportion of secondary roots, since comparisons are not usually made within root classes (Eissenstat *et al.* 2000; Forde and Lorenzo 2001). Evidence for increased, reduced, or unchanged SRL can be found in the literature, but most reports do not consider variation in tissue density or variation in SRL within root classes, and are therefore not direct measurements of root diameter (Ryser and Lambers 1995; Gahoonia and Nielsen 2004). Careful studies of the effect of nutrient stress on root diameter within root classes and orders are needed to determine whether root etiolation could be an adaptive trait.

Under low P availability, root elongation is emphasized at the expense of lateral branching (Borch *et al.* 1999) and secondary growth (Eshel *et al.* 1995). There have been a few reports on increased diameter of specific root classes under high nutrient availability, including nitrate (Hackett 1972; Drew and Saker 1978; Ryser and Lambers 1995) and P (Xie and Yu 2003; Zhu and Lynch 2004). Bean basal roots show increased root diameter under high P, primarily in the older parts of the root (Figure 5.8). The larger diameter of the older parts of basal roots grown in high P was largely a result of a greater area of the stele, both in absolute area and relative to total root area (Fan *et al.* 2003). Similarly, barley roots grown with high nitrate showed an increase in stele diameter (Drew and Saker 1978), so this response is not restricted to dicots. In our study of maize genotypes with contrasting P efficiencies, we found that lateral root SRL and diameter varied among genotypes, and that smaller diameter and greater SRL of lateral roots was associated with faster lateral root growth, which in turn was associated with higher shoot growth and P efficiency (Zhu and Lynch 2004). Furthermore, there was genetic variation in plasticity of this trait, i.e. its response to P availability.

Particular root types may be more likely to alter their diameter in response to nutrient stress. In studies of barley, high nitrate increased the diameter of first and second order lateral roots, but not seminal roots (Drew and Saker 1978). In our study of primary root elongation in Arabidopsis, no difference in diameter could be discerned between high and low P treatments (Ma *et al.* 2003). Timing and extent of etiolation may vary with root class, order, and extent of the nutrient stress. ‘Root etiolation’ is presumably adaptive by reducing the metabolic costs of root extension into new soil domains that may have greater P availability. This phenomenon deserves further study.

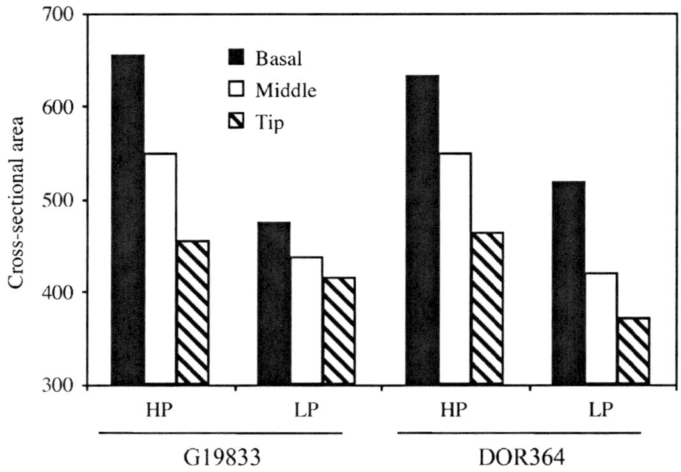

Fig. 5.8 Cross-sectional area of common bean (*Phaseolus vulgaris* L.) basal roots grown for six weeks with high (1 mM, HP) or low (1 μM, LP) phosphorus. Total cross-sectional area was measured from segments of the most basal (2 cm from point of origin), central, and apical (2 cm from root tip) portions of one basal root from each of six plants per genotype and treatment. Values shown are means, error bars = SEM. (Graph created from data in Fan *et al.* (2003). With permission.)

Like shoot etiolation, root etiolation increases exploration at the expense of mechanical strength. Finer roots may be able to penetrate smaller pores in soil, but have less ability to push soil particles aside, so roots grown in soils with high bulk density tend to have a larger diameter and reduced branching (Bennie 1991). In experiments on the effects of co-occurring soil compaction and P-deficiency, roots increased their diameter with increasing bulk density only when supplied with P (Hoffmann and Jungk 1995). Root etiolation may also have negative tradeoffs in terms of turnover rates, desiccation tolerance, susceptibility to herbivory and other characteristics (Eissenstat *et al.* 2000).

Root hairs

Root hairs are subcellular protrusions of root epidermal cells that are important for the acquisition of relatively immobile nutrients such as P (Clarkson 1985; Peterson and Farquhar 1996; Jungk 2001). Several lines of evidence indicate that root hairs contribute to Pi acquisition. First, mathematical modeling indicates that root hairs substantially increase Pi acquisition from the soil solution, by expanding the soil volume subject to Pi depletion through diffusion to the root surface (Bouldin 1961). Indirect evidence from autoradiography demonstrated that root hairs increase the size of Pi depletion zones around roots (Lewis and Quirk 1967; Bhat and Nye 1974). The inclusion of root hairs improved estimates of crop Pi uptake in simulation models (Itoh and Barber 1983a,b). More recently, direct evidence was provided for Pi uptake by root hairs (Gahoonia and Nielsen 1998).

Root hairs may also assist the dispersion of exudates such as organic acids throughout the rhizosphere, which improve P bioavailability in many soils (Hinsinger 2001; Ryan *et al.* 2001). Mutants of Arabidopsis and barley lacking root hairs have severely impaired Pi uptake (Bates and Lynch 2000a,b; Gahoonia and Nielsen 2003) and in the case of Arabidopsis, reduced competitiveness in low P soil (Bates and Lynch 2001). Both root hair length (Bates and Lynch 1996) and root hair density (Ma *et al.* 2001a) are highly regulated by P availability, which suggests that they have value to plants in low P soil. Geometric modeling indicated that responses of root hairs to P availability interact synergistically to improve P acquisition (Ma *et al.* 2001b). Variation among species in root hair length is correlated with P acquisition (Itoh and Barber 1983b; Föhse *et al.* 1991; Gahoonia *et al.* 1999), as is intraspecific variation among genotypes of white clover (Caradus 1981), barley and wheat (Gahoonia *et al.* 1997, 2001; Gahoonia and Nielsen 1997), common bean (Miguel 2004; Yan *et al.* 2004), and turfgrass (Green *et al.* 1991).

Genotypic variation in root hair length and density in maize and common bean is controlled by several major Quantitative Trait Loci (QTL; Yan *et al.* 2004; Zhu *et al.* 2005c), suggesting that this trait could be selected in crop breeding programs through marker aided selection (MAS), as well as through direct phenotypic screening. Root hairs are particularly important for P acquisition in non-mycorrhizal plants, since mycorrhizal hyphae fulfill some of the same functions as root hairs. However, genotypic variation in root hair length and density is important for PAE regardless of the mycorrhizal status of the plant (Figure 5.9; Miguel 2004). Root hairs are attractive targets for crop breeding programs because there is large genotypic variation, substantial effect of this variation on PAE (regardless of mycorrhizal status), relatively simple genetic control, and opportunities for direct phenotypic selection (Gahoonia and Nielsen 2004; Lynch 2007).

Root turnover

Root senescence, or turnover, could have positive or negative effects on the efficiency of Pi acquisition. Negative effects would result if roots were lost in fertile soil domains, resulting in loss of prior metabolic investment in those roots, as well as the opportunity costs of P that is unexploited, or worse, exploited by a competitor. Positive effects could result from the pruning of roots in infertile soil domains, thereby avoiding ongoing maintenance costs of unproductive organs, which is important, since maintenance costs rapidly overtake construction cost in most roots (e.g. Table 5.1; Peng *et al.* 1993; Nielsen *et al.* 1998). It has also been proposed that greater root turnover or "root renewal" could enhance P acquisition by increasing soil exploration and by replacing older roots with younger ones more active in Pi

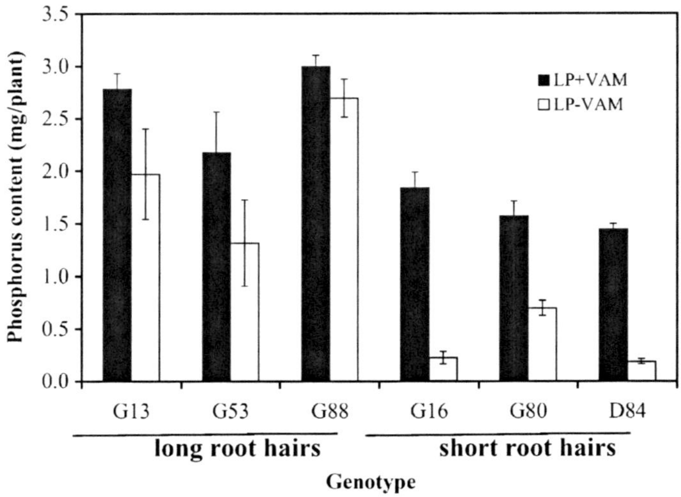

Fig. 5.9 Longer root hairs improve phosphorus acquisition in the presence and absence of mycorrhizal inoculation in common bean (*Phaseolus vulgaris* L.). Plants were grown for 28 days in low-P soil with (+VAM) or without (-VAM) mycorrhizal inoculum. Genotypes are recombinant inbred lines having long or short root hairs. Each bar is the mean of four replicates, error bars = SEM. (Miguel 2004.)

uptake (Steingrobe *et al.* 2001). Regulated senescence of roots would permit the remobilization of root resources, including carbohydrates and nutrients, to other plant activities, notably to reproductive growth in annual plants. In common bean, there is no evidence that roots in infertile soil domains are preferentially senesced (Snapp and Lynch 1996), or that programmed root senescence occurs during reproductive development (Fisher *et al.* 2002). It appears that significant root turnover observed in the field is the result of biotic and abiotic stress rather than programmed plant responses (Eissenstat and Yanai 1997; Fisher *et al.* 2002). This is supported by the observation that P availability was positively associated with soil biological activity and fine root turnover in a Hawaiian montane forest (Ostertag 2001). A report of increased root turnover with lower P availability in barley late in the season (Steingrobe *et al.* 2001) may have been confounded by P effects on phenology (see below). Therefore, traits that affect root lifespan, such as defense chemistry or tissue composition, may have only indirect effects on low P adaptation.

TARGETING P MOBILIZATION IN THE RHIZOSPHERE

Several root traits contribute to Pi acquisition by increasing the bioavailability of P in the rhizosphere. Exudation of carboxylates, such as citrate and malate, is particularly important for Pi acquisition from P-fixing soils. Deprotonated carboxylates chelate Al^{3+}, Fe^{3+} and Ca^{2+}, which results in mobilization of Pi from bound forms (Hinsinger 2001). This activity is complemented in neutral and alkaline soils by rhizosphere acidification, which results in increased solubility of Ca-phosphates (Hinsinger 2001). Although carboxylate release from roots is accentuated under P-deficiency conditions in many species, recent evidence showed that this activity is constitutive in three genotypes of chickpea (Wouterlood *et al.* 2004; Wouterlood *et al.* 2005). The subject of organic acid excretion and its importance for release of Pi from inorganic forms has been discussed extensively in several recent reviews (Hinsinger 2001; Kochian *et al.* 2004). Organic acid excretion is also important for aluminum tolerance, which is related to P efficiency traits, since excess aluminum availability coincides with P deficiency in many acid soils (Kochian *et al.* 2005). Overexpression of enzymes responsible for organic acid production in roots improves plant growth in soils with excess aluminum or low P availability (Koyama *et al.* 2000; Lopez-Bucio *et al.* 2000; Tesfaye *et al.* 2001).

Since a considerable proportion of P can occur in organic forms, plants can increase P availability in the rhizosphere by secreting phosphohydrolases to mineralize Pi from organic compounds (Marschner 1995; Abel *et al.* 2002; Vance *et al.* 2003). Secreted acid phosphatases have been shown to be upregulated under P deficiency (Marschner 1995; Vance *et al.* 2003; Tomscha *et al.* 2004). Their significance for P nutrition under P-limiting conditions has been demonstrated (Barrett-Lennard *et al.* 1993; Li *et al.* 2003, 2004; Tomscha *et al.* 2004), although their importance seems to vary with species, cropping system, and forms of organic P in the soil (Yun and Kaeppler 2001; Li *et al.* 2003, 2004; George *et al.* 2005). Exudate metabolism and production by rhizosphere microflora introduce significant complexity in the relationship between root release of exudates and the effectiveness of exudates in improving Pi acquisition by the root.

The strong interactions of root exudates with the chemical and biological characteristics of the rhizosphere highlight the important role of root architecture in placing exudates in specific soil domains. Many well-developed soils such as Spodosols, Mollisols, Andisols, Ultisols, Alfisols, and Gelisols, together comprising some 39% of the earth's ice-free land surface, have highly differentiated horizons that vary sharply in chemical, physical, and biological properties (Soil Survey Staff 1999). Phosphatases are useful for mobilization of organic phosphate esters, which are found in organic-matter rich surface horizons but, in most soils, are scarce in subsurface horizons several centimeters away. Carboxylates are especially useful in liberating Pi adsorbed to oxide surfaces, which are common in subsurface horizons but less so in the topsoil. Root architectures that deploy roots to surface or subsurface horizons should therefore have a significant impact on the functional importance of exudates for P acquisition. This largely unexplored topic is an example of trait synergy (see below).

Cluster roots

Cluster roots are zones of tightly packed, short, hairy rootlets that occur widely in Proteaceae (where they are called proteoid roots) and in several other plant families (Barber 1995; Diem *et al.* 2000; Neumann and Martinoia 2002; Lamont 2003;

Shane and Lambers 2005; Lambers *et al.* 2006). Cluster roots provide a unique mechanism for acquiring P in extremely P-poor environments by concentrating the P-mobilizing mechanisms described above into a small volume of soil. Cluster root formation and attendant secretion of organic acids, H^+ ions, and acid phosphatases are promoted by P-deficiency and in some species under other conditions such as Fe and N deficiency (Skene 2001; Neumann and Martinoia 2002; Lamont 2003; Vance *et al.* 2003).

Only a few crop species form cluster roots, notably white lupin and some Cucurbitaceae (Waters and Blevins 2000; Neumann and Martinoia 2002). While white lupin has been studied extensively, the impact of cluster roots on other crops has received little attention. One report on cucurbits implicates cluster root formation in Fe(III) reduction (Waters and Blevins 2000). We have failed to observe cluster root formation in P-stressed cucurbits (Postma and Lynch 2006; unpublished data). This adaptation may not be significant for Pi acquisition for crops other than white lupin.

The fact that so few species employ this strategy, and the Proteaceae are endemic to soils with extremely low P availability, indicates that there are significant tradeoffs to this strategy. One is that cluster roots have a high metabolic requirement. Lambers and colleagues estimate that over half of all photosynthetic carbohydrate production is required for the growth, respiration, and exudate production by cluster roots of one species (Lambers *et al.* 2006). Another potential liability to localized rhizosphere acidification is the release of toxic metals including aluminum or heavy metals. A comparison of cluster forming vs. non-cluster forming lupins showed that the cluster forming species took up more Cd, a toxic contaminant of P fertilizer sources (Brennan and Bolland 2003). The risk of metal toxicity may be one reason that cluster-rooted species are often found on sandy soils with low metal content. There may be additional tradeoffs associated with concentration of root foraging activity in limited domains, such as reduced acquisition of more mobile and dispersed nutrients, as well as water.

MYCORRHIZAL SYMBIOSES

The majority of higher plant species have mycorrhizal symbioses with fungi that assist nutrient acquisition (Smith and Read 1997). Ectomycorrhizas enhance P acquisition via mobilization of sparingly soluble P complexes, whereas both ectomycorrhizas and the vesicular-arbuscular mycorrhizas common in many annuals and hardwood species enhance Pi acquisition because they increase the volume of soil explored beyond the depletion zone surrounding the root itself. In exchange for Pi supplied to the plant, the fungal symbiont obtains reduced carbon. Therefore, the carbon cost to the plant of mycorrhizal symbioses is one component of the cost of Pi acquisition in most species. In bean, mycorrhizal colonization increased root Pi acquisition, but the resulting increase in shoot photosynthesis did not result in increased plant growth because of greater root respiration (Nielsen *et al.* 1998). At high P supply, mycorrhizal colonization reduced the growth of citrus seedlings because of greater root carbon cost (Peng *et al.* 1993). In general, the costs of the mycorrhizal symbiosis in various herbaceous and woody species ranges from 4% to 20% of daily net photosynthesis (Koch and Johnson 1984; Harris and Paul 1987; Douds *et al.* 1988; Jakobsen and Rosendahl 1990; Eissenstat *et al.* 1993; Nielsen *et al.* 1998). The greater metabolic burden of mycorrhizal roots may contribute to the non-beneficial or even parasitic role that mycorrhizal fungi play in agroecosystems (Ryan and Graham 2002).

Mycorrhizal symbioses have attracted a great deal of attention by researchers in the past 30 years. The importance of mycorrhizal symbioses for Pi acquisition has led some mycorrhizal researchers to the belief that root traits are secondary or trivial in importance for Pi acquisition compared to fungal-assisted Pi acquisition (Smith *et al.* 2003). In this context it is useful to consider the strong correlations observed between Pi uptake and root traits such as root hair length (Figure 5.10; Miguel 2004; see also references cited above) and root shallowness (Lynch and Brown 2001) even in the presence of mycorrhizas. This could signify that mycorrhizal foraging is incomplete and can be supplemented by direct root foraging. Alternatively, extraradical hyphae could be restricted to the volume of soil near the root (Owusu-Bennoah *et al.* 2002), so that root architectural patterns have a strong influence on foraging patterns by the fungal symbiont. In our research with maize, soybean, and common bean, we have observed similar genotypic rankings for plant growth in low P soil in the field where mycorrhizas are formed and in controlled environments without mycorrhizas (Bonser *et al.* 1996; Miguel 2004; Ho *et al.* 2005). This suggests that for these annual crops, mycorrhizal symbiosis changes the effective fertility status of the soil environment but does not represent a selection criterion (either through natural selection or in plant breeding) among genotypes, possibly because it is ubiquitous.

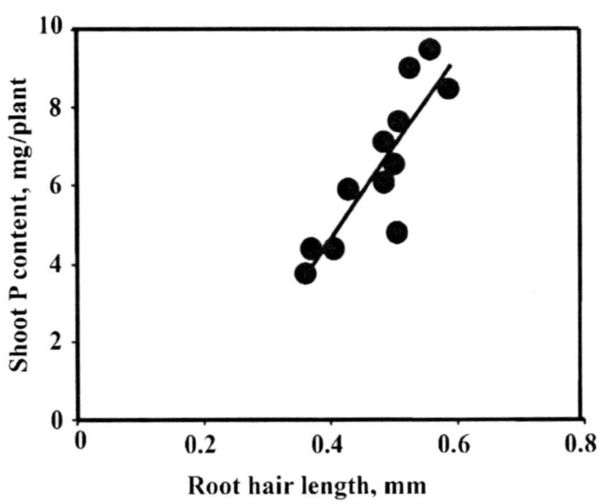

Fig. 5.10 Effect of root hair length on phosphorus content of common bean (*Phaseolus vulgaris* L.) genotypes. Plants were grown for 35 days in low-P soil in the field in Costa Rica. Each point is the mean of four replicates of one genotype; the set of genotypes are six recombinant inbred lines having long or short root hairs. (Miguel 2004)

PHENOLOGY

Some annual plants respond to P stress by delayed maturation (Rossiter 1978; Chauhan *et al.* 1992; Ma *et al.* 2002). This could be adaptive for P acquisition by permitting continued root growth, and by extending the period of time in which existing roots acquire Pi. Time is particularly important for Pi acquisition, since Pi diffusion through soil is slow, as is recharge of P-depleted soil (Tinker and Nye 2000). We call this phenomenon 'root foraging duration' by analogy with leaf area duration. We have observed that root foraging duration (the integral of root length over time) is highly correlated with Pi acquisition in Arabidopsis genotypes of contrasting phenology (Figure 5.11). In addition to possible benefits for Pi acquisition, an extended growing season would also increase the metabolic utility of acquired P, for example by extending the time leaf P could be employed to generate photosynthates. In other words, the utility of P to the plant is dependent on the length of time the P is used by the plant, which in general would be greater with an extended growing season. Phenology is responsive to P availability in some plants and there is a range of maturities available within crop species. If it is demonstrated that delayed maturation is a positive adaptation to low P availability, genotypic variation for this trait may have value in crop breeding programs, especially in tropical agroecosystems where temperature and moisture availability do not limit the effective growing season.

TRAIT SYNERGY

Several root strategies for Pi acquisition may have functional interactions with each other or with other plant traits. These interactions could be positive or synergistic in improving P efficiency traits, or they may be antagonistic. An example of trait

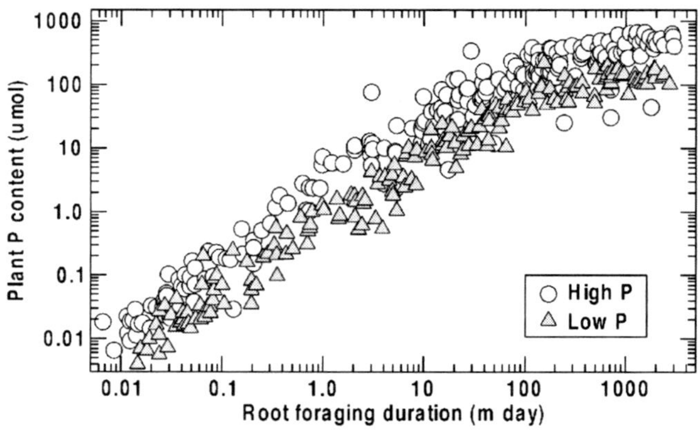

Fig. 5.11 Plant phosphorus content and root foraging duration for seven genotypes of *Arabidopsis thaliana* grown in soils with high and low phosphorus availability. Total phosphorus acquired is strongly correlated with root foraging duration, with plants grown in high phosphorus soil acquiring more phosphorus at a given level of root foraging duration. (Nord and Lynch, unpublished, 2007)

synergy in Pi acquisition is the interaction of four distinct root hair traits; root hair length, root hair density, the distance from the root tip to the first appearance of root hairs, and the pattern of root hair bearing epidermal cells (trichoblasts) among non hair bearing cells (atrichoblasts). Low P availability causes coordinated increases in root hair length and density in many species (Brewster *et al.* 1976; Foehse and Jungk 1983; Bates and Lynch 1996; Ma *et al.* 2001a). In Arabidopsis, low P availability also shortens the distance from the first root hair to the root tip, and changes the geometry of trichoblasts by increasing the number of trichoblast files, caused by cortical reorganization (Ma *et al.* 2001a, 2003). Geometric modeling showed that the combined effect of these four traits on Pi acquisition was 371% greater than their additive effects, demonstrating substantial morphological synergy (Ma *et al.* 2001b). Synergism among root hair traits may account for their coordinated regulation.

Traits of individual root axes such as root hairs and root exudates may have synergism with root architectural traits, which locate root axes in soil domains with varying P availability. For example, longer root hairs would be expected to provide greater benefit to the plant if they were positioned in P-replete topsoil as opposed to P-deficient subsoil. Phosphatases that mobilize Pi from soil organic matter would be more useful if exuded by shallow roots than by deep roots, since in most soils organic matter decreases with depth. In contrast, organic acids that mobilize Pi from Fe and Al oxides may be more useful when released into deeper soil horizons where these forms of P predominate. Root architectural traits may themselves display interactions, by altering the extent of inter-root competition, which is an important component of overall root foraging efficiency (Ge *et al.* 2000; Rubio *et al.* 2001, 2003). For example, root systems combining deep rooting (through, for example, lateral branching from the deeper parts of the primary and basal roots) with shallow rooting (through adventitious roots or shallow basal roots) would be expected to be more complementary than root systems in which distinct root classes competed for the same soil niche (Walk *et al.* 2006). This is especially relevant in the context of drought, since in many environments water is a deep soil resource while P is a shallow resource (see discussion below). We know very little about the interaction of traits related to P acquisition, despite the importance of trait interactions for whole plant performance. This is pertinent to plant breeding, since traits under distinct genetic control could be combined to maximize positive synergy.

TRADEOFFS

The utility of a trait for plants in low P environments must take into account potential tradeoffs of the trait for other plant processes. The most obvious tradeoff for many traits is simply the opportunity cost resulting from diversion of plant resources from other functions. For example, the production of adventitious roots reduces the development of basal roots, which in certain soils can be detrimental to overall plant P acquisition (Walk *et al.* 2006). Since soil resource distribution is heterogeneous, architectural tradeoffs can also occur when the exploitation of one soil domain reduces exploitation of another soil domain with its attendant resources.

An important tradeoff or opportunity cost to topsoil foraging is increased sensitivity to drought stress, since water is a deep soil resource in many environments. A comparison of deep-rooted and shallow-rooted bean genotypes showed that while shallower genotypes had superior growth under P stress, deep-rooted genotypes had superior growth under water stress (Figure 5.12; Ho *et al.* 2005). These results are consistent with economic optimization modeling of the relationship between root architecture and multiple resource acquisition, particularly water and P (Ho *et al.* 2004). The general solution of the model states that a plant will locate its roots at a soil depth where the marginal benefit of water and P acquisition will exactly equal the marginal cost of inter-root competition (Ho *et al.* 2004). Indeed, bean genotypes that are best adapted to low P environments, where P is localized in the surface soil, tend to have a shallower basal root angle, whereas genotypes that are adapted to terminal drought environments have deeper root systems (Ho *et al.* 2005). This example illustrates the importance of considering tradeoffs in assessing the adaptive importance of specific root traits, especially in crop breeding for distinct environments.

The large genotypic variation for root traits that appear to be positive adaptations for nutrient acquisition may be caused or maintained by tradeoffs incurred by certain phenotypes. For example, long, dense root hairs improve Pi acquisition at minimal metabolic cost (see references above), yet a large proportion of crop genotypes have few, sparse root hairs, and many genotypes display plasticity in root hair traits, so that under high fertility, root hairs are suppressed. This could suggest that there are potential costs to root hairs, such as increased susceptibility to root pathogens. Similarly, cortical aerenchyma appears to reduce the metabolic costs of soil exploration (see references above), yet substantial intraspecific variation for constitutive aerenchyma formation exists, and aerenchyma formation is suppressed under high fertility. This suggests that there are potential costs to aerenchyma formation, such as reduced radial transport of water (Fan *et al.* 2007) and nutrients, or reduced mycorrhizal habitat. Such questions are largely unresolved.

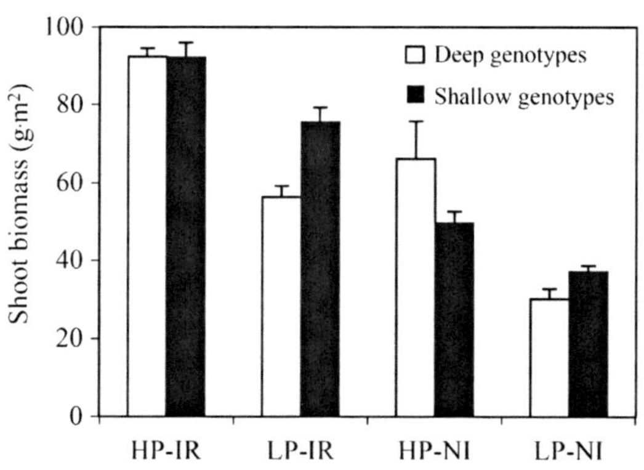

Fig. 5.12 Shoot biomass at 44 days after planting for three shallow-rooted and three deep-rooted common bean (*Phaseolus vulgaris* L.) genotypes in the field. HP = high phosphorus availability, LP = low phosphorus availability, IR = irrigated, NI = non-irrigated. Each bar is the mean of four replicates, error bars = SEM. (From Ho *et al.* (2005). With permission)

RESPONSES TO HETEROGENEOUS P AVAILABILITY

In addition to the variability in P availability with soil depth (discussed above), P availability may be heterogeneous in space and time as a result of organic matter decomposition, variation in soil composition, competition with the same or other root systems, water availability, temperature, etc. (Jackson and Caldwell 1993). Many plants have the ability to respond to these patches of higher nutrient availability in ways that are expected to increase their ability to compete for these nutrients. The responses of root systems to heterogeneous nutrient distribution have been reviewed recently (Hodge 2004), so this topic will be discussed here only in the context of agroecology.

When P-stressed plants encounter a patch of higher P availability, one advantageous response is to proliferate roots to enhance Pi acquisition from the patch. Root proliferation has been observed in nutrient patches and includes increased number, length, and branching of lateral roots. The extent to which this occurs varies among species, some showing very dramatic effects (e.g. barley; Drew 1975), while others show little or no response (Campbell *et al.* 1991; Farley and Fitter 1999). To complicate matters further, plants may alter root development in nutrient patches when roots of another plant (even of the same species) are competing within the patch (Robinson *et al.* 1999; Gersani *et al.* 2001; Maina *et al.* 2002). The available data justifies the conclusion that root proliferation in nutrient patches is likely to be useful for plants grown in intercrop systems, as is the case for many crops grown in poor soils in the tropics and subtropics.

INTERPLANT COMPETITION

The utility of traits for P efficiency will be most evident in competitive environments, including those experienced by wild plants, crops in subsistence agroecosystems, and in the high-density genetic monocultures typical of commercial agriculture. Traits influencing P efficiency will affect plant productivity, and thereby competitive performance, under P-stress. An example of this is the positive effect of root hairs on plant performance in mixed stands of Arabidopsis at low P but not at high P (Bates and Lynch 2001). Traits influencing Pi acquisition can directly affect interplant competition by removing soil P that could be accessed by competitors. For example, bean genotypes with shallow basal roots out-compete genotypes with deep basal roots in low P fields (Rubio *et al.* 2003), because of enhanced topsoil exploitation and reduced competition among roots of the same plant (Rubio *et al.* 2001).

At the population level, competition among root systems can be important in determining the utility of root traits for P efficiency. This appears to be the case for plasticity of basal root shallowness, for which genetic variation exists, i.e. some genotypes respond to P-stress by becoming more shallow, whereas others are unaffected or become deeper (Bonser *et al.* 1996; Ho *et al.* 2004, 2005). Plasticity

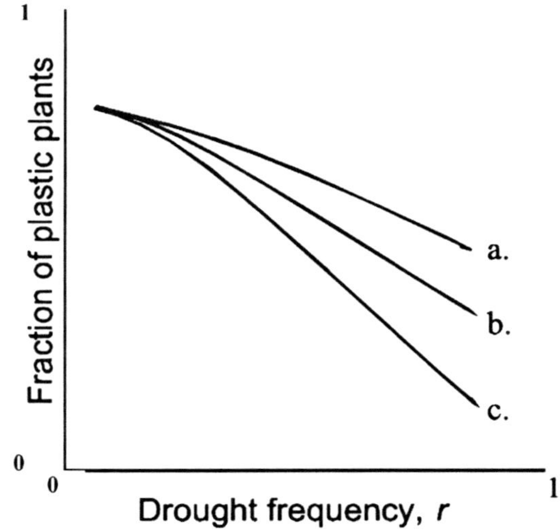

Fig. 5.13 The theoretical relationship of the equilibrium steady-state fraction of plastic plants in a population and drought frequency, where spatial competition is considered, under (a) low, (b) intermediate or (c) high drought intensity. Plastic plants in this model respond to low P availability by increasing topsoil foraging, which incurs costs in conditions of terminal drought, and when neighboring individuals have the same phenotype. (Ho 2004)

of root shallowness would generally be considered a useful trait, since plasticity would permit a plant to modify its root architecture to adapt to the prevailing edaphic stress. However, if all plants in a population were equally plastic, and therefore had the same root architecture, greater interplant competition would occur than if distinct root phenotypes existed in a population, thereby permitting complementary exploitation of distinct soil domains. Theoretical modeling showed that interplant competition could be important in determining an optimal balance of plastic and non-plastic root phenotypes under conditions of P-stress and combined P and water stress (Figure 5.13; Ho 2004). This suggests that genetic mixtures or multilines may have better performance in low P agroecosystems than genetic monocultures, especially in drought-prone environments.

ECOSYSTEM ISSUES

A better understanding of plant adaptations to P-stress is critically needed for two of the greatest challenges facing humanity in the 21st century: eliminating world hunger and understanding how natural and managed ecosystems will respond to global climate change.

The development of crops with superior growth in low P soil and with better responsiveness to applied P inputs would have tremendous value in many developing countries, where yields are limited by low soil fertility and fertilizer use is minimal (World Bank 2004). Since genotypic variation for PAE is much larger than variation for P use efficiency in crop plants, development of P-efficient crops is likely to have a great impact on agricultural productivity in these agroecosystems (Lynch and Beebe 1995). Although such genotypes would extract more P from the soil than conventional genotypes, they may actually enhance soil fertility in the

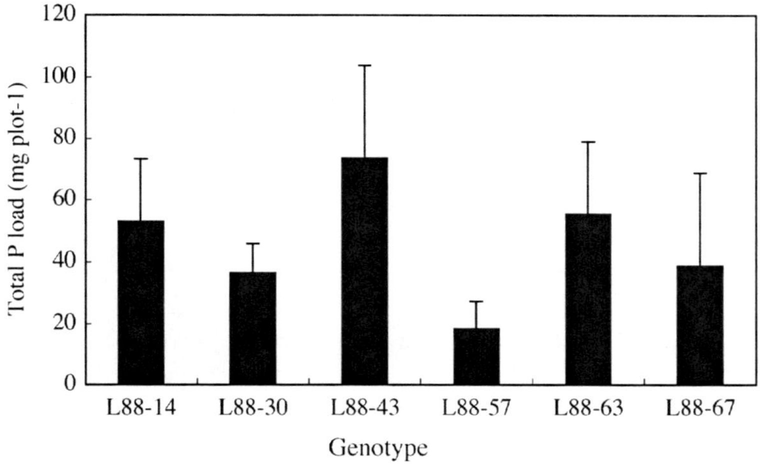

Fig. 5.14 Effect of genotype on phosphorus runoff from an unfertilized on-farm site in Costa Rica during one growing season. Beans were planted at densities typical of local practice. (Henry and Lynch, unpublished, 2007)

long term through beneficial effects on soil erosion and nutrient cycling, as well as benefits they accrue to farm income and thereby the use of fertility amendments (Lynch and Deikman 1998). For example, P runoff from common bean plots in a farmer's unfertilized field varied substantially among genotypes (Figure 5.14; Henry and Lynch 2007; unpublished).

Several genetic traits have been identified with potential utility in breeding P efficient crops, as discussed above, including root exudates, root hair traits, cortical aerenchyma, topsoil foraging through basal or adventitious rooting, and the use of multiline mixtures of root phenotypes. Deployment of these traits through plant breeding programs is resulting in progress in several crops including common bean (CIAT 1999) and soybean (Yan 2005). The success of this effort would constitute a second 'Green Revolution', benefiting the resource-poor farmers who were largely left behind by the first Green Revolution, and who represent the single largest human labor occupation (Lynch and Deikman 1998). A better understanding of the biology of traits associated with P efficiency, especially how these traits combine and their tradeoffs in specific production environments, is needed to guide plant breeding efforts.

We will not be able to understand or manage ecosystem responses to global change unless we learn more about how global change variables such as CO_2, temperature, and ozone interact with the edaphic stresses prevalent in most terrestrial ecosystems (Lynch and St. Clair 2004). The vast majority of research on plant response to global change has focused on leaf responses and has not considered edaphic stresses other than water and possibly nitrogen with any rigor, despite the fact that plant responses to edaphic stresses are primary limitations to plant productivity in most forests and managed systems. Plants limited by low soil P availability may respond to elevated CO_2 by producing more exudates and by altered root growth and architecture, which may partially alleviate P stress, but interactions with other global change variables such as drought are likely to be detrimental, as discussed above. This topic merits research.

CONCLUSIONS

Low soil P availability is a primary constraint to plant growth on earth. Accordingly, plants express a wide array of root strategies that improve Pi acquisition, including increased biomass allocation to roots and to specific root classes within the root system, root architectural traits that enhance topsoil foraging, including basal root gravitropism, adventitious rooting, and lateral root branching, reduced metabolic costs of soil exploration, via formation of cortical aerenchyma, the formation of finer roots and possibly root etiolation, root hairs, P-solubilizing root exudates, mycorrhizal symbioses, phenological plasticity, and morphological plasticity. Ecological tradeoffs and interactions among these traits are poorly understood but are likely to be important in determining the functional utility of these traits, especially in competitive environments. A better understanding of these traits is needed to guide the development of more P efficient crops for developing nations, and to understand how ecosystems will respond to global climate change.

REFERENCES

Abel S, Ticconi CA, Delatorre CA (2002) Phosphate sensing in higher plants. Physiol Plant 115: 1–8

Abelson PH (1999) A potential phosphate crisis. Science 283: 2015–2015

Aguilar EA, Turner DW, Sivasithamparam K (1999) Aerenchyma formation in roots of four banana (*Musa* spp.) cultivars. Sci Hort 80: 57–72

Al-Ghazi Y, Muller B, Pinloche S, Tranbarger TJ, Nacry P, Rossignol M, Tardieu F, Doumas P (2003) Temporal responses of Arabidopsis root architecture to phosphate starvation: evidence for the involvement of auxin signalling. Plant Cell Environ 26: 1053–1066

Amthor JS (2000) The McCree-de Wit-Penning de Vries-Thornley respiration paradigms: 30 years later. Ann Bot 86: 1–20

Anderson G (1980) Assessing organic phosphorus in soils. In: Khasawneh, FE, Sample EC, Kamprath EJ (eds), The Role of Phosphorus in Agriculture. ASA/CSSA/SSSA, Madison, WI, pp 263–310

Barber S (1962) A diffusion and mass flow concept of nutrient availability. Soil Sci 93: 39–49

Barber SA (1995) Soil Nutrient Bioavailability: A Mechanistic Approach. Wiley, New York

Barrett-Lennard EG, Dracup M, Greenway H (1993) Role of extracellular phosphatases in the phosphorus-nutrition of clover. J Exp Bot 44: 1595–1600

Bates TR, Lynch JP (1996) Stimulation of root hair elongation in *Arabidopsis thaliana* by low phosphorus availability. Plant Cell Environ 19: 529–538

Bates TR, Lynch JP (2000a) Plant growth and phosphorus accumulation of wild type and two root hair mutants of *Arabidopsis thaliana* (Brassicaceae). Am J Bot 87: 958–963

Bates TR, Lynch JP (2000b) The efficiency of *Arabidopsis thaliana* (Brassicaceae) root hairs in phosphorus acquisition. Am J Bot 87: 964–970

Bates TR, Lynch JP (2001) Root hairs confer a competitive advantage under low phosphorus availability. Plant Soil 236: 243–250

Bennie A (1991) Growth and mechanical impedance. In: Waisel Y, Eshel A, Kafkafi U (eds), Plant Roots: The Hidden Half. Marcel Dekker, New York, pp 393–414

Bhat KKS, Nye PH (1974) Diffusion of phosphate to plant roots in soil III. Depletion around onion roots without root hairs. Plant Soil 41: 383–394

Bloom A, Chapin F, Mooney H (1985) Resource limitation in plants - an economic analogy. Ann Rev Ecol Syst 16: 363–392

Bonser AM, Lynch J, Snapp S (1996) Effect of phosphorus deficiency on growth angle of basal roots in *Phaseolus vulgaris*. New Phytol 132: 281–288

Borch K, Bouma TJ, Lynch JP, Brown KM (1999) Ethylene: a regulator of root architectural responses to soil phosphorus availability. Plant Cell Environ 22: 425–431

Bouldin D (1961) Mathematical description of diffusion process in the soil. Soil Sci Soc Am Proc 25: 476–480

Bouma TJ, Broekhuysen AGM, Veen BW (1996) Analysis of root respiration of *Solanum tuberosum* as related to growth, ion uptake and maintenance of biomass. Plant Physiol Biochem 34: 795–806

Bouranis DL, Chorianopoulou SN, Siyiannis VF, Protonotarios VE, Hawkesford MJ (2003) Aerenchyma formation in roots of maize during sulphate starvation. Planta 217: 382–391

Brennan RF, Bolland MDA (2003) *Lupinus luteus* cv. Wodjil takes up more phosphorus and cadmium than *Lupinus angustifolius* cv. Kalya. Plant Soil 248: 167–185

Brewster J, Bhat K, Nye P (1976) The possibility of predicting solute uptake and plant growth response from independently measured soil and plant characteristics. IV. The growth and uptake of rape in solutions of different phosphorus concentrations. Plant Soil 44: 279–293

Campbell BD, Grime JP, Mackey JML (1991) A trade-off between scale and precision in resource foraging. Oecologia 87: 532–538

Caradus J (1981) Effect of root hair length on white clover growth over a range of soil phosphorus levels. New Zeal J Agr Res 24: 359–364

Chauhan YS, Johansen C, Venkataratnam N (1992) Effects of phosphorus deficiency on phenology and yield components of short-duration pigeonpea. Trop Agr 69: 235–238

Chu W, Chang S (1966) Surface activity of inorganic soil phosphorus. Soil Sci 101: 459–464

CIAT (1999) Bean Project 1998 Annual Report. CIAT, Cali, Colombia

Clarkson D (1985) Factors affecting mineral nutrient acquisition by plants. Ann Rev Plant Physiol 36: 77–115

Comerford N (1998) Soil phosphorus bioavailability. In: Lynch J, Deikman J (eds), Phosphorus in Plant Biology: Regulatory Roles in Molecular, Cellular, Organismic, and Ecosystem Processes. ASPP, Rockville, MD, pp 136–147

Constable J, Longstreth DJ (1994) Aerenchyma carbon dioxide can be assimilated in *Typha latifolia* L. leaves. Plant Physiol 106: 1065–1072

Constable JVH, Grace JB, Longstreth DJ (1992) High carbon dioxide concentrations in aerenchyma of *Typha latifolia*. Am J Bot 79: 415–418

Diem HG, Duhoux E, Zaid H, Arahou M (2000) Cluster roots in Casuarinaceae: role and relationship to soil nutrient factors. Ann Bot 85: 929–936

Douds DD, Johnson CR, Koch KE (1988) Carbon cost of the fungal symbiont relative to net leaf-P accumulation in a split-root VA mycorrhizal symbiosis. Plant Physiol 86: 491–496

Drew MC (1975) Comparison of the effects of a localized supply of phosphate, nitrate, ammonium and potassium on the growth of the seminal root system, and the shoot, in barley. New Phytol 75: 479–490

Drew M, He C, Morgan P (1989) Decreased ethylene biosynthesis, and induction of aerenchyma, by nitrogen- or phosphate-starvation in adventitious roots of *Zea mays* L. Plant Physiol 91: 266–271

Drew MC, Saker LR (1978) Nutrient supply and the growth of the seminal root system in barley. J Exp Bot 29: 435–451
Eissenstat DM, Yanai RD (1997) The ecology of root lifespan. Adv Ecol Res 27: 1–60
Eissenstat DM, Graham JH, Syvertsen JP, Drouillard DL (1993) Carbon economy of sour orange in relation to mycorrhizal colonization and phosphorus status. Ann Bot 71: 1–10
Eissenstat DM, Wells CE, Yanai RD, Whitbeck JL (2000) Building roots in a changing environment: implications for root longevity. New Phytol 147: 33–42
Esau K (1977) Anatomy of Seed Plants. Wiley, New York
Eshel A, Nielsen K, Lynch J (1995) Response of bean root systems to low level of P. In: Plant Roots - From Cells to Systems. 14th Long Ashton International Symposium. IACR-Long Ashton Research Station, Bristol, UK
Fan MS, Zhu JM, Richards C, Brown KM, Lynch JP (2003) Physiological roles for aerenchyma in phosphorus-stressed roots. Funct Plant Biol 30: 493–506
Fan M, Bai R, Zhao X, Zhang J (2007) Aerenchyma formed under phosphorus deficiency contributes to the reduced root hydraulic conductivity in maize roots. J Integr Plant Biol 49: 598–604
Farley RA, Fitter AH (1999) The responses of seven co-occurring woodland herbaceous perennials to localized nutrient-rich patches. J Ecol 87: 849–859
Fisher MCT, Eissenstat DM, Lynch JP (2002) Lack of evidence for programmed root senescence in common bean (*Phaseolus vulgaris*) grown at different levels of phosphorus supply. New Phytol 153: 63–71
Foehse D, Jungk A (1983) Influence of phosphate and nitrate supply on root hair formation of rape, spinach and tomato plants. Plant Soil 74: 359–368
Föhse D, Claassen N, Jungk A (1991) Phosphorus efficiency of plants II. Significance of root radius, root hairs and cation-anion balance for phosphorus influx in seven plant species. Plant Soil 132: 261–272
Forde B, Lorenzo H (2001) The nutritional control of root development. Plant Soil 232: 51–68
Gahoonia TS, Nielsen NE (1997) Variation in root hairs of barley cultivars doubled soil phosphorus uptake. Euphytica 98: 177–182
Gahoonia TS, Nielsen NE (1998) Direct evidence on participation of root hairs in phosphorus (^{32}P) uptake from soil. Plant Soil 198: 147–152
Gahoonia TS, Nielsen NE (2003) Phosphorus (P) uptake and growth of a root hairless barley mutant (bald root barley, *brb*) and wild type in low- and high-P soils. Plant Cell Environ 26: 1759–1766
Gahoonia TS, Nielsen NE (2004) Root traits as tools for creating phosphorus efficient crop varieties. Plant Soil 260: 47–57
Gahoonia TS, Care D, Nielsen NE (1997) Root hairs and phosphorus acquisition of wheat and barley cultivars. Plant Soil 191: 181–188
Gahoonia TS, Nielsen NE, Lyshede OB (1999) Phosphorus (P) acquisition of cereal cultivars in the field at three levels of P fertilization. Plant Soil 211: 269–281
Gahoonia TS, Nielsen NE, Joshi PA, Jahoor A (2001) A root hairless barley mutant for elucidating genetic of root hairs and phosphorus uptake. Plant Soil 235: 211–219
Garthwaite AJ, von Bothmer R, Colmer TD (2003) Diversity in root aeration traits associated with waterlogging tolerance in the genus *Hordeum*. Funct Plant Biol 30: 875–889
Ge Z, Rubio G, Lynch JP (2000) The importance of root gravitropism for inter-root competition and phosphorus acquisition efficiency: results from a geometric simulation model. Plant Soil 218: 159–171
George TS, Richardson AE (2008) Potential and limitations to improving crops for enhanced phosphorus utilization. In: White PJ, Hammond JP (eds), The Ecophysiology of Plant-Phosphorus Interactions. Springer, Dordrecht, The Netherlands, pp 247–270
George TS, Simpson RJ, Hadobas PA, Richardson AE (2005) Expression of a fungal phytase gene in *Nicotiana tabacum* improves phosphorus nutrition of plants grown in amended soils. Plant Biotechnol J 3: 129–140

Gersani M, Brown JS, O'Brien EE, Maina GM, Abramsky Z (2001) Tragedy of the commons as a result of root competition. J Ecol 89: 660–669

Gillespie IMM, Deacon JW (1988) Effects of mineral nutrients on senescence of the cortex of wheat roots and root pieces. Soil Biol Biochem 20: 525–531

Green RL, Beard JB, Oprisko MJ (1991) Root hairs and root lengths in nine warm-season turfgrass genotypes. J Am Soc Hortic Sci 116: 965–969

Gutschick V (1993) Nutrient-limited growth rates: roles of nutrient-use efficiency and of adaptations to increase uptake rate. J Exp Bot 44: 41–51

Hackett C (1972) A method of applying nutrients locally to roots under controlled conditions, and some morphological effects of locally applied nitrate on the branching of wheat roots. Aust J Biol Sci 25: 1169–1180

Halsted M, Lynch J (1996) Phosphorus responses of C-3 and C-4 species. J Exp Bot 47: 497–505

Hansen CW, Lynch J, Ottosen CO (1998) Response to phosphorus availability during vegetative and reproductive growth of chrysanthemum: I. Whole-plant carbon dioxide exchange. J Am Soc Hortic Sci 123: 215–222

Harris D, Paul E (1987) Carbon requirements of vesicular-arbuscular mycorrhizae. In: Safir GR (ed), Ecophysiology of VA Mycorrhizae. CRC, Boca Raton, FL, pp 93–105

Harrison MJ (1999) Molecular and cellular aspects of the arbuscular mycorrhizal symbiosis. Annu Rev Plant Physiol Plant Mol Biol 50: 361–389

Hartemink AE (2003) Soil Fertility Decline in the Tropics. CABI, Wageningen, The Netherlands

Hayes JE, Richardson AE, Simpson RJ (1999) Phytase and acid phosphatase activities in extracts from roots of temperate pasture grass and legume seedlings. Aust J Plant Physiol 26: 801–809

He CJ, Morgan PW, Drew MC (1992) Enhanced sensitivity to ethylene in nitrogen-starved or phosphate-starved roots of *Zea mays* l. During aerenchyma formation. Plant Physiol 98: 137–142

Hinsinger P (2001) Bioavailability of soil inorganic P in the rhizosphere as affected by root-induced chemical changes: a review. Plant Soil 237: 173–195

Hinsinger P, Plassard C, Tang CX, Jaillard B (2003) Origins of root-mediated pH changes in the rhizosphere and their responses to environmental constraints: a review. Plant Soil 248: 43–59

Hinsinger P, Gobran GR, Gregory PJ, Wenzel WW (2005) Rhizosphere geometry and heterogeneity arising from root mediated physical and chemical processes. New Phytol 168: 293–303

Ho MD (2004) Effects of root architecture, plasticity, and tradeoffs on water and phosphorus acquisition in heterogeneous environments. Dissertation, Penn State University

Ho MD, McCannon BC, Lynch JP (2004) Optimization modeling of plant root architecture for water and phosphorus acquisition. J Theor Biol 226: 331–340

Ho M, Rosas J, Brown K, Lynch J (2005) Root architectural tradeoffs for water and phosphorus acquisition. Funct Plant Biol 32: 737–748

Hodge A (2004) The plastic plant: root responses to heterogeneous supplies of nutrients. New Phytol 162: 9–24

Hoffmann C, Jungk A (1995) Growth and phosphorus supply of sugar-beet as affected by soil compaction and water tension. Plant Soil 176: 15–25

Huang BR, Johnson JW, Nesmith DS, Bridges DC (1994) Root and shoot growth of wheat genotypes in response to hypoxia and subsequent resumption of aeration. Crop Sci 34: 1538–1544

Itoh S, Barber SA (1983a) A numerical solution of whole plant nutrient uptake for soil root systems with root hairs. Plant Soil 70: 403–413

Itoh S, Barber SA (1983b) Phosphorus uptake by six plant species as related to root hairs. Agron J 75: 457–461

Jackson MB, Armstrong W (1999) Formation of aerenchyma and the processes of plant ventilation in relation to soil flooding and submergence. Plant Biol 1: 274–287

Jackson RB, Caldwell MM (1993) The scale of nutrient heterogeneity around individual plants and its quantification with geostatistics. Ecology 74: 612–614

Jakobsen I, Rosendahl L (1990) Carbon flow into soil and external hyphae from roots of mycorrhizal cucumber plants. New Phytol 115: 77–83

Jeschke WD, Holobrada M, Hartung W (1997) Growth of *Zea mays* l. Plants with their seminal roots only. Effects on plant development, xylem transport, mineral nutrition and the flow and distribution of abscisic acid (ABA) as a possible shoot- to-root signal. J Exp Bot 48: 1229–1239

Joner EJ, Magid J, Gahoonia TS, Jakobsen I (1995) P depletion and activity of phosphatases in the rhizosphere of mycorrhizal and nonmycorrhizal cucumber (*Cucumis sativus* L). Soil Biol Biochem 27: 1145–1151

Jones DL (1998) Organic acids in the rhizosphere - a critical review. Plant Soil 205: 25–44

Jungk A (2001) Root hairs and the acquisition of plant nutrients from soil. J Plant Nutr Soil Sci 164: 121–129

Kirkby EA, Johnston AE (2008) Soil and fertilizer phosphorus in relation to crop nutrition. In: White PJ, Hammond JP (eds), The Ecophysiology of Plant-Phosphorus Interactions. Springer, Dordrecht, The Netherlands, pp 177–223

Koch KE, Johnson CR (1984) Photosynthate partitioning in split root citrus seedlings with mycorrhizal and non-mycorrhizal root systems. Plant Physiol 75: 26–30

Kochian LV, Hoekenga OA, Piñeros MA (2004) How do crop plants tolerate acid soils? Mechanisms of aluminum tolerance and phosphorus efficiency. Annu Rev Plant Biol 55: 459–493

Kochian LV, Piñeros MA, Hoekenga OA (2005) The physiology, genetics and molecular biology of plant aluminum resistance and toxicity. Plant Soil 274: 175–195

Koide R, Elliott G (1989) Cost, benefit and efficiency of the vesicular-arbuscular mycorrhizal symbiosis. Funct Ecol 3: 252–255

Koide RT, Goff MD, Dickie IA (2000) Component growth efficiencies of mycorrhizal and non-mycorrhizal plants. New Phytol 148: 163–168

Konings H, Verschuren G (1980) Formation of aerenchyma in roots of *Zea mays* in aerated solutions, and its relation to nutrient supply. Physiol Plant 49: 265–279

Koyama H, Kawamura A, Kihara T, Hara T, Takita E, Shibata D (2000) Overexpression of mitochondrial citrate synthase in *Arabidopsis thaliana* improved growth on a phosphorus-limited soil. Plant Cell Physiol 41: 1030–1037

Lambers H, Atkin O, Millenaar FF (2002) Respiratory patterns in roots in relation to their functioning. In: Waisel Y, Eshel A, Kafkaki K (eds), Plant Roots: The Hidden Half. Marcel Dekker, New York, pp 521–552

Lambers H, Shane MW, Cramer MD, Pearse SJ, Veneklaas EJ (2006) Root structure and functioning for efficient acquisition of phosphorus: matching morphological and physiological traits. Ann Bot 98: 693–713

Lamont BB (2003) Structure, ecology and physiology of root clusters - a review. Plant Soil 248: 1–19

Lascaris D, Deacon JW (1991) Relationship between root cortical senescence and growth of wheat as influenced by mineral nutrition, *Idriella bolleyi* (sprague) von Arx and pruning of leaves. New Phytol 118: 391–396

Lewis DG, Quirk JP (1967) Phosphate diffusion in soil and uptake by plants. Plant Soil 26: 445–453

Li L, Tang C, Rengel Z, Zhang F (2003) Chickpea facilitates phosphorus uptake by intercropped wheat from an organic phosphorus source. Plant Soil 248: 297–303

Li SM, Li L, Zhang F, Tang C (2004) Acid phosphatase role in chickpea/maize intercropping. Ann Bot 94: 297–303

Liao H, Rubio G, Yan XL, Cao AQ, Brown KM, Lynch JP (2001) Effect of phosphorus availability on basal root shallowness in common bean. Plant Soil 232: 69–79

Liao H, Yan X, Rubio G, Beebe SE, Blair MW, Lynch JP (2004) Genetic mapping of basal root gravitropism and phosphorus acquisition efficiency in common bean. Funct Plant Biol 31: 959–970

Lizaso JI, Melendez LM, Ramirez R (2001) Early flooding of two cultivars of tropical maize. I. Shoot and root growth. J Plant Nutr 24: 979–995

Lopez-Bucio J, Nieto-Jacobo MF, Ramirez-Rodriguez V, Herrera-Estrella L (2000) Organic acid metabolism in plants: from adaptive physiology to transgenic varieties for cultivation in extreme soils. Plant Sci 160: 1–13

Lopez-Bucio J, Hernandez-Abreu E, Sánchez-Calderon L, Nieto-Jacobo MF, Simpson J, Herrera-Estrella L (2002) Phosphate availability alters architecture and causes changes in hormone sensitivity in the Arabidopsis root system. Plant Physiol 129: 244–256

Lu Y, Wassmann R, Neue HU, Huang C (1999) Impact of phosphorus supply on root exudation, aerenchyma formation and methane emission of rice plants. Biogeochemistry 47: 203–218

Lynch JP (1995) Root architecture and plant productivity. Plant Physiol 109: 7–13

Lynch J (2005) Root architecture and nutrient acquisition. In: BassiriRad H (ed), Nutrient Acquisition by Plants. An Ecological Perspective. Springer, Berlin/Heidelberg/New York, pp 147–184

Lynch J (2007) Roots of the second green revolution. Aust J Bot 55: 493–512

Lynch JP, Beebe SE (1995) Adaptation of beans (*Phaseolus vulgaris* L.) to low phosphorus availability. HortScience 30: 1165–1171

Lynch J, Brown K (1998) Root architecture and phosphorus acquisition efficiency in common bean. In: Lynch J, Deikman J (eds), Phosphorus in Plant Biology: Regulatory Roles in Molecular, Cellular, Organismic, and Ecosystem Processes. ASPP, Rockville, MD, pp 148–156

Lynch JP, Brown KM (2001) Topsoil foraging - an architectural adaptation of plants to low phosphorus availability. Plant Soil 237: 225–237

Lynch J, Brown K (2006) Whole plant adaptations to low phosphorus availability. In: Huang B (ed), Plant-Environment Interactions (3rd edition). Taylor & Francis, New York, pp 209–242

Lynch J, Deikman J (1998) Phosphorus in Plant Biology: Regulatory Roles in Molecular, Cellular, Organismic, and Ecosystem Processes. ASPP, Rockville, MD

Lynch JP, Ho MD (2005) Rhizoeconomics: carbon costs of phosphorus acquisition. Plant Soil 269: 45–56

Lynch JP, St. Clair SB (2004) Mineral stress: the missing link in understanding how global climate change will affect plants in real world soils. Field Crop Res 90: 101–115

Lynch J, Lauchli A, Epstein E (1991) Vegetative growth of the common bean in response to phosphorus nutrition. Crop Sci 31: 380–387

Ma QF, Longnecker N, Atkins C (2002) Varying phosphorus supply and development, growth and seed yield in narrow-leafed lupin. Plant Soil 239: 79–85

Ma Z, Bielenberg DG, Brown KM, Lynch JP (2001a) Regulation of root hair density by phosphorus availability in *Arabidopsis thaliana*. Plant Cell Environ 24: 459–467

Ma Z, Walk TC, Marcus A, Lynch JP (2001b) Morphological synergism in root hair length, density, initiation and geometry for phosphorus acquisition in *Arabidopsis thaliana*: a modeling approach. Plant Soil 236: 221–235

Ma Z, Baskin TI, Brown KM, Lynch JP (2003) Regulation of root elongation under phosphorus stress involves changes in ethylene responsiveness. Plant Physiol 131: 1381–1390

Maina GG, Brown JS, Gersani M (2002) Intra-plant versus inter-plant root competition in beans: avoidance, resource matching or tragedy of the commons. Plant Ecol 160: 235–247

Mano Y, Omori F, Takamizo T, Kindiger B, Bird RM, Loaisiga CH (2006) Variation for root aerenchyma formation in flooded and non-flooded maize and teosinte seedlings. Plant Soil 281: 269–279

Mano Y, Omori F, Takamizo T, Kindiger B, Bird RM, Loaisiga CH, Takahashi H (2007) QTL mapping of root aerenchyma formation in seedlings of a maize x rare teosinte "*Zea nicaraguensis*" cross. Plant Soil 295: 103–113

Manske GGB, Ortiz-Monasterio JI, Van Grinkel M, González R, Rajaram S, Molina E, Vlek PLG (2000) Traits associated with improved P-uptake efficiency in CIMMYT's semidwarf spring bread wheat grown on an acid Andisol in Mexico. Plant Soil 221: 189–204

Marschner H (1995) Mineral Nutrition of Higher Plants (2nd edition). Academic, London

Miguel M (2004) Genotypic variation in root hairs and phosphorus efficiency in common bean (*Phaseolus vulgaris* L.). Dissertation, Penn State University

Miller CR, Ochoa I, Nielsen KL, Beck D, Lynch JP (2003) Genetic variation for adventitious rooting in response to low phosphorus availability: potential utility for phosphorus acquisition from stratified soils. Funct Plant Biol 30: 973–985

Mollier A, Pellerin S (1999) Maize root system growth and development as influenced by phosphorus deficiency. J Exp Bot 50: 487–497

Neumann G, Martinoia E (2002) Cluster roots - an underground adaptation for survival in extreme environments. Trends Plant Sci 7: 162–167

Nielsen KL, Bouma TJ, Lynch JP, Eissenstat DM (1998) Effects of phosphorus availability and vesicular-arbuscular mycorrhizas on the carbon budget of common bean (*Phaseolus vulgaris*). New Phytol 139: 647–656

Nielsen KL, Eshel A, Lynch JP (2001) The effect of phosphorus availability on the carbon economy of contrasting common bean (*Phaseolus vulgaris* L.) genotypes. J Exp Bot 52: 329–339

Niklas K (1994) Plant Allometry: The Scaling of Form and Process. University of Chicago Press, Chicago, IL

Ochoa I, Blair M, Lynch J (2006) QTL analysis of adventitious root formation in common bean (*Phaseolus vulgaris* L.) under contrasting phosphorus availability. Crop Sci 46: 1609–1621

Oldeman L, Hakkeling R, Sombroek W (1991) World Map of the Status of Human-Induced Soil Degradation: An Explanatory Note (2nd revised edition). ISRIC, Wageningen, The Netherlands

Oldroyd GED, Harrison MJ, Udvardi M (2005) Peace talks and trade deals. Keys to long-term harmony in legume-microbe symbioses. Plant Physiol 137: 1205–1210

Ostertag R (2001) Effects of nitrogen and phosphorus availability on fine-root dynamics in Hawaiian montane forests. Ecology 82: 485–499

Owusu-Bennoah E, Zapata F, Fardeau JC (2002) Comparison of greenhouse and ^{32}P isotopic laboratory methods for evaluating the agronomic effectiveness of natural and modified rock phosphates in some acid soils of Ghana. Nutr Cycl Agroecosys 63: 1–12

Pellerin S, Mollier A, Plenet D (2000) Phosphorus deficiency affects the rate of emergence and number of maize adventitious nodal roots. Agron J 92: 690–697

Peng SB, Eissenstat DM, Graham JH, Williams K, Hodge NC (1993) Growth depression in mycorrhizal citrus at high-phosphorus supply - analysis of carbon costs. Plant Physiol 101: 1063–1071

Peterson RL, Farquhar ML (1996) Root hairs: specialized tubular cells extending root surfaces. Bot Rev 62: 1–40

Pothuluri J, Kissel D, Whitney D, Thien S (1986) Phosphorus uptake from soil layers having different soil test phosphorus levels. Agron J 78: 991–994

Radin J, Eidenbock M (1984) Hydraulic conductance as a factor limiting leaf expansion of phosphorus-deficient cotton plants. Plant Physiol 75: 372–377

Ray JD, Kindiger B, Dewald CL, Sinclair TR (1998) Preliminary survey of root aerenchyma in Tripsacum. Maydica 43: 49–53

Ray JD, Kindiger B, Sinclair TR (1999) Introgressing root aerenchyma into maize. Maydica 44: 113–117

Riedell WE, Reese RN (1999) Maize morphology and shoot CO_2 assimilation after root damage by western corn rootworm larvae. Crop Sci 39: 1332–1340

Robinson D (2005) Integrated root responses to variations in nutrient supply. In: BassiriRad H (ed), Nutrient Acquisition by Plants. An Ecological Perspective. Springer, Berlin/Heidelberg/New York, pp 43–62

Robinson D, Hodge A, Griffiths BS, Fitter AH (1999) Plant root proliferation in nitrogen-rich patches confers competitive advantage. Proc Roy Soc Lond B 266: 431–435

Roman-Aviles B, Snapp SS, Kelly JD (2004) Assessing root traits associated with root rot resistance in common bean. Field Crop Res 86: 147–156

Rossiter R (1978) Phosphorus deficiency and flowering time in subterranean clover *Trifolium subterraneum*. Ann Bot 42: 325–330

Rubio G, Walk T, Ge ZY, Yan XL, Liao H, Lynch JP (2001) Root gravitropism and below-ground competition among neighbouring plants: a modelling approach. Ann Bot 88: 929–940
Rubio G, Liao H, Yan XL, Lynch JP (2003) Topsoil foraging and its role in plant competitiveness for phosphorus in common bean. Crop Sci 43: 598–607
Ryan MH, Graham JH (2002) Is there a role for arbuscular mycorrhizal fungi in production agriculture? Plant Soil 244: 263–271
Ryan PR, Delhaize E, Jones DL (2001) Function and mechanism of organic anion exudation from plant roots. Annu Rev Plant Physiol Plant Mol Biol 52: 527–560
Ryser P, Lambers H (1995) Root and leaf attributes accounting for the performance of fast-growing and slow-growing grasses at different nutrient supply. Plant Soil 170: 251–265
Setter TL, Waters I (2003) Review of prospects for germplasm improvement for waterlogging tolerance in wheat, barley and oats. Plant Soil 253: 1–34
Shane MW, Lambers H (2005) Cluster roots: a curiosity in context. Plant Soil 274: 101–125
Skene KR (2001) Cluster roots: model experimental tools for key biological problems. J Exp Bot 52: 479–485
Smith S, Read D (1997) Mycorrhizal Symbiosis. Academic, San Diego, CA
Smith SE, Smith FA, Jakobsen I (2003) Mycorrhizal fungi can dominate phosphate supply to plants irrespective of growth responses. Plant Physiol 133: 16–20
Snapp SS, Lynch JP (1996) Phosphorus distribution and remobilization in bean plants as influenced by phosphorus nutrition. Crop Sci 36: 929–935
Snapp S, Koide R, Lynch J (1995) Exploitation of localized phosphorus-patches by common bean roots. Plant Soil 177: 211–218
Snapp S, Kirk W, Roman-Aviles B, Kelly J (2003) Root traits play a role in integrated management of Fusarium root rot in snap beans. HortScience 38: 187–191
Soil Survey Staff (1999) Soil Taxonomy. A Basic System of Soil Classification for Making and Interpreting Soil Surveys (2nd edition). USDA/NRCS
Steen I (1998) Phosphorus availability in the 21st century. Management of a non-renewable resource. Phosphorus Potassium 217: 25–31
Steingrobe B, Schmid H, Claassen N (2001) Root production and root mortality of winter barley and its implication with regard to phosphate acquisition. Plant Soil 237: 239–248
Striker GG, Insausti P, Grimoldi AA, Leon RJC (2006) Root strength and trampling tolerance in the grass *Paspalum dilatatum* and the dicot *Lotus glaber* in flooded soil. Funct Ecol 20: 4–10
Striker GG, Insausti I P, Grimoldi AA, Vega AS (2007) Trade-off between root porosity and mechanical strength in species with different types of aerenchyma. Plant Cell Environ 30: 580–589
Tesfaye M, Temple SJ, Allan DL, Vance CP, Samac DA (2001) Overexpression of malate dehydrogenase in transgenic alfalfa enhances organic acid synthesis and confers tolerance to aluminum. Plant Physiol 127: 1836–1844
Tesfaye M, Liu J, Allan DL, Vance CP (2007) Genomic and genetic control of phosphate stress in legumes. Plant Physiol 144: 594–603
Ticconi CA, Abel S (2004) Short on phosphate: plant surveillance and countermeasures. Trends Plant Sci 9: 548–555
Tinker P, Nye P (2000) Solute Movement in the Rhizosphere. Oxford University Press, New York
Tomscha JL, Trull MC, Deikman J, Lynch JP, Guiltinan MJ (2004) Phosphatase under-producer mutants have altered phosphorus relations. Plant Physiol 135: 334–345
Van der Werf A, Welschen R, Lambers H (1992) Respiratory losses increase with decreasing inherent growth rate of a species and with decreasing nitrate supply: a search for explanations for these observations. In: Lambers H, Van der Plas L (eds), Molecular, Biochemical, and Physiological Aspects of Plant Respiration. SPB Academic Publishing, The Hague, pp 421–432
Vance CP (2008) Plants without arbuscular mycorrhizae. In: White PJ, Hammond JP (eds), The Ecophysiology of Plant-Phosphorus Interactions. Springer, Dordrecht, The Netherlands, pp 117–142

Vance CP, Uhde-Stone C, Allan DL (2003) Phosphorus acquisition and use: critical adaptations by plants for securing a nonrenewable resource. New Phytol 157: 423–447

Vartapetian BB, Jackson MB (1997) Plant adaptations to anaerobic stress. Ann Bot 79: 3–20

Visser EJW, Colmer TD, Blom C, Voesenek L (2000) Changes in growth, porosity, and radial oxygen loss from adventitious roots of selected mono- and dicotyledonous wetland species with contrasting types of aerenchyma. Plant Cell Environ 23: 1237–1245

Walk T, Jaramillo R, Lynch J (2006) Architectural tradeoffs between adventitious and basal roots for phosphorus acquisition. Plant Soil 279: 347–366

Walker T (1965) The significance of phosphorus in pedogenesis. In: Hallsworth E, Crawford D (eds), Experimental Pedology. Butterworths, London, pp 295–316

Waters BM, Blevins DG (2000) Ethylene production, cluster root formation, and localization of iron (III) reducing capacity in Fe deficient squash roots. Plant Soil 225: 21–31

White PJ, Hammond JP (2008) Phosphorus nutrition of terrestrial plants. In: White PJ, Hammond JP (eds), The Ecophysiology of Plant-Phosphorus Interactions. Springer, Dordrecht, The Netherlands, pp 51–81

Whiteaker G, Gerloff G, Gabelman W, Lindgren D (1976) Intraspecific differences in growth of beans at stress levels of phosphorus. J Am Soc Hortic Sci 101: 472–475

Williamson LC, Ribrioux SPCP, Fitter AH, Leyser HMO (2001) Phosphate availability regulates root system architecture in Arabidopsis. Plant Physiol 126: 875–890

Wood S, Sebastian K, Scherr S (2000) Pilot Analysis of Global Ecosystems: Agroecosystems. World Resources Institute, Washington, DC

World Bank (2004) World Development Indicators. Washington, DC

Wouterlood M, Cawthray GR, Scanlon TT, Lambers H, Veneklaas EJ (2004) Carboxylate concentrations in the rhizosphere of lateral roots of chickpea (*Cicer arietinum*) increase during plant development, but are not correlated with phosphorus status of soil or plants. New Phytol 162: 745–753

Wouterlood M, Lambers H, Veneklaas EJ (2005) Plant phosphorus status has a limited influence on the concentration of phosphorus-mobilising carboxylates in the rhizosphere of chickpea. Funct Plant Biol 32: 153–159

Xie YJ, Yu D (2003) The significance of lateral roots in phosphorus (P) acquisition of water hyacinth (*Eichhornia crassipes*). Aquat Bot 75: 311–321

Yan X (2005) The roots of P-efficient soybean: Theories and practices. In: Li CJ *et al.* (eds), Plant Nutrition for Food Security. Human Health and Environmental Protection, Tsinghia University Press, Beijing, China, pp 36–37

Yan X, Liao H, Beebe SE, Blair MW, Lynch JP (2004) QTL mapping of root hair and acid exudation traits and their relationship to phosphorus uptake in common bean. Plant Soil 265: 17–29

Yun SJ, Kaeppler SM (2001) Induction of maize acid phosphatase activities under phosphorus starvation. Plant Soil 237: 109–115

Zhao J, Fu J, Liao H, He Y, Nian H, Hu Y, Qiu L, Dong Y, Yan X (2004) Characterization of root architecture in an applied core collection for phosphorus efficiency of soybean germplasm. Chinese Science Bull 49: 1611–1620

Zhu J, Lynch JP (2004) The contribution of lateral rooting to phosphorus acquisition efficiency in maize (*Zea mays*) seedlings. Funct Plant Biol 31: 949–958

Zhu J, Kaeppler SM, Lynch JP (2005a) Mapping of QTL for lateral root branching and length in maize (*Zea mays* L.) under differential phosphorus supply. Theor Appl Genet 111: 688–695

Zhu J, Kaeppler SM, Lynch JP (2005b) Topsoil foraging and phosphorus acquisition efficiency in maize (*Zea mays*). Funct Plant Biol 32: 749–762

Zhu J, Kaeppler SM, Lynch JP (2005c) Mapping of QTL controlling root hair length in maize (*Zea mays* L.) under phosphorus deficiency. Plant Soil 270: 299–310

Chapter 6
PLANTS WITHOUT ARBUSCULAR MYCORRHIZAE*

Carroll P. Vance

INTRODUCTION

Although mycorrhizal symbioses (described elsewhere in this volume) are the most important adaptation for angiosperms to acquire scarce phosphorus (P), many plant families contain species that either do not form or rarely form this pivotal association (Skene 1998; Miller *et al.* 1999; Cripps and Eddington 2005; Miller 2005). This review will address adaptations and mechanisms for acquisition and use of scarce P in plants lacking effective mycorrhizal symbioses. The primary focus will be on root adaptations in species that develop specialized-complex roots (cluster and dauciform) in response to P deficiency. Although not producing cluster or dauciform roots in response to P deficiency, Arabidopsis will also be considered because it does not form mycorrhizal symbiosis and is a model species for evaluating plant adaptation to P deficiency.

Plants have evolved two broad strategies for improved P acquisition and use in nutrient-limiting environments: (1) those aimed at conservation of use; and (2) those directed toward enhanced acquisition or uptake (Vance *et al.* 2003; Ticconi and Abel 2004; Misson *et al.* 2005; Morcuende *et al.* 2007). Processes that conserve the use of P involve decreased growth rate, increased growth per unit of P uptake, remobilization of internal P, modifications in carbon (C) metabolism that bypass P-requiring steps, alternative respiratory pathways, and alterations in membrane biosynthesis requiring less P (Plaxton and Carswell 1999; Uhde-Stone *et al.* 2003a,b; Wasaki *et al.* 2003; Misson *et al.* 2005; Lambers *et al.* 2006). In comparison, processes that lead to enhanced P uptake include modified root architecture and greater root growth, prolific development of root hairs leading to expanded root surface area, enhanced expression of Pi transporter genes, and increased production and exudation of phosphatases and organic acids (Marschner *et al.* 1986; López-Bucio *et al.* 2002; Shane and Lambers 2005). These numerous adaptive responses to P-deficiency are not mutually exclusive and all may occur within a single species.

P.J. White, J.P. Hammond (eds.) *The Ecophysiology of Plant-Phosphorus Interactions*,

ROOT ARCHITECTURE

Root architecture (Dinkelaker *et al.* 1995; Lynch and Brown 2001; Williamson *et al.* 2001; López-Bucio *et al.* 2003; Lambers *et al.* 2006) refers to the complexity of root system spatial configurations that arise in response to soil conditions. It includes root morphology, topology, and distribution patterns. Soil P limitation is a primary effector of root architecture (Dinkelaker *et al.* 1995; Borch *et al.* 1999; Williamson *et al.* 2001; López-Bucio *et al.* 2003; Lambers *et al.* 2006) and is known to impact all aspects of root growth and development. Phenotypic and genotypic adaptations to P deficiency involve changes in root architecture that facilitate acquisition of P from the topsoil (Lamont 1982; Lynch and Brown 2001; López-Bucio *et al.* 2003). Adaptations that enhance acquisition of P from topsoil are important because of the relative immobility of P in soil, with the highest concentrations usually found in the topsoil and little movement of P into the lower soil profiles. Lynch and Brown (2001) refer to P deficiency induced modification of root architecture as adaptations for topsoil foraging. Root characteristics associated with improved topsoil foraging for scarce P are a more horizontal basal-root growth angle resulting in more shallow roots, increased adventitious root formation, enhanced lateral root proliferation, and increased root hair density and length. Such modifications in root architecture in response to P deficiency are well documented in Arabidopsis and in those species forming either cluster or dauciform roots.

ROOT DEVELOPMENT AND FUNCTION ARE ALTERED BY P DEFICIENCY

Cluster root species

Distribution

While Arabidopsis is a compelling model species for the study of plant adaptation to P deficiency, a wide group of species respond to the scarcity of P with profound changes in root architecture resulting in cluster roots (Dinkelaker *et al.* 1995; Diem *et al.* 2000; Pate and Watt 2001; Neumann and Martinoia 2002; Lamont 2003; Vance *et al.* 2003; Shane and Lambers 2005; Lambers *et al.* 2006). Phosphorus-deficiency induced cluster root formation involves the localized formation of massive numbers of determinate secondary/tertiary rootlets having prolific root hair formation (Figures 6.1, 6.2). In the literature they can be referred to as cluster, proteoid, bottlebrush, and/or dauciform roots (Lamont 2003; Vance *et al.* 2003; Lambers *et al.* 2006). Originally thought to be present in only native species of the Proteaceae in Australia and South Africa, cluster roots are now known to have widespread occurrence (Table 6.1) in members of the Betulaceae, Moraceae, Casuarinaceae, Myricaceae, Eleagnaceae, Cucurbitaceae, Cyperaceae, Restionaceae, and Fabaceae

Fig. 6.1 Cluster roots of members of the family Protaceae (**a**, **b**, **c**) and dauciform roots of the Cyperaceae (**d**). **a**. Compound cluster root of *Banksia cirsioides*. This compound cluster root is 6 cm in length (Permission and courtesy of Byron Lamont). **b**. Young cluster root of *Leucodendron laureolum* prior to root hair development on rootlets (Scanning electron microscopy. Permission and courtesy of Byron Lamont). **c**. Cluster root of *L. laureolum* emerging through epidermis (Scanning electron microscopy. Permission and courtesy of Byron Lamont). **d**. Dauciform roots of *Caustis* sp (Cyperaceae) with prolific root hair development. (Permission and courtesy of Michael Shane.)

(Lamont 1993, 2003; Skene 1998; Lambers *et al.* 2006). Many species in these families are particularly adapted to growth in nutrient-impoverished soils through the evolution of cluster roots for acquisition of scarce nutrients. Members of the Cyperaceae and Restionaceae are frequently primary colonizers of marshes and wetland. All cluster root species are well adapted to low P growth environments (Lamont 1993, 2003; Lambers *et al.* 2006). Beside mycorrhizal associations, cluster roots are regarded as one of the major adaptations for P acquisition (Dinkelaker *et al.* 1995; Skene 1998; Diem *et al.* 2000; Adams and Pate 2002; Lamont 2003; Vance *et al.* 2003; Shane and Lambers 2005).

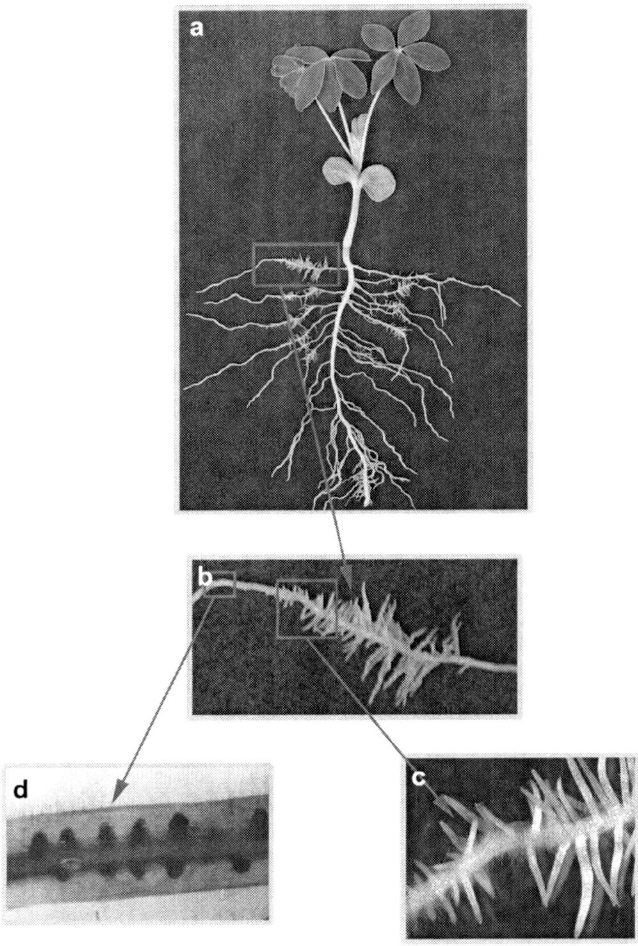

Fig. 6.2 White lupin plant grown under P-deficient conditions 14 days after emergence. **a**. Note cluster root zones on first-order lateral roots. **b**. Cluster root zone boxed is enlarged to show developmental zones. Premature (**c**) and pre-emergent (**d**) zones are shown in boxes at higher magnification. (Prepared by Dr. B. Bucciarelli.)

Development

Recent reviews (Lamont 2003; Shane and Lambers 2005; Lambers *et al.* 2006) have documented the striking diversity in cluster root morphology. Cluster roots form through synchronized development of densely packed determinate rootlets. They can comprise single clusters formed on parent axis, as found in members of the Proteaceae, such as *Hakeae* spp., *Leucadendron* spp., *Grevillea robusta*, and in the Fabaceae, such as white lupin (Lamont 1982, 2003; Skene *et al.* 1996; Lambers *et al.* 2006; Figure 6.1). However, some species from the Proteaceae, like *Banksia*

Table 6.1 Plant families having specialized compound (cluster or dauciform) root development and the prevalence of either nitrogen (N_2) fixing or mycorrhizal symbioses. (Adapted from: Dinkelaker *et al.* 1995; Skene 1998; Waters and Blevins 2000; Lamont 2003; Shane and Lambers 2005; Lambers *et al.* 2006.)

		Symbiosis		
Family	Root	Mycorrhizal	N_2 fixation	
			Rhizobial	Actinorrhizal
Fabaceae[a]	Cluster	–/+	+	–
Moraceae	Cluster	–	–	–
Betulaceae[a]	Cluster	–/+	–	+
Casuarinaceae[a]	Cluster	–/+	–	+
Eleagnaceae	Cluster	+	–	+
Myricaceae[a]	Cluster	–/+	–	+
Proteaceae	Cluster	–	–	–
Cucurbitaceae[a]	Cluster	+	–	–
Cyperaceae[a]	Dauciform	–/+	–	–

[a]Families noted as –/+ may be mycorrhizal but those species forming cluster roots are either non-mycorrhizal– or weakly mycorrhizal (+ = mycorrhizal/–/+ = weakly mycorrhizal).

grandis, form more complex compound clusters that form embedded mats in the soil profile (Lamont 1993, 2003; Adams and Pate 2002; Shane and Lambers 2005). Several features of dicot cluster root development and morphology are distinguished from that of typical dicot lateral roots. First, lateral roots are initiated usually in an alternating pattern from the pericycle of primary roots near the zone of metaxylem differentiation (Celenza *et al.* 1995; Charlton 1996), while cluster rootlets are initiated in waves along the axis of second and third order lateral roots (Skene and James 2000; Pate and Watt 2001; Lamont 2003). Second, typical lateral roots are initiated singularly opposite a protoxylem point, unlike cluster roots which are initiated in multiples opposite every protoxylem point within the wave of differentiation (Figures 6.1, 6.2). Third, cluster rootlets produce a superabundance of root hairs due to an apparent loss of regulation of trichoblast differentiation, while root hair development in typical roots is highly regulated and occurs from a discrete number of trichoblasts (Dolan 2001; Müller and Schmidt 2004). The length of cluster root segments varies from a few millimeters to 2–4 cm. Under P-deficient conditions cluster roots can comprise more than 70% of the plant root mass (Lamont 1982, 2003; Pate and Watt 2001; Lambers *et al.* 2006). The frequency of cluster roots in P-deficient soil and the accompanying increase in root hair density results in an increase in root surface area of greater than 100-fold (Lamont 2003; Vance *et al.* 2003; Shane and Lambers 2005). Another notable feature of cluster rootlets is that their growth is determinate, ceasing shortly after emergence, as contrasted to the indeterminate growth of lateral roots (Neumann and Martinoia 2002). This highly synchronous developmental pattern reflects that cluster root formation is an elegant, finely tuned process. Moreover, because root pericycle cells are arrested in the G2 phase of the cell cycle (Skene 1998; Skene and James 2000), cluster root initiation likely involves a hormone mediated concerted release of multiple pericycle cells from the G2 phase in a wave-like pattern along second-order lateral roots.

Root architectural changes, classified as cluster root types, also occur in the sedges (Cyperaceae) and the rushes (Restionaceae). Lamont (1974) characterized the cluster root morphology found in the Cyperaceae as dauciform due to the carrot-like developmental pattern along the root axis. As dauciform roots mature they develop large numbers of long (3–5 mm) root hairs (Figure 6.1; Shane and Lambers 2005; Shane *et al.* 2005). The development of dauciform roots in *Schoenus unispiculatus* was directly related to P availability, with their formation being suppressed as P availability increased (Shane *et al.* 2005). Recently Shane *et al.* (2006) have shown that dauciform roots of *S. unispiculatus* are structurally distinct from typical cluster roots, but functionally analogous to them. In the Restionaceae, the cluster root morphology has been characterized as capillaroid (Lamont 1982) due to the sponge-like, water-holding capacity of the root-soil aggregate. Capillaroid species have root clumps with exceptionally long root hairs (see Figure 6.3; Lambers *et al.* 2006).

Cluster root function

Similar to mycorrhizal association, cluster roots increase the root surface area and soil volume exploited for mining nutrients. In contrast to mycorrhizae, which grow over much of the entire root surface, the hairy and densely packed lateral rootlets in cluster root zones bind tightly to trapped soil particles and organic matter in localized soil volumes. Cluster root aggregates are most prominent in the upper layers of the soil profile where P is most abundant (Lamont 2003; Shane and Lambers 2005;

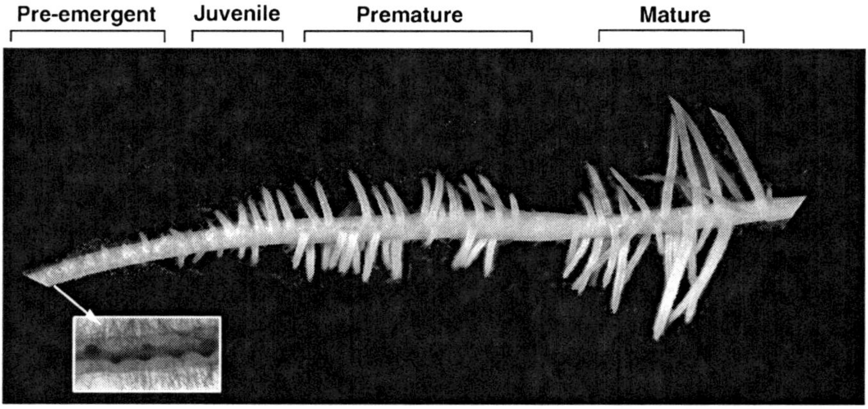

Fig. 6.3 Cluster root segment of P-deficient white lupin. Rootlet development proceeds from left to right. Pre-emergent zone contains rootlet confined to the root cortex and not yet emerged. Juvenile, premature, and mature zones contain emerged rootlets as described by Neumann and Martinoia (2002). Mature cluster roots have ceased elongating and have determinate development. Inset shows root hairs in pre-emergent zone. (Prepared by Dr. B. Bucciarelli.)

Lambers *et al.* 2006). The dense aggregation of cluster roots and soil facilitates the acquisition and uptake of nutrients, particularly P (Lamont 2003; Vance *et al.* 2003; Shane and Lambers 2005; Lambers *et al.* 2006). Enhanced capacity for nutrient acquisition and uptake by cluster roots occurs by effectively concentrating root exudates and plant nutrient uptake systems in localized patches of soil (Grierson and Attiwill 1989). Cluster roots exude organic acids (Gardner *et al.* 1983; Dinkelaker *et al.* 1995; Johnson *et al.* 1996a,b; Shane *et al.* 2006), protons (H^+; Dinkelaker *et al.* 1995; Neumann and Martinoia 2002; Yan *et al.* 2002; Shane *et al.* 2006), phenolics (Neumann *et al.* 1999; Weisskopf *et al.* 2006a,b), acid phosphatases (Dinkelaker *et al.* 1995; Gilbert *et al.* 1999; Miller *et al.* 2001), and iron chelate reductase (Dinkelaker *et al.* 1995; Neumann and Martinoia 2002). Cluster roots have enhanced expression of phosphate (Pi) transporters (Liu *et al.* 2001) and increased Pi uptake (Keerthisinghe *et al.* 1998; Neumann *et al.* 2000; Sousa *et al.* 2007). Exudation of carboxylates, H^+, phenolics, and acid phosphatases releases P bound to inorganic and organic particles making it available for rapid uptake by Pi transporters. Cluster root development accompanied by corresponding changes in cluster root metabolism and membrane uptake systems reflect an elegant, highly coordinated molecular response to P deficiency.

Although the development of cluster roots in Proteaceae and Cyperaceae has received growing attention, white lupin has served as the model for analysis of biochemical and molecular adaptations contributing to enhanced P acquisition and use by cluster roots (Dinkelaker *et al.* 1995; Watt and Evans 1999; Neumann and Martinoia 2002; Uhde-Stone *et al.* 2003a,b; Liu *et al.* 2005). As P deficiency occurs in white lupin, a cascade of changes in gene expression occurs resulting in synchronous development of: cluster roots having prolific root hair density; exudation of carboxylates, H^+, and enzymes; enhanced expression of membrane transporters; and apparent heightened sensitivity in roots to hormonal signals. Neumann *et al.* (2000) have staged cluster root development into four phases distinguished by rootlet emergence and biochemical response (Figure 6.3). In the juvenile and premature stages, cluster rootlets have emerged from the cortex and are actively elongating. At these stages rootlets are exuding malate and citrate in fairly equal amounts (300–700 nmol h^{-1} g^{-1} fw) accompanied by uniform rhizosphere pH (5–6) and little extrusion of protons. As the mature stage is attained, rootlet elongation stops, citrate exudation exceeds malate by four- to five-fold, copious amounts of acid phosphatase and phenolics are exuded accompanied by H^+ extrusion and a reduction in rhizosphere pH (Neumann and Martinoia 2002). Mature cluster root zones have increased transcript abundance for P transporters (Liu *et al.* 2001), acid phosphatase (Miller *et al.* 2001), sugar metabolism genes (Uhde-Stone *et al.* 2003a), and MATE-transporters (Uhde-Stone *et al.* 2005). It is noteworthy that total RNA concentration decreases as cluster root zones progress through development into the later phase of maturity, suggesting an increased turnover of nucleic acids to provide P for remobilization (Johnson *et al.* 1996a,b). The burst of exudation occurring in mature cluster root zones occurs over a three to four day period followed by the senescent stages of differentiation in which cluster roots turn brown and physically deteriorate (Watt and Evans 1999). Recent studies with cluster roots of *Hakea*, Proteaceae (Shane *et al.* 2004; Shane and Lambers 2005) and with dauciform roots of *Schoenus*, Cyperaceae (Shane *et al.* 2006) show developmental stages and biochemical modifications analogous to those that occur in white lupin.

Phosphate acquisition

Phosphate acquisition from soil is enhanced in cluster root species (Gardner *et al.* 1982; Lamont 1982; Grierson 1992; Dinkelaker *et al.* 1995; Neumann *et al.* 2000; Keerthisinghe *et al.* 1998; Sousa *et al.* 2007). Enhanced acquisition of soil P occurs not only because of the biochemical and architectural changes that occur in cluster roots, which facilitate increased surface area for explorations and release of exudates to solubilize P, but also because Pi uptake is enhanced in cluster roots. Increased Pi uptake has been noted in cluster root zones of both white lupin (Keerthisinghe *et al.* 1998; Neumann *et al.* 2000) and *Hakea* (Sousa *et al.* 2007). The Pi concentration at which Pi uptake is half-maximal (Km) in cluster root species ranges from 1.0 μM for the high affinity uptake in low P soils to 40 μM under high P conditions. Sousa *et al.* (2007) showed a biphasic Pi uptake for *Hakea serica* indicating a high and low affinity system. Interestingly, Liu *et al.* (2001) isolated and characterized two Pi-transporter genes from white lupin cluster roots. The white lupin phosphate transporter 1 (*LaPT1*) was specifically expressed in cluster roots under P-deficiency and probably corresponds to a high affinity transporter. In contrast, *LaPT2* is expressed in all tissues in fairly high abundance and likely represents a low affinity transporter. We have noted the expression abundance of numerous transporter genes in P deficiency induced cluster roots of white lupin (Uhde-Stone *et al.* 2003b) ranging from Pi transporters to sulfate transporters, amino acid, and sugar transporters. The abundance of these transporters in white lupin cluster roots suggests the cluster root system is geared up for effective transport of many nutrients.

Modified carbon metabolism and partitioning is required for adaptation to P deficiency

Under normal growth and developmental conditions, plant roots exude a wide variety of organic compounds including: simple sugars, organic acids, amino acids, phenolics, quinones, (iso)-flavonoids, growth hormones, proteins, and polysaccharides (Marschner *et al.* 1986; Kochian *et al.* 2004). Exudation of organic compounds from roots can alter rhizosphere chemistry, soil microbial populations, competition, and plant growth. Exuded compounds are functionally diverse and can be involved in a wide array of processes ranging from signaling in plant-microbe interactions, to allelopathy and nutrient acquisition (Marschner *et al.* 1986; Hinsinger 2001; Ryan *et al.* 2001; Kochian *et al.* 2004). During P deficiency, roots show enhanced accumulation of sugars, increased synthesis and exudation of carboxylates, and a dependence upon sugars or current phloem transport for P-stress induced gene expression in roots (Johnson *et al.* 1996a; Keerthisinghe *et al.* 1998; Watt and Evans 1999; Neumann and Martinoia 2002; Shane *et al.* 2004; Liu *et al.* 2005).

Convincing evidence exists for exudation of malate and citrate as a principal mechanism for relieving the edaphic stress of P deficiency. The release of carboxylates allows for the chelation of Al^{3+}, Fe^{3+}, and Ca^{2+} and the subsequent displacement of Pi from bound or precipitated forms (Gardner *et al.* 1982; Dinkelaker *et al.* 1995; Ryan *et al.* 2001; Vance *et al.* 2003; Shane and Lambers 2005) and may also make organic P more susceptible to acid phosphatase activity. Tricarboxylates, such as citrate, are more effective than dicarboxylates, such as malate, at displacing bound P due to their greater affinity for Fe^{3+}, Al^{3+}, and Ca^{2+} (Hinsinger 2001; Ryan *et al.* 2001; Kochian *et al.* 2004). While many plant species exude carboxylates under P-deficient conditions (Marschner *et al.* 1986; Ryan *et al.* 2001; Lambers *et al.* 2006), this trait achieves maximum effectiveness in cluster root species (Johnson *et al.* 1996b; Keerthisinghe *et al.* 1998; Neumann and Martinoia 2002; Kihara *et al.* 2003; Shane *et al.* 2004, 2006). When grown under P-deficient conditions, cluster roots exude 20- to 40-fold more citrate and malate than P-sufficient roots. Shane *et al.* (2004) reported that a significant accumulation of carboxylates (75 μmol g^{-1} fw) in *Hakea* cluster roots was coincident with the maximum rate of organic acid exudation. Likewise, Watt and Evans (1999) showed that organic acid exudation from mature lupin cluster roots peaked at 34 nmol min^{-1} g^{-1} fw, compared to near non-detectable levels in the early stages of their development.

Malate is frequently detected as the primary component of exudates in juvenile and premature cluster roots, while citrate predominates at peak exudation in mature cluster roots (Johnson *et al.* 1994, 1996b; Watt and Evans 1999; Neumann *et al.* 2000; Shane *et al.* 2004, 2006). Peak exudation has been referred to as the exudative burst. The amount of carbon exuded in citrate and malate can constitute from 10% to greater than 25% of the apparent current photosynthate of the plant (Johnson *et al.* 1994, 1996a; Neumann *et al.* 2000). Cluster roots and P-stressed roots in general are strong sinks for photosynthate (Johnson *et al.* 1996a; Watt and Evans 1999; Neumann *et al.* 2000; Shane *et al.* 2004; Hernández *et al.* 2007; Morcuende *et al.* 2007).

The striking change that occurs in the organic acid exudation of P-deficient cluster roots is invariably reflected in concurrent changes in RNA expression and activity of enzymes involved in C metabolism (Johnson *et al.* 1994; Neumann *et al.* 2000; Massonneau *et al.* 2001; Uhde-Stone *et al.* 2003a; Shane *et al.* 2004). Measurements of shoot and root CO_2 fixation with $^{14}CO_2$, accompanied by measurements of C partitioning and exudation, show that organic acids exuded from lupin cluster roots are derived from both photosynthetic CO_2 fixation and root dark (anapleurotic) CO_2 fixation (Johnson *et al.* 1994, 1996a,b). Some 60–65% of C exuded from roots is shoot derived while 35–40% is root derived. Exudation of carboxylates from roots is accompanied by an increase in the activities of sucrose synthase (SS), enzymes of glycolysis, and organic acids, particularly phospho*enol*pyruvate carboxylase (PEPC) and malate dehydrogenase (MDH; Johnson *et al.*

1994; Massonneau *et al.* 2001; Uhde-Stone *et al.* 2003a; Peñaloza *et al.* 2005; Figure 6.4). Similar changes in C metabolism have been noted for P-deficient roots of bean (Ciereszko *et al.* 1998; Hernández *et al.* 2007) and Arabidopsis (Morcuende *et al.* 2007).

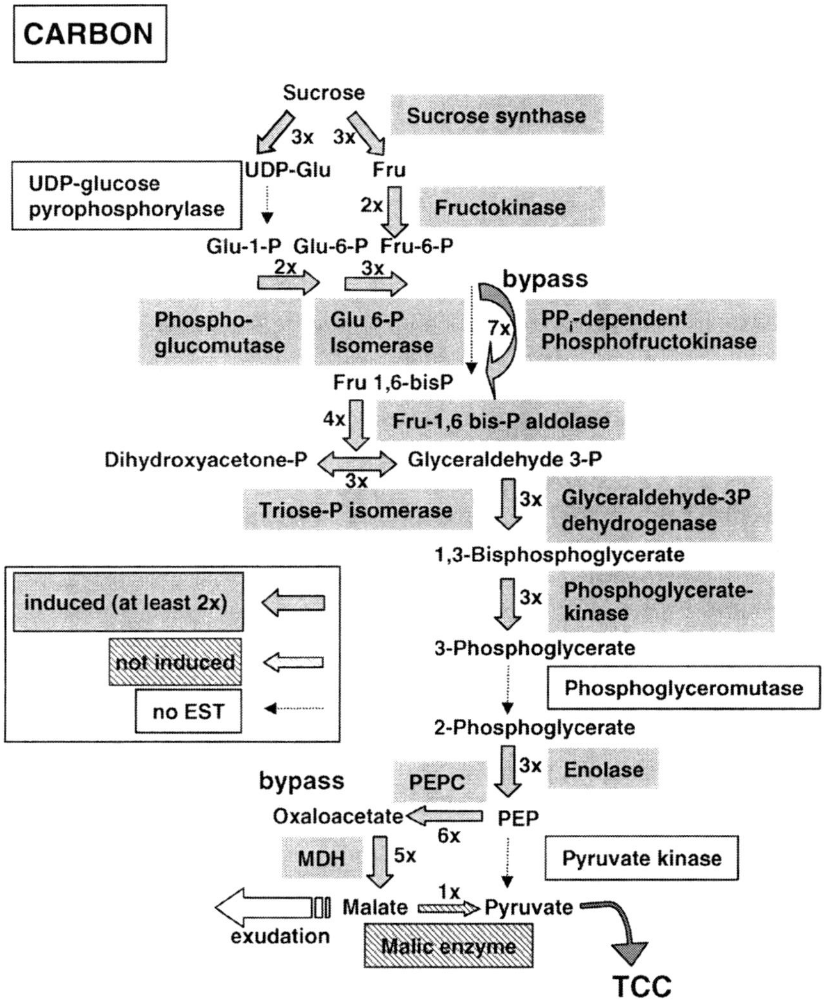

Fig. 6.4 Schematic representation of sucrose (carbon) metabolism through glycolysis with the formation of malate for exudation. ESTs with induced expression in P-deficient proteoid roots, compared to P-sufficient normal roots are represented in grey boxes. The average gene induction determined by two independent macroarrays is indicated at the corresponding arrows (e.g. 2x). Enzymes of the glycolytic pathway that were not found in the collection of 1,250 ESTs are shown in white boxes and are represented by dotted arrows. The majority of ESTs with possible function in the glycolytic pathway displayed increased expression in P-deficient proteoid roots, compared to P-sufficient normal roots. (From Uhde-Stone *et al.* 2003a.)

Interestingly, even though cluster roots are strong sinks for C there is not a striking increase in respiration. Cluster root C metabolism and respiration appear to be relatively efficient due to adaptive changes that occur in response to low P. For example, P deficiency can limit the activity of pyruvate kinase (PK), an enzyme requiring Pi and ADP. The PEPC, MDH, NAD-malic enzyme pathway can bypass the PK entryway into the TCA cycle, thereby maintaining the flow of C while avoiding the use of ADP and generating Pi (Plaxton and Carswell 1999; Morcuende *et al.* 2007). Likewise, the reduction in cellular ADP and Pi in P deficient plants can result in reduced efficiency of respiration by inhibiting the cytochrome pathway of electron transport. Non-phosphorylating pathways that bypass energy-requiring steps, like the alternative oxidase (AOX) system, can maintain cellular metabolic integrity (Theodorou and Plaxton 1993; Day and Wiskich 1995; Vance *et al.* 2003). Recent studies show that AOX is enhanced in *Hakea* cluster roots (Shane *et al.* 2004), Arabidopsis (Morcuende *et al.* 2007), and bean (Hernández *et al.* 2007).

A series of elegant studies have now conclusively demonstrated a tight link between sugars and plant adaptation to P deficiency (Liu *et al.* 2005; Hernández *et al.* 2007; Karthikeyan *et al.* 2007; Morcuende *et al.* 2007; Müller *et al.* 2007). Sugars are required for the expression of many P deficiency induced genes in both roots and shoots. Liu *et al.* (2005) and Tesfaye *et al.* (2007) showed that the expression of a phosphate transporter (Pt), an acid phosphatase (APase), a multidrug toxin (MATE) protein, sucrose synthase (SS), hexokinase (HXK), and fructokinase (FK), in P deficiency induced cluster roots required sugar or phloem transport (Figure 6.5). Dark adaptation or blocking phloem transport reduced the expression of these genes to non-detectable levels in 24 hours, but re-exposure to light activated their transcription within a few hours. The interplay of sugars and P deficiency induced gene expression observed in white lupin has also been validated with Arabidopsis. Karthikeyan *et al.* (2007) and Müller *et al.* (2007) have shown that P deficiency induced gene expression and modified root architecture in Arabidopsis is tightly linked to sugar signaling, probably through HXK.

Hormonal regulation of cluster root development

Since cluster root development involves the synchronized initiation and growth of a large number of tertiary lateral roots in distinct wavelike patterns originating from primary and secondary lateral roots, it would not be surprising that the balance of hormones plays a role in this adaptive response to low P availability (Gilbert *et al.* 2000; Skene and James 2000). Many hormonally controlled developmental responses occurring in P-deficient Arabidopsis plants, which give rise to modified root architecture, appear also to be involved in modifying cluster root architecture. We have detected a number of plant hormone related ESTs during sequencing from cDNA libraries made to P-deficiency induced cluster roots (Table 6.2). As described later for Arabidopsis lateral roots, substantial support for the role of auxin in cluster root formation comes from observations showing: exogenous application of auxin

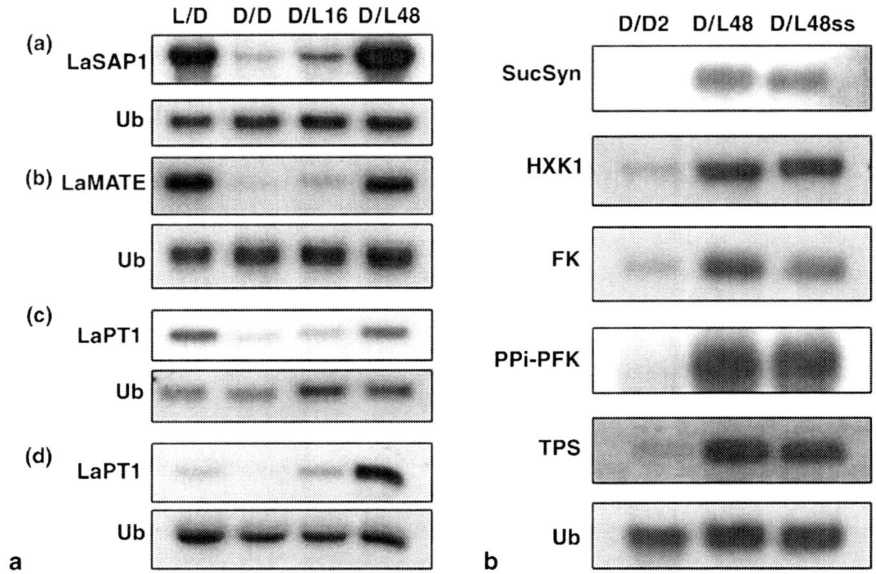

Fig. 6.5 RNA blot of lupin P-stressed induced genes that are co-regulated by products of photosynthesis. **a**. Total RNA from P-deficient proteoid roots and leaves were separated on agarose gels. Polyubiquitin probe (UB) was used as a control for loading and RNA quality. Light/dark regime: L/D 16/8 hour photoperiod; D/D dark-adapted for 24 hours prior to harvest; D/L18 or D/L48 plant transferred to continuous light for 16 or 48 hours after dark adaptation. Dark adaptation of P-deficient plants results in loss of RNAs encoding root expressed (a, b, c) LaSAP (lupin secreted acid phosphatase), LaMATE (lupin multidrug-toxin-efflux protein), and LaPT1 (lupin phosphate transporter). Dark adaptation of P-deficient plants results in loss of RNAs encoding leaf expressed (d) LaPT1. Upon returning plants to light mRNA abundance of P-deficient induced genes recovers (from Liu *et al.* 2005). **b**. Phosphorus deficiency and light affect expression of sugar-related genes in proteoid roots of white lupin. Total RNA isolated from proteoid roots of P-deficient white lupin plants at 14 days after emergence under different light regimes. Treatments: D/D2, P-deficient plants shaded in continuous dark for 48 hours; D/L48, dark-treated P-deficient plants re-exposed to continuous light for 48 hours; D/L48ss, one-half of the shoot of dark-grown plants re-exposed to light and the other one-half remaining in the dark. SucSyn, Suc synthase; HXK1, hexokinase; FK, fructokinase; PPi-PFK, PPi-dependent phosphofructokinase-1; TPS, trehalose-6-P synthase; Ub, polyubiquitin. Dark adaptation of P-deficient plants results in the loss of RNAs encoding genes of sugar metabolism. Upon returning P-deficient plants to light, P deficiency induced gene expression recovers. (From Tesfaye *et al.* 2007.)

stimulates cluster root formation in P-sufficient white lupin and *Protea* species (Gilbert *et al.* 2000; Skene and James 2000); auxin transport inhibitors block cluster root formation in P-deficient plants; and many genes involved in auxin synthesis and signaling are abundantly expressed in developing cluster roots of white lupin (Vance *et al.* 2003; Vance CP, Uhde-Stone C, Yamagishi M 2007; unpublished). These data clearly show a significant component of P-deficiency induced cluster root formation is due to auxin synthesis and transport. Currently, experiments are underway to assess auxin signaling in cluster roots through the transformation of cluster roots with the auxin reporter construct DR5-GUS (Ulmasov *et al.* 1997).

Table 6.2 Expressed sequence tags (ESTs) related to plant hormone balance isolated from phosphorus deficient white lupin cluster roots

EST Annotation	Number of ESTs[a]
Auxin influx/efflux	2
Auxin binding	3
Auxin response/repressed	10
IAA hydrolyase	3
Cytokinin oxidase	6
Methionine synthase	8
Aminocyclopropane carboxylic acid oxidase (ACC)	5
Giberrillin induced	14

[a]Number of ESTs detected in each category out of a total of 3,000 sequenced.

Although the role of ethylene in P-deficiency induced cluster root architecture is not clear, the fact that ethylene plays a role in root hair length, density, and development, as well as a role in lateral root emergence in P-deficient Arabidopsis (Ticconi and Abel 2004), suggests a similar role in cluster roots. As noted previously, cluster roots and dauciform roots have densely packed, unusually long root hairs. Studies with P deficiency-induced cluster roots of white lupin (Gilbert *et al.* 2000), *Casurina* (Zaid *et al.* 2003), and squash (Waters and Blevins 2000) have correlated ethylene production with cluster root formation. Gilbert *et al.* (2000) demonstrated a two- to three-fold increase in ethylene as cluster roots develop in P-deficient white lupin. Zaid *et al.* (2003) showed that Fe-deficiency stimulated cluster roots in *Casurina* and inhibition of ethylene synthesis reduced cluster root formation. Waters and Blevins (2000) noted a correlation between ethylene production, root iron reduction, and the formation of cluster roots in squash. We have found a number of genes involved in ethylene biosynthesis to be over-represented (Table 6.2) in our sequencing of ESTs derived from P-deficiency induced cluster root cDNA libraries, further suggesting a role for ethylene in cluster root architecture.

In their classical study of the physiology of cluster roots, Neumann *et al.* (2000) found that addition of cytokinins to lupin significantly reduced cluster root formation and cluster rootlet elongation. They also found elevated levels of cytokinin in four-week-old P-deficient white lupin roots, compared to P-sufficient roots. They postulated that auxin stimulates emergence of cluster rootlets in P-deficient plants, which results in increased production of cytokinin due to the numerous emerged root tips. In mature segments of P deficiency induced cluster roots we have found numerous ESTs with strong homology to cytokinin oxidase (Table 6.2), suggesting that cytokinins may be subject to degradation as cluster rootlets mature. Cytokinin oxidase is the key enzyme implicated in cytokinin degradation (Morris *et al.* 1999). Enhanced degradation of cytokinins in cluster root formation and/or development might be expected because low cytokinin levels favor root growth, and P deficiency reportedly results in reduced xylem sap cytokinin levels (Salama and Wareing 1979; Martín *et al.* 2000; Emery and Atkins 2002). Alternatively, *in planta* regulation of potentially large quantities of cytokinins, which could be released by the mass induction of cluster root meristems, may require cytokinin oxidase. As described later, cytokinins reduce lateral root initiation in Arabidopsis.

While strong correlative physiological and gene expression data suggest a critical role for auxins, cytokinins, and ethylene in P deficiency induced cluster root development, definitive genetic and biochemical experiments have yet to be performed. Salient questions to be addressed include: What is/are the internal signal(s) that initiate(s) the cascade of developmental biochemical and genetic changes resulting in cluster roots? How is determinacy in cluster roots regulated? Are reactive oxygen and programmed cell death part of the cluster root developmental phenomenon? Can gene knock-down and overexpression studies be harnessed to definitively answer questions regarding the role of growth hormones in cluster root development and function? Can the genetic control mechanism(s) for cluster root formation be identified and used to enhance P-uptake and use efficiency in other plant species? Recent advances in genetic transformation of root tissue via *Agrobacterium rhizogenes* (Boisson-Dernier *et al.* 2001; Limpens *et al.* 2004) facilitate high throughput experiments to evaluate gene function with either knock-down or overexpression approaches. Coupled with metabolomics and proteomics, new insights will be gained into the role of P deficiency in the development of complex root architecture.

ARABIDOPSIS

P deficiency mediates determinate growth of the primary root

Phosphorus availability has a marked effect on the root system architecture of Arabidopsis (Narang *et al.* 2000; Williamson *et al.* 2001; López-Bucio *et al.* 2002; Al-Ghazi *et al.* 2003; Chevalier *et al.* 2003; Müller and Schmidt 2004; Nacry *et al.* 2005). Growth under P-limiting conditions results in determinate growth of the primary root and redistribution of root growth from the primary root to lateral roots (Ticconi and Abel 2004). Reduced primary root growth under low P is accompanied by increased lateral root density, along with increased root hair length and number. Arabidopsis root biomass is concentrated near the soil surface, suggesting topsoil foraging by roots. Moreover, accessions with enhanced P acquisition also appear to have strengthened root penetration capacity (Narang *et al.* 2000).

Decreased primary root elongation during P deficiency is thought to involve localized sensing of P in the rhizosphere by root meristem and cap cells, which results in cessation of cell division in the primary root meristem (Sánchez-Calderón *et al.* 2005; Svistoonoff *et al.* 2007). Ticconi *et al.* (2004) identified an Arabidopsis mutant *pdr2* (Pi deficiency response) that has a disruption in P-sensing. This mutant shows a short-root phenotype that is specific for P deficiency. The short-root phenotype of *pdr2* is the result of inhibition of primary root cell division at P concentrations below 0.1 mM external P. Ticconi *et al.* (2004) proposed that *pdr2* monitors external environmental P and regulates primary root meristem activity to adjust root system architecture to maximize P acquisition. More recently, Svistoonoff *et al.*

(2007) identified an Arabidopsis gene that mediates the signaling between the root tip and contact with low P. The gene conditions reprogramming of plant root architecture. Previous studies by this group identified an Arabidopsis QTL designated low phosphate root (*LPR*) associated with arrest in primary root growth under low P (Reymond *et al.* 2006). Svistoonoff *et al.* (2007) then employed map-based cloning to identify the gene associated with the *LPR* QTL. They found that *LPR* encoded a multicopper oxidase (MCO) protein. Transcripts of the *LPR* MCO were most abundant in the root meristem and cap. *LPR* mutants lost the determinate primary root arrest phenotype under low P. They proposed that as the primary root tip came in contact with a low P area, MCO activity modifies either the amount or distribution of hormone-like compounds that signal determinacy in primary root growth and stimulates lateral root development.

Root hairs

The abundant development of lateral roots following cessation of primary root growth in low P environments is almost invariably accompanied by increased root hair density and length. Root-hairs are tubular-shaped cells specialized for nutrient uptake (Peterson and Farquhar 1996; Gahoonia and Nielsen 1998; Gilroy and Jones 2000). They arise from root epidermal cells known as trichoblasts and undergo tip growth, thereby extending the root surface area in contact with the soil matrix (Ridge 1995; Dolan 2001). Root hairs can form as much as 77% of the root surface area of field crops (Peterson and Farquhar 1996). For plants lacking mycorrhizal associations, they are the primary site of nutrient uptake (Gahoonia and Nielsen 1998; Gilroy and Jones 2000; Jungk 2001). Root hair formation and growth is regulated largely by the supply of mineral nutrients, particularly P and NO_3^- (Bates and Lynch 1996, 2000; Gilroy and Jones 2000; Ma *et al.* 2001).

In Arabidopsis, trichoblasts are located over the junction of two underlying cortical cells. Trichoblasts can be distinguished by the late stages of embryogenesis. In recent years, root hair formation in Arabidopsis has become a model system for evaluating cell fate and hormonal interactions. Grierson *et al.* (2001) report that at least 40 genes in Arabidopsis affect root hair initiation and development. Five loci involved in root hair formation encode transparent testa glabra (*TIG*), glabra2 (*GL2*), constitutive triple response1 (*CTR1*), root hair defective6 (*RHD6*), and auxin resistant2 (*AXR2*; Masucci and Schiefelbein 1996). Analysis of these genes in Arabidopsis mutants demonstrated that a network of hormone interactions involving auxin and ethylene regulate root epidermal cell fate and root hair initiation (Tanimoto *et al.* 1995). Likewise, Parker *et al.* (2000) found eight genes whose loss-of-function mutants showed root hair defects. Although the function of these genes was not established, they were mapped to Arabidopsis chromosomes. More recently, Jones *et al.* (2006) showed that in addition to previously reported genes, some 606 novel genes had enhanced expression in Arabidopsis root hairs. They identified several gene families that appeared to be important in root hair morphogenesis including cell wall synthesis enzymes, glycosylphosphatidylinositol (GPI) anchored proteins associated with lipid rafts, armadillo-repeat proteins, and leucine-rich receptor kinases. Importantly, a basic helix-loop-helix transcription factor has been shown to be the ancestral regulator of root hair development in Arabidopsis (Menand *et al.* 2007).

Phosphate availability regulates root hair elongation (Bates and Lynch 1996) and root hair density (Ma *et al.* 2001). The average length of root hairs on P-deficient plants was three-fold greater and root hair density was five-fold greater than in P-sufficient plants. Trichoblast files increased from 8 to 12 on P-deficient plants. Analysis of P acquisition in Arabidopsis wild type and root hair mutants, *rhd6* and *rhd2,* showed that wild type plants were more efficient than root hair mutants in obtaining P when plants were grown under low P conditions. However, there was no difference in growth when grown under high P conditions (Bates and Lynch 2000). Phosphorus deficiency altered radical cell pattern formation in roots by increasing cortical number and reduced epidermal cell elongation resulting in increased trichoblasts (Ma *et al.* 2001, 2003; Müller and Schmidt 2004). Modified root hair patterning was most evident in mutant genotypes having defects in first stage of epidermal cell differentiation, suggesting P deficiency signals are perceived very early in epidermal cell development. Modified root cell patterning in P-deficient plants is regulated by auxin-ethylene interactions (Dolan 2001; Stepanova *et al.* 2007).

Lateral roots

The prototype root architecture modification to occur in Arabidopsis in response to P deficiency is the promotion of lateral root development at the expense of primary root growth (Williamson *et al.* 2001; López-Bucio *et al.* 2002). Phosphorus deficiency appears to impair cell division in the primary root meristem, but stimulate elongation of initiated lateral root primordia and newly emerged lateral roots (Al-Ghazi *et al.* 2003; Malamy 2005; Nacry *et al.* 2005; Sánchez-Calderón *et al.* 2005). The root architecture changes induced by P deficiency are strikingly similar to those induced by either auxin addition or overproduction (Al-Ghazi *et al.* 2003; Malamy 2005; Nacry *et al.* 2005). Utilizing Arabidopsis mutants with lesions in auxin and ethylene signaling, a growing body of evidence indicates that P deficiency induced root architectural changes result from modifications in auxin and ethylene synthesis and/or signaling. The current working hypothesis is that P deficiency induces an increase in auxin synthesis and sensitivity in the primary root meristem, resulting in a modification or cessation of cell division. The modification in auxin signaling may result from P deficiency induced ethylene effects on auxin distribution (López-Bucio *et al.* 2002; Ma *et al.* 2003; Ortega-Martínez *et al.* 2007; Ruzicka *et al.* 2007; Swarup *et al.* 2007). Transcript profiling (Himanen *et al.* 2004; Laskowski *et al.* 2006) shows that more than 900 genes are differentially regulated in early lateral root initiation. Numerous genes related to auxin and ethylene signaling, cell wall modification, and the cell cycle are rapidly induced as lateral root primordia are formed.

The plant hormone auxin (primarily indole-3-acetic acid [IAA]) has long been known to stimulate lateral root formation (Boerjan *et al.* 1995; Reed *et al.* 1998; Casimiro *et al.* 2001; Marchant *et al.* 2002). Nacry *et al.* (2005) showed that during P deficiency, IAA increases in the whole primary root and young lateral roots of Arabidopsis. Without IAA, only primary root growth was observed in Arabidopsis (Karthikeyan *et al.* 2007). Regardless of root P status, exogenous auxin application enhanced lateral root formation while suppressing primary root elongation in Arabidopsis (Al-Ghazi *et al.* 2003; Nacry *et al.* 2005; De Smet *et al.* 2007; Stepanova *et al.* 2007). Auxin transport inhibitors suppress lateral root formation in many species including Arabidopsis. As noted earlier, auxin has a striking effect on cluster root formation in white lupin and other cluster root species (Gilbert *et al.* 2000; Skene and James 2000).

Root architectural changes in response to P deficiency have been shown to be related to ethylene in Arabidopsis (López-Bucio *et al.* 2002; Ma *et al.* 2003). Low P conditions appear to enhance the roots' sensitivity to ethylene. Ethylene is known to be an effector of primary and lateral root cell elongation. Several recent reports document the interrelationship between auxin and ethylene in regulating root growth and development (Ruzicka *et al.* 2007; Stepanova *et al.* 2007; Swarup *et al.* 2007). Ethylene appears to stimulate auxin biosynthesis and increase auxin transport. Moreover, auxin must be present for ethylene inhibition of cell expansion. Ethylene may also interact with auxin to release lateral root meristems for growth. Thus, low P availability may alter root architecture by stimulating multicopper oxidase activity (Svistoonoff *et al.* 2007), which modulates the auxin-ethylene balance, resulting in determinate root growth and enhanced lateral root initiation.

Traditionally, cytokinins are thought to be negative regulators of root growth while having positive effects on shoot growth (Christianson and Warnick 1985; Werner *et al.* 2001, 2003; Aloni *et al.* 2006). Both N and P deficiency result in decreased cytokinin content (López-Bucio *et al.* 2003) accompanied by increased lateral root formation. Application of cytokinin inhibits root development and abolishes the auxin effect of increased lateral root formation. Cytokinin inhibition of primary root growth appears to occur through its effect on regulating root meristem size (Dello Ioio *et al.* 2007). Arabidopsis plants that overexpress cytokinin oxidase (CKX), the key enzyme regulating cytokinin degradation, have reduced cytokinin concentrations accompanied by increased lateral and adventitious root formation (Werner *et al.* 2003; Lohar *et al.* 2004). Li *et al.* (2006) have recently shown that cytokinins inhibit lateral root formation by blocking the pericycle founder cells transition at the G_2 phase. It is quite apparent that the modification of root architecture in P-deficient Arabidopsis is regulated by the subtle interaction of auxin, ethylene, and cytokinin. Localized changes in hormone transport, synthesis and cellular concentration shut down primary root growth, activate lateral root development, and increase trichoblast frequency giving rise to the root architecture favorable for topsoil foraging for P.

Phosphorus deficiency induced gene expression

Several studies of Arabidopsis (Hammond *et al.* 2003, 2004; Wu *et al.* 2003; Misson *et al.* 2005; Morcuende *et al.* 2007) and other species (Wang *et al.* 2001; Wasaki *et al.* 2003; Graham *et al.* 2006; Hernández *et al.* 2007) have now addressed whole genome changes in gene expression in response to P deficiency. Each study has noted that scores of genes respond with a two-fold change in expression. Hammond *et al.* (2004) noted that these changes can be classified as either early genes, which respond rapidly and frequently non-specifically to P deficiency or late genes, which lead to modifications in root architecture and metabolism. Invariably these studies have shown that early response genes include transporters particularly Pi transporters. However numerous other transporters also respond rapidly to P deficiency including aquaporins, sulfate transporters, sugar transporters, and amino acid transporters (Wang *et al.* 2001; Misson *et al.* 2005; Morcuende *et al.* 2007). Transcription factor genes also respond quickly to P deficiency, along with cell wall synthesis genes, PR genes, and acid phosphatases (Franco-Zorrilla *et al.* 2004). Genes induced in late responses include those encoding enzymes of glycolysis and the carboxylate cycle. Internal remobilization of P was affected by late response genes including: glucose-6P/Pi translocator, nucleotide pyrophosphatases, and glycerol-P-permeases (Misson *et al.* 2005). A large number of genes involved in phospholipid degradation and galacto- and sulfolipid synthesis also have enhanced expression after prolonged P deficiency.

Activation of these lipid-related genes probably reflects a shift in membrane biogenesis from incorporation of phospholipids to the use of galacto- and sulfo-lipids as membrane components (Essigmann *et al.* 1998; Misson *et al.* 2005). Morcuende *et al.* (2007) have completed an elegant study of Arabidopsis in which they followed the recovery of plants from P-stress over a 24-hour period. Plants subjected to P deficiency were supplemented with sufficient P to induce the recovery of genes suppressed under low P. They also evaluated metabolites and enzyme activities of several proteins. This analysis allowed for the identification of genes, proteins, and metabolites specifically involved in recovery from P deficiency. In addition to genes and pathways noted above, they found that phosphoenolpyruvate carboxylase (PEPC) and PEPC kinase were integrally involved in the shift in C metabolism that occurs during P deficiency. They also showed that C metabolism shifted toward starch synthesis under low P conditions and towards starch catabolism under high P conditions. Interestingly, Morcuende *et al.* (2007) found a striking increase in glycerophosphodiesterases (GPDEs) gene expression. GPDE genes have been recently shown to be important in root hair development (Jones *et al.* 2006) and the phenomenon of increased root hair length and density is a common response to P deficiency.

Whole genome studies have also led to the discovery of many transcription factors that appear to be important in response to P deficiency. Current evidence suggests that control of signaling P deficiency is complex, involving both positive and negative regulation of gene expression. Some of the transcription factors proven to be involved in P deficiency signal transduction include: MYBs, WRKYs, bHLH, Zn-finger proteins, and HD-Zip factors (Rubio *et al.* 2001; Hammond *et al.* 2004; Yi *et al.* 2005; Devaiah *et al.* 2007; Tesfaye *et al.* 2007).

SYNOPSIS

Phosphorus is second to N as the most limiting element for plant growth. Plants have evolved a number of effective strategies to acquire P and grow in P limited environments. Physiological, biochemical, and molecular studies of plant adaptations to P deficiency, which occur in non-mycorrhizal species, have provided striking new insights into how plants respond to P deficiency and which traits appear important in acquiring P. Because inexpensive P fertilizer is a limited resource and because many farmers of the world cannot afford P fertilizer (Steen 1997), it is imperative that plant biologists discover strategies and approaches that will enhance efficiency of P acquisition and use. The fact that we know modified root architecture, increased root hairs, and exuded carboxylates are key components in P acquisition offers targets to improve or modify in efforts toward increasing efficiency of P acquisition and use. A targeted approach is borne out by the recent identification of an Arabidopsis QTL gene affecting primary root extension and low P signal recognition as a multicopper oxidase (Svistoonoff *et al.* 2007). Utilizing a QTL approach for improving P acquisition in bean (Beebe *et al.* 2006) has resulted in beans with enhanced top-soil foraging, but the gene controlling this trait has not been cloned. With the wide array of genomic and bioinformatics research platforms now available, coupled to QTL and whole genome analysis, research into plant responses to P deficiency is moving toward an exciting phase, centered around signal transduction and regulation of root developmental plasticity and gene function, aimed at increasing efficiency of use. Sustainable cropping systems in the developed world and the need for efficient P acquisition in low P conditions of the developing world, require that plant research identify and exploit mechanisms to improve P use efficiency. Efforts that improve soil P availability to plants are urgently needed to maintain and increase economically and environmentally sustainable crop agriculture.

Acknowledgments The author wishes to thank Drs M. Shane, B. Lamont, and B. Bucciarelli for images of cluster and dauciform roots. Preparation of this contribution was supported in part by U.S. Department of Agriculture (USDA) Agricultural Research Service CRIS 3640-21000-024-00D, USDA-CSREES Grant # 2005-35100-16002, and the Minnesota Agricultural Experiment Station. Mention of trade names or commercial products in this publication is solely for the purpose of providing specific information and does not imply recommendation or endorsement by the U.S. Department of Agriculture.

REFERENCES

Adams MA, Pate JS (2002) Phosphorus sources and availability modify growth and distribution of root clusters and nodules of native Australian legumes. Plant Cell Environ 26: 837–850

Al-Ghazi Y, Muller B, Pinloche S, Tranbarger TJ, Nacry P, Rossignol M, Tardieu F, Doumas P (2003) Temporal responses of *Arabidopsis* root architecture to phosphate starvation: evidence for the involvement of auxin signalling. Plant Cell Environ 26: 1053–1066

Aloni R, Aloni E, Langhans M, Ullrich CI (2006) Role of cytokinin and auxin shaping root architecture: regulating vascular differentiation, lateral root initiation, root apical dominance and root gravitropism. Ann Bot 97: 883–893

Bates TR, Lynch JP (1996) Stimulation of root hair elongation in *Arabidopsis thaliana* by low phosphorus availability. Plant Cell Environ 19: 529–538

Bates TR, Lynch JP (2000) The efficiency of *Arabidopsis thaliana* (Brassicaceae) root hairs in phosphorus acquisition. Am J Bot 87: 964–970

Beebe SE, Rojas-Pierce M, Yan X, Blair MW, Pedraza F, Muñoz F, Tohme J, Lynch JP (2006) Quantitative trait loci for root architecture traits correlated with phosphorus acquisition in common bean. Crop Sci 46: 413–423

Boerjan W, Cervera MT, Delarue M, Beeckman T, Dewitte W, Bellini C, Caboche M, Van Onckelen H, Van Montagu M, Inzé D (1995) *superroot*, a recessive mutation in Arabidopsis, confers auxin overproduction. Plant Cell 7: 1405–1419

Boisson-Dernier A, Chabaud M, Garcia F, Bécard G, Rosenberg C, Barker DG (2001) *Agrobacterium rhizogenes* - transformed roots of *Medicago truncatula* for the study of nitrogen-fixing and endomycorrhizal symbiotic associations. Mol Plant Microbe Interact 14: 695–700

Borch K, Bouma TJ, Lynch JP, Brown KM (1999) Ethylene: a regulator of root architectural responses to soil phosphorus availability. Plant Cell Environ 22: 425–431

Casimiro I, Marchant A, Bhalerao RP, Beeckman T, Dhooge S, Swarup R, Graham N, Inzé D, Sandberg G, Casero PJ, Bennett MJ (2001) Auxin transport promotes Arabidopsis lateral root initiation. Plant Cell 13: 843–852

Celenza JL Jr., Grisafi PL, Fink GR (1995) A pathway for lateral root formation in *Arabidopsis thaliana*. Genes Dev 9: 2131–2142

Charlton WA (1996) Lateral root initiation. In: Waisel Y, Eshel A, Kafkafi U (eds), Plant Roots: The Hidden Half (2nd edition). Marcel Dekker, New York

Chevalier F, Pata M, Nacry P, Doumas P, Rossignol M (2003) Effects of phosphate availability on the root system architecture: large-scale analysis of the natural variation between *Arabidopsis* accessions. Plant Cell Environ 26: 1839–1850

Christianson ML, Warnick DA (1985) Temporal requirements for phytohormone balance in the control of organogenesis in vitro. Dev Biol 112: 494–497

Ciereszko I, Zambrzycka A, Rychter A (1998) Sucrose hydrolysis in bean roots (*Phaseolus vulgaris* L.) under phosphate deficiency. Plant Sci 133: 139–144

Cripps CL, Eddington LH (2005) Distribution of mycorrhizal types among alpine vascular plant families on the Beartooth Plateau, Rocky Mountains, U.S.A., in reference to large-scale patterns in arctic-alpine habitats. Arct Antarct Alp Res 37: 177–188

Day DA, Wiskich JT (1995) Regulation of alternative oxidase activity in higher plants. J Bioenerg Biomembr 27: 379–385

De Smet I, Tetsurmura T, De Rybel B, Frey NF, Laplaze L, Casimiro I, Swarup R, Naudts M, Vanneste S, Audenaert D, Inzé D, Bennett MJ, Beeckman T (2007) Auxin-dependent regulation of lateral root positioning in the basal meristem of *Arabidopsis*. Development 134: 681–690

Dello Ioio R, Linhares FS, Scacchi E, Casamitjana-Martinez E, Heidstra R, Costantino P, Sabatini S (2007) Cytokinins determine *Arabidopsis* root-meristem size by controlling cell differentiation. Curr Biol 17: 678–682

Devaiah BN, Karthikeyan AS, Raghothama KG (2007) *WRKY75* transcription factor is modulated by phosphate acquisition and root development in *Arabidopsis*. Plant Physiol 143: 1789–1801

Diem HG, Duhoux E, Zaid H, Arahou M (2000) Cluster roots in Casuarinaceae: role and relationship to soil nutrient factors. Ann Bot 85: 929–936

Dinkelaker B, Hengeler C, Marschner H (1995) Distribution and function of proteoid roots and other root clusters. Bot Acta 108: 183–200

Dolan L (2001) The role of ethylene in root hair growth in *Arabidopsis*. J Plant Nutr Soil Sci 164: 141–145

Emery NRJ, Atkins CA (2002) Roots and cytokinins. In: Waisel Y, Eshel A, Kafkafi U (eds), Plant Roots: The Hidden Half (3rd edition). Marcel Dekker, New York, pp 417–434

Essigmann B, Güler S, Narang RA, Linke D, Benning C (1998) Phosphate availability affects the thylakoid lipid composition and the expression of *SQD1*, a gene required for sulfolipid biosynthesis in *Arabidopsis thaliana*. Proc Natl Acad Sci USA 95: 1950–1955

Franco-Zorrilla JM, González E, Bustos R, Linhares F, Leyva A, Paz-Ares J (2004) The transcriptional control of plant responses to phosphate limitation. J Exp Bot 55: 285–293

Gahoonia TS, Nielsen NE (1998) Direct evidence on participation of root hairs in phosphorus (^{32}P) uptake from soil. Plant Soil 198: 147–152

Gardner WK, Parbery DG, Barber DA (1982) The acquisition of phosphorus by *Lupinus albus* L. I. Some characteristics of the soil/root interface. Plant Soil 68: 19–32

Gardner WK, Barber DA, Parbery DG (1983) The acquisition of phosphorus by *Lupinus albus* L. III. The probable mechanism by which phosphorus movement in the soil/root interface is enhanced. Plant Soil 70: 107–124

Gilbert GA, Knight JD, Vance CP, Allan DL (1999) Acid phosphatase activity in phosphorus-deficient white lupin roots. Plant Cell Environ 22: 801–810

Gilbert GA, Knight JD, Vance CP, Allan DL (2000) Proteoid root development of phosphorus deficient lupin is mimicked by auxin and phosphonate. Ann Bot 85: 921–928

Gilroy S, Jones DL (2000) Through form to function: root hair development and nutrient uptake. Trends Plant Sci 5: 56–60

Graham MA, Ramírez M, Valdes-López O, Lara M, Tesfaye M, Vance CP, Hernandez G (2006) Identification of candidate phosphorus stress induced genes in *Phaseolus vulgaris* through clustering analysis across several plant species. Funct Plant Biol 33: 789–797

Grierson PF (1992) Organic acids in the rhizosphere of *Banksia integrifolia* L.f. Plant Soil 144: 259–265

Grierson PF, Attiwill PM (1989) Chemical characteristics of the proteoid root mat of *Banksia integrifolia* L. Aust J Bot 37: 137–143

Grierson CS, Parker JS, Kemp AC (2001) *Arabidopsis* genes with roles in root hair development. J Plant Nutr Soil Sci 164: 131–140

Hammond JP, Bennett MJ, Bowen HC, Broadley MR, Eastwood DC, May ST, Rahn C, Swarup R, Woolaway KE, White PJ (2003) Changes in gene expression in *Arabidopsis* shoots during phosphate starvation and the potential for developing smart plants. Plant Physiol 132: 578–596

Hammond JP, Broadley MR, White PJ (2004) Genetic responses to phosphorus deficiency. Ann Bot 94: 323–332

Hernández G, Ramirez M, Valdés-López O, Tesfaye M, Graham MA, Czechowski T, Schlereth A, Wandrey M, Erban A, Cheung F, Wu HC, Lara M, Town CD, Kopka J, Udvardi MK, Vance CP (2007) Phosphorus stress in common bean: root transcript and metabolic responses. Plant Physiol 144: 752–767

Himanen K, Vuylsteke M, Vanneste S, Vercruysse S, Boucheron E, Alard P, Chriqui D, Van Montagu M, Inzé D, Beeckman T (2004) Transcript profiling of early lateral root initiation Proc Natl Acad Sci USA 101: 5146–5151

Hinsinger P (2001) Bioavailability of soil inorganic P in the rhizosphere as affected by root-induced chemical changes: a review. Plant Soil 237: 173–195

Johnson JF, Allan DL, Vance CP (1994) Phosphorus stress-induced proteoid roots show altered metabolism in *Lupinus albus*. Plant Physiol 104: 657–665

Johnson JF, Allan DL, Vance CP, Weiblen G (1996a) Root carbon dioxide fixation by phosphorus-deficient *Lupinus albus*. Contribution to organic acid exudation by proteoid roots. Plant Physiol 112: 19–30

Johnson JF, Vance CP, Allan DL (1996b) Phosphorus deficiency in *Lupinus albus*. Altered lateral root development and enhanced expression of phospho*enol*pyruvate carboxylase. Plant Physiol 112: 31–41

Jones MA, Raymond MJ, Smirnoff N (2006) Analysis of the root-hair morphogenesis transcriptome reveals the molecular identity of six genes with roles in root-hair development. Plant J 45: 83–100

Jungk A (2001) Root hairs and the acquisition of plant nutrients from soil. J Plant Nutri Soil Sci 164: 121–129

Karthikeyan AS, Varadarajan DK, Jain A, Held MA, Carpita NC, Raghothama KG (2007) Phosphate starvation responses are mediated by sugar signaling in *Arabidopsis*. Planta 225: 907–918.

Keerthisinghe G, Hocking PJ, Ryan PR, Delhaize E (1998) Effect of phosphorus supply on the formation and function of proteoid roots of white lupin (*Lupinus albus* L.). Plant Cell Environ 21: 467–478

Kihara T, Wada T, Suzuki Y, Hara T, Koyama H (2003) Alteration of citrate metabolism in cluster roots of white lupin. Plant Cell Physiol 44: 901–908

Kochian LV, Hoekenga OA, Piñeros MA (2004) How do crop plants tolerate acid soils? Mechanisms of aluminum tolerance and phosphorus efficiency. Annu Rev Plant Biol 55: 459–493

Lambers H, Shane MW, Cramer MD, Pearse SJ, Veneklaas EJ (2006) Root structure and functioning for efficient acquisition of phosphorus: matching morphological and physiological traits. Ann Bot 98: 693–713

Lamont B (1974) The biology of dauciform roots in the sedge *Cyathochaete avenacea*. New Phytol 73: 985–996

Lamont B (1982) Mechanisms for enhancing nutrient uptake in plants, with particular reference to Mediterranean, South Africa and Australia. Bot Rev 48: 597–689

Lamont BB (1993) Why are hairy root clusters so abundant in the most nutrient-impoverished soils of Australia? Plant Soil 155/156: 269–272

Lamont BB (2003) Structure, ecology and physiology of root clusters - a review. Plant Soil 48: 1–19

Laskowski M, Biller S, Stanley K, Kajstura T, Prusty R (2006) Expression profiling of auxin-treated *Arabidopsis* roots: toward a molecular analysis of lateral root emergence. Plant Cell Physiol 47: 788–792

Li X, Mo X, Shou H, Wu P (2006) Cytokinin-mediated cell cycling arrest of pericycle founder cells in lateral root initiation of *Arabidopsis*. Plant Cell Physiol 47: 1112–1123

Limpens E, Ramos J, Franken C, Raz V, Compaan B, Franssen H, Bisseling T, Geurts R (2004) RNA interference in *Agrobacterium rhizogenes*-transformed roots of *Arabidopsis* and *Medicago truncatula*. J Exp Bot 55: 983–992

Liu J, Uhde-Stone C, Li A, Vance C, Allan D (2001) A phosphate transporter with enhanced expression in proteoid roots of white lupin (*Lupinus albus* L.). Plant Soil 237: 257–266

Liu JQ, Samac DA, Bucciarelli B, Allan DL, Vance CP (2005) Signaling of phosphorus deficiency-induced gene expression in white lupin requires sugar and phloem transport. Plant J 41: 257–268

Lohar DP, Schaff JE, Laskey JG, Kieber JJ, Bilyeu KD, Bird DM (2004) Cytokinins play opposite roles in lateral root formation, and nematode and rhizobial symbioses. Plant J 38: 203–214

López-Bucio J, Hernández-Abreu E, Sánchez-Calderón L, Nieto-Jacobo MF, Simpson J, Herrera-Estrella L (2002) Phosphate availability alters architecture and causes changes in hormone sensitivity in the Arabidopsis root system. Plant Physiol 129: 244–256

López-Bucio J, Cruz-Ramírez A, Herrera-Estrella L (2003) The role of nutrient availability in regulating root architecture. Curr Opini Plant Biol 6: 280–287

Lynch JP, Brown KM (2001) Topsoil foraging - an architectural adaptation of plants to low phosphorus availability. Plant Soil 237: 225–237

Ma Z, Bielenberg DG, Brown KM, Lynch JP (2001) Regulation of root hair density by phosphorus availability in *Arabidopsis thaliana*. Plant Cell Environ 24: 459–467

Ma Z, Baskin TI, Brown KM, Lynch JP (2003) Regulation of root elongation under phosphorus stress involves changes in ethylene responsiveness. Plant Physiol 131: 1381–1390

Malamy JE (2005) Intrinsic and environmental response pathways that regulate root system architecture. Plant Cell Environ 28: 67–77

Marchant A, Bhalerao R, Casimiro I, Eklot J, Casero PJ, Bennett M, Sandberg G (2002) AUX1 promotes lateral root formation by facilitating indole-3-acetic acid distribution between sink and source tissue in Arabidopsis seedling. Plant Cell 14: 589–597

Marschner H, Römheld V, Horst WJ, Martin P (1986) Root induced changes in the rhizosphere: importance for the mineral nutrition of plants. Z Pflanz Bodenkunde 149: 441–456

Martín AC, del Pozo JC, Iglesias J, Rubio V, Solano R, de la Peña A, Leyva A, Paz-Ares J (2000) Influence of cytokinins on the expression of phosphate starvation responsive genes in *Arabidopsis*. Plant J 24: 559–567

Massonneau A, Langlade N, Léon S, Smutny J, Vogt E, Neumann G, Martinoia E (2001) Metabolic changes associated with cluster root development in white lupin (*Lupinus albus* L.): relationship between organic acid excretion, sucrose metabolism and energy status. Planta 213: 534–542

Masucci JD, Schiefelbein JW (1996) Hormones act downstream of *TTG* and *GL2* to promote root hair outgrowth during epidermis development in the Arabidopsis root. Plant Cell 8: 1505–1517

Menand B, Yi K, Jouanic S, Hoffmann L, Ryan E, Linstead P, Schaefer DG, Dolan L (2007) An ancient mechanism controls the development of cells with a rooting function in land plants. Science 316: 1477–1480

Miller RM (2005) The nonmycorrhizal root - a strategy for survival in nutrient-impoverished soils. New Phytol 165: 655–658

Miller RM, Smith CI, Jastrow JD, Bever JD (1999) Mycorrhizal status of the genus *Carex* (Cyperaceae). Am J Bot 86: 547–553

Miller SS, Liu J, Allan DL, Menzhuber CJ, Fedorova M, Vance CP (2001) Molecular control of acid phosphatase secretion into the rhizosphere of proteoid roots from phosphorus-stressed white lupin. Plant Physiol 127: 594–606

Misson J, Raghothama KG, Jain A, Jouhet J, Block MA, Bligny R, Ortet P, Creff A, Somerville S, Rolland N, Doumas P, Nacry P, Herrerra-Estrella L, Nussaume L, Thibaud M-C (2005) A genome-wide transcriptional analysis using *Arabidopsis thaliana* Affymetrix gene chips determined plant responses to phosphate deprivation. Proc Natl Acad Sci USA 102: 11934–11939

Morcuende R, Bari R, Gibon Y, Zheng WM, Pant BD, Blasing O, Usadel B, Czechowski T, Udvardi MK, Stitt M, Scheible WR (2007) Genome-wide reprogramming of metabolism and regulatory networks of *Arabidopsis* in response to phosphorus. Plant Cell Environ 30: 85–112

Morris RO, Bilyeu KD, Laskey JG, Cheikh NN (1999) Isolation of a gene encoding a glycosylated cytokinin oxidase from maize. Biochem Biophys Res Commun 255: 328–333

Müller M, Schmidt W (2004) Environmentally induced plasticity of root hair development in *Arabidopsis*. Plant Physiol 134: 409–419

Müller R, Morant M, Jarmer H, Nilsson L, Nielsen TH (2007) Genome-wide analysis of the *Arabidopsis* leaf transcriptome reveals interaction of phosphate and sugar metabolism. Plant Physiol 143: 156–171

Nacry P, Canivenc G, Muller B, Azmi A, Van Onckelen H, Rossignol M, Doumas P (2005) A role for auxin redistribution in the responses of the root system architecture to phosphate starvation in Arabidopsis. Plant Physiol 138: 2061–2074

Narang RA, Bruene A, Aitmann T (2000) Analysis of phosphate acquisition efficiency in Arabidopsis accessions. Plant Physiol 124: 1786–1799

Neumann G, Martinoia E (2002) Cluster root - an underground adaptation for survival in extreme environments. Trends Plant Sci 7: 162–167

Neumann G, Massonneau A, Martinoia E, Römheld V (1999) Physiological adaptations to phosphorus deficiency during proteoid root development in white lupin. Planta 208: 373–382

Neumann G, Massonneau A, Langlade N, Dinkelaker B, Hengeler C, Römheld V, Martinoia E (2000) Physiological aspects of cluster root function and development in phosphorus-deficient white lupin (*Lupinus albus* L.). Ann Bot 85: 909–919

Ortega-Martínez O, Pernas M, Carol RJ, Dolan L (2007) Ethylene modulates stem cell division in the *Arabidopsis thaliana* root. Science 317: 507–510

Parker JS, Cavell AC, Dolan L, Roberts K, Grierson CS (2000) Genetic interactions during root hair morphogenesis in Arabidopsis. Plant Cell 12: 1961–1974

Pate J, Watt M (2001) Roots of *Banksia* spp. (Proteacea) with special reference to functioning of their specialized root clusters. In: Waisel Y, Eshel A, Kafkafi U (eds), Plant Roots: The Hidden Half (3rd edition). Marcel Dekker, New York, pp 989–1006

Peñaloza E, Muñoz G, Salvo-Garrido H, Silva H, Luis J, Corcuera LJ (2005) Phosphate deficiency regulates phospho*enol*pyruvate carboxylase expression in proteoid root clusters of white lupin. J Exp Bot 56: 145–153

Peterson RL, Farquhar ML (1996) Root hairs: specialized tubular cells extending root surfaces. Bot Gaz 62: 1–40

Plaxton WC, Carswell MC (1999) Metabolic aspects of the phosphate starvation response in plants. In: Lerner HR (ed), Plant Responses to Environmental Stresses: From Phytohormones to Genome Reorganisation. Dekker, New York, pp 349–372

Reed RC, Brady SR, Muday GK (1998) Inhibition of auxin movement from the shoot into the root inhibits lateral root development in Arabidopsis. Plant Physiol 118: 1369–1378

Reymond M, Svistoonoff S, Loudet O, Nussaume L, Desnos T (2006) Identification of QTL controlling root growth response to phosphate starvation in *Arabidopsis thaliana*. Plant Cell Environ 29: 115–125

Ridge RW (1995) Recent developments in the cell and molecular biology of root hairs. J Plant Res 108: 399–405

Rubio V, Linhares F, Solano R, Martín AC, Iglesias J, Leyva A, Paz-Ares J (2001) A conserved MYB transcription factor involved in phosphate starvation signaling both in vascular plants and in unicellular algae. Genes Dev 15: 2122–2133

Ruzicka K, Ljung K, Vanneste S, Podhorska R, Beeckman T, Frimi J, Benkova E (2007) Ethylene regulates root growth through effects on auxin biosynthesis and transport auxin distribution. Plant Cell 19: 2197–2212

Ryan PR, Delhaize E, Jones DL (2001) Function and mechanism of organic anion exudation from plant roots. Annu Rev Plant Physiol Plant Mol Biol 52: 527–560

Salama AMSE-DA, Wareing PF (1979) Effects of mineral nutrition on endogenous cytokinins in plants of sunflower (*Helianthus annuus* L.). J Exp Bot 30: 971–981

Sánchez-Calderón L, López-Bucio J, Chacón-López A, Cruz-Ramirez A, Nieto-Jacobo F, Dubrovsky JG, Herrera-Estrella L (2005) Phosphate starvation induces a determinate developmental program in the roots of *Arabidopsis thaliana*. Plant Cell Physiol 46: 174–184

Shane MW, Lambers H (2005) Cluster roots: a curiosity in context. Plant Soil 274: 101–125

Shane MW, Cramer MD, Funayama-Noguchi S, Cawthray GR, Millar AH, Day DA, Lambers H (2004) Developmental physiology of cluster-root carboxylate synthesis and exudation in harsh Hakea. Expression of phospho*enol*pyruvate carboxylase and the alternative oxidase. Plant Physiol 135: 549–560

Shane MW, Dixon KW, Lambers H (2005) The occurrence of dauciform roots amongst Western Australian reeds, rushes and sedges, and the impact of phosphorus supply on dauciform-root development in *Schoenus unispiculatus* (Cyperaceae). New Phytol 165: 887–898

Shane MW, Cawthray GR, Cramer MD, Kuo J, Lambers H (2006) Specialized 'dauciform' roots of Cyperaceae are structurally distinct but functionally analogous with 'cluster' roots. Plant Cell Environ 29: 1989–1999

Skene KR (1998) Cluster roots: some ecological considerations. J Ecol 86: 1060–1064

Skene KR, James WM (2000) A comparison of the effects of auxin on cluster root initiation and development in *Grevillea robusta* Cunn. ex R. Br. (Proteaceae) and the genus *Lupinus* (Leguminosae). Plant Soil 219: 221–229

Skene KR, Kierans M, Sprent J, Raven JA (1996) Structural aspects of cluster root development and their possible significance for nutrient acquisition in *Grevillea robusta* (Proteacea). Ann Bot 77: 443–451

Sousa MF, Façanha AR, Tavares RM, Lino-Neto T, Gerós H (2007) Phosphate transport by proteoid roots of *Hakea sericea*. Plant Sci 173: 550–558

Steen I (1997) Phosphorus availability in the 21st century. Management of a non-renewable resource. Phosphorus Potassium 217: 25–31

Stepanova AN, Yun J, Likhacheva AV, Alonso JM (2007) Multilevel interactions between ethylene and auxin in *Arabidopsis* roots. Plant Cell 19: 2169–2185

Svistoonoff S, Creff A, Reymond M, Sigoillot-Claude C, Ricaud L, Blanchet A, Nussaume L, Desnos T (2007) Root tip contact with low phosphate media reprograms plant root architecture. Nat Genet 19: 792–796

Swarup R, Perry P, Hagenbeck D, Van Der Straeten D, Beemster GTS, Sandberg G, Bhalerao R, Ljung K, Bennett MJ (2007) Ethylene upregulates auxin biosynthesis in *Arabidopsis* seedlings to enhance inhibition of root elongation. Plant Cell 19: 2186–2196

Tanimoto M, Roberts K, Dolan L (1995) Ethylene is a positive regulator of root hair development in *Arabidopsis thaliana*. Plant J 8: 943–948

Tesfaye M, Liu J, Allan DL, Vance CP (2007) Genomic and genetic control of phosphate stress in legumes. Plant Physiol 144: 594–603

Theodorou ME, Plaxton WC (1993) Metabolic adaptations of plant respiration to nutritional phosphate deprivation. Plant Physiol 101: 339–344

Ticconi CA, Abel S (2004) Short on phosphate: plant surveillance and countermeasures. Trends Plant Sci 9: 548–555

Ticconi CA, Delatorre CA, Lahner B, Salt DE, Abel S (2004) *Arabidopsis pdr2* reveals a phosphate-sensitive checkpoint in root development. Plant J 37: 801–814

Uhde-Stone C, Gilbert G, Johnson JMF, Litjens R, Zinn KE, Temple SJ, Vance CP, Allan DL (2003a) Acclimation of white lupin to phosphorus deficiency involves enhanced expression of genes related to organic acid metabolism. Plant Soil 248: 99–116

Uhde-Stone C, Zinn KE, Ramirez-Yáñez M, Li A, Vance CP, Allan DL (2003b) Nylon filter arrays reveal differential gene expression in proteoid roots of white lupin in response to phosphorus deficiency. Plant Physiol 131: 1064–1079

Uhde-Stone C, Liu J, Zinn KE, Allan DL, Vance CP (2005) Transgenic proteoid roots of white lupin: a vehicle for characterizing and silencing root genes involved in adaptation to P stress. Plant J 44: 840–853

Ulmasov T, Murfett J, Hagen G, Guilfoyle T (1997) Aux/IAA proteins repress expression of reporter genes containing natural and highly active synthetic auxin response elements. Plant Cell 9: 1963–1971

Vance CP, Uhde-Stone C, Allan DL (2003) Phosphorus acquisition and use: critical adaptations by plants for securing a nonrenewable resource. New Phytol 157: 423–447

Wang Y-H, Garvin DF, Kochian LV (2001) Nitrate-induced genes in tomato roots. Array analysis reveals novel genes that may play a role in nitrogen nutrition. Plant Physiol 127: 345–359

Wasaki J, Yonetani R, Kuroda S, Shinano T, Yazaki J, Fujii F, Shimbo K, Yamamoto K, Sakata K, Sasaki T, Kishimoto N, Kikuchi S, Yamagishi M, Osaki M (2003) Transcriptomic analysis of metabolic changes by phosphorus stress in rice plant roots. Plant Cell Environ 26: 1515–1523

Waters BM, Blevins DG (2000) Ethylene production, cluster root formation, and localization of iron (III) reducing capacity in Fe deficient squash roots. Plant Soil 225: 21–31

Watt M, Evans JR (1999) Linking development and determinacy with organic acid efflux from proteoid roots of white lupin grown with low phosphorus and ambient or elevated atmospheric CO_2 concentration. Plant Physiol 120: 705–716

Weisskopf L, Abou-Mansour E, Fromin N, Tomasi N, Santelia D, Edelkott I, Neumann G, Aragno M, Tabacchi R, Martinoia E (2006a) White lupin has developed a complex strategy to limit microbial degradation of secreted citrate required for phosphate acquisition. Plant Cell Environ 29: 919–927

Weisskopf L, Tomasi N, Santelia D, Martinoia E, Langlade NB, Tabacchi R, Abou-Mansour E (2006b) Isoflavonoid exudation from white lupin roots is influenced by phosphate supply, root type and cluster-root stage. New Phytol 171: 657–668

Werner T, Motyka V, Strand M, Schmülling T (2001) Regulation of plant growth by cytokinin. Proc Natl Acad Sci USA 98: 10487–10492

Werner T, Motyka V, Laucou V, Smets R, Van Onckelen H, Schmülling T (2003) Cytokinin-deficient transgenic Arabidopsis plants show multiple developmental alterations indicating opposite functions of cytokinins in the regulation of shoot and root meristem activity. Plant Cell 15: 2532–2550

Williamson LC, Ribrioux SPCP, Fitter AH, Leyser HMO (2001) Phosphate availability regulates root system architecture in Arabidopsis. Plant Physiol 126: 875–890

Wu P, Ma L, Hou X, Wang M, Wu Y, Liu F, Deng XW (2003) Phosphate starvation triggers distinct alterations of genome expression in Arabidopsis roots and leaves. Plant Physiol 132: 1260–1271

Yan F, Zhu Y, Müller C, Schubert S (2002) Adaptation of H^+-pumping and plasma membrane H^+ ATPase activity in proteoid roots of white lupin under phosphate deficiency. Plant Physiol 129: 50–63

Yi K, Wu Z, Zhou J, Du L, Guo L, Wu Y, Wu P (2005) *OsPTF1*, a novel transcription factor involved in tolerance to phosphate starvation in rice. Plant Physiol 138: 2087–2096

Zaid H, El Morabet R, Diem HG, Arahou M (2003) Does ethylene mediate cluster root formation under iron deficiency? Ann Bot 92: 673–677

Chapter 7
MYCORRHIZAL SYMBIOSES

José-Miguel Barea, Nuria Ferrol, Concepción Azcón-Aguilar, and Rosario Azcón

INTRODUCTION

Phosphorus (P) is essential for plants, and its availability is one of the most influential factors limiting their growth and development in both natural and agricultural systems. Even though the total concentration of P in soil is high, the concentration of inorganic phosphate (Pi) in the soil solution, the primary source of P taken up by plants, is usually very low, ranging from 1 to 10 μM (Bieleski 1973). This restricts the diffusion of Pi to plant roots. The low solubility of Pi minerals, Pi adsorption to soil particles and formation of organic P complexes are the primary reasons for the low concentrations of Pi in the soil solution (Marschner 1995). In addition, the rate of Pi uptake by plant roots often exceeds its rate of diffusion in the soil solution, which results in the generation of a Pi-depletion zone surrounding the root system (Smith 2002). Consequently, to cope with P limitations plants have evolved strategies to increase Pi uptake and/or improve the efficiency of tissue P utilization. The establishment of mycorrhizal (fungus-root) symbioses is considered to be one of the most successful and widespread strategies to maximize the access of plant roots to available Pi. Through mycorrhizal associations, the extraradical fungal mycelia function as an additional absorptive surface area for the plant, which increases its capacity to forage for nutrients beyond the Pi depletion zone surrounding the roots. Thus, mycorrhizas enable plants to cope with low Pi availability in natural ecosystems and benefit crop P nutrition in agriculture.

The aim of this Chapter is to review and integrate current knowledge of the impact of mycorrhizal symbioses on plant functioning and adaptation with specific emphasis on P acquisition at various levels of cellular organization, from the molecular, biochemical and physiological to the whole plant. Accordingly, the available information will be structured as follows: (i) mycorrhizas as a plant strategy for P acquisition; (ii) fungal-plant integration to establish functional arbuscular mycorrhizal (AM) symbiosis; (iii) functional biology of Pi uptake by AM plants; (iv) ecological impact of AM symbiosis on plant community structure and productivity; and (v) mycorrhizosphere interactions and ecosystem P cycling.

P.J. White, J.P. Hammond (eds.) *The Ecophysiology of Plant-Phosphorus Interactions*,

MYCORRHIZA AS A PLANT STRATEGY FOR PHOSPHORUS ACQUISITION

Mycorrhizas are symbiotic, generally mutualistic and balanced, associations established between soil fungi and most vascular plants where both partners exchange nutrients and energy (Brundrett 2002). Basically, the host plant receives mineral nutrients via the fungal mycelium (mycotrophism), while the heterotrophic fungus obtains carbon compounds from the host's photosynthesis (Harley and Smith 1983). It is universally accepted that mycorrhizal symbioses, which can be found in almost all ecosystems worldwide, are fundamental to improve plant fitness and soil quality (Smith and Read 1997).

The mycorrhizal fungi colonize the root cortex of most plant species and, once biotrophically established into the root tissues, develop an extraradical mycelium which overgrows the soil surrounding plant roots. This hyphal net is a structure specialized for the acquisition of mineral nutrients from the soil, particularly those whose ionic forms have poor mobility or are present in low concentration in the soil solution, as it is the case with P (Barea 1991). It provides the plant with an adaptive strategy for P acquisition in soils with low P availability.

There are two main types of mycorrhiza, ecto- and endomycorrhiza, which have considerable differences in their structure and physiological relationships between symbionts. Since the structural and functional features of the different mycorrhizal types have been reviewed in detail elsewhere (Smith and Read 1997), only a brief overview will be given here. In ectomycorrhizas, the fungus develops a sheath or mantle around the feeder roots. The mycelium penetrates the root and develops between the cortical cells forming the so-called 'Hartig net' that constitutes the site of nutrient exchange between partners. About 3% of higher plants, mainly forest trees in the Fagaceae, Betulaceae, Pinaceae, *Eucalyptus*, and some woody legumes, form ectomycorrhiza. The fungi involved are mostly Basidiomycetes and Ascomycetes. In endomycorrhizas, the fungi colonize the root cortex both intercellularly and intracellularly. Some endomycorrhizal types are restricted to species in the Ericaceae ("ericoid" mycorrhiza) or Orchidaceae ("orchid" mycorrhiza), whilst the common arbuscular mycorrhizal (AM) type is widely distributed throughout the plant kingdom. An intermediate mycorrhizal type, the ectendomycorrhiza, is formed by plants in families other than the Ericaceae, but in the Ericales, and in the Monotropaceae. In these mycorrhizal associations the fungi form both a sheath and intracellular penetrations.

Species from most of the major plant families form AM symbioses. This widespread and ubiquitous mycorrhizal type is characterized by the tree-like symbiotic structures, termed "arbuscules", that the fungus develops within the root cortical cells, and where most of the nutrient exchange between the fungus and the plant is thought to occur (Smith and Read 1997). These AM associations are especially important for P acquisition by plants growing in both agricultural and natural systems (Barea 1991).

The discovery of well-preserved fossil plants in the Early Devonian Rhynie Chert, which was critical for our knowledge of the origin and early evolution of land plants (Kenrick and Crane 1997), also served to reveal the existence of fossil mycorrhizal associations (Stubblefield *et al.* 1987; Remy *et al.* 1994; Kenrick and Crane 1997). Fungal structures, such as hyphae and spores, resembling those of extant AM fungi, were found colonizing the thallus and rhizomes in fossil records of minute 400-million-year-old plants (Kenrick 2003). Later on, certain "arbuscular" structures, very similar to those present in AM symbiosis of the present day, were clearly observed in the root cortical cells of Triassic plants (Phipps and Taylor 1996). Fossilized spores from the Ordovician (about 460 million years old), similar to those of modern AM fungi (Schüβler *et al.* 2001a,b), indicate that these micro-organisms were present at very early stages of plant evolution and probably even predated land plants (Redecker *et al.* 2000a). These paleobotanical observations were subsequently validated by phylogenetic studies based on DNA sequence data from living taxa (Simon *et al.* 1993; Redecker *et al.* 2000b). Since algae and cyano-bacteria able to fix atmospheric carbon (C) and nitrogen (N) were present in the primitive soil before colonization of the land by plants (Barghoorn 1974; Kar and Saxena 1976; Awramik 1981; Schopf 1983; Gehrig *et al.* 1996; Evans and Johansen 1999), a biological system capable of facilitating P acquisition for plant nutrition appears to have been critical to the early evolution of land plants. It is clear that this role was played by AM fungi from the very beginning of land colonization by plants (Pirozynski and Malloch 1975). Therefore, primitive roots developed in association with AM fungi and co-evolved with them to build the mycorrhizal root systems of extant vascular plants (Brundrett 2002). Thus, the establishment of AM symbioses represents an important innovation of plant adaptation, and can be envisaged as the first mechanism plants evolved to cope with low P availability in natural ecosystems. Accordingly, AM fungi have been suggested to have played a key role in the colonization of the land by plants and, subsequently, on the evolution of land plants (Pirozynski and Malloch 1975; Simon *et al.* 1993). Later on, during the Cretaceous, when the angiosperm radiation took place, other types of mycorrhiza evolved independently from AM symbiosis through parallel evolution (Brundrett 2002; Wang and Qiu 2006). The long co-evolution of plants and AM fungi originated diverse inter-dependencies between the symbionts, one of which is the obligate biotrophic character of AM fungi, which renders them unable to complete their life cycle in the absence of an appropriate host plant.

Since the AM association is the most widespread mycorrhizal symbiosis and influences the P nutrition of land plants considerably, this Chapter will focus only on AM symbiosis. The next section will describe how the fungus and plant establish symbiotic interfaces that allow the exchange of P and C in a functional AM symbiosis.

FUNGAL-PLANT INTEGRATION TO ESTABLISH FUNCTIONAL AM SYMBIOSIS

Development of the AM symbiosis starts when fungal hyphae arising from resting spores or from previously colonized roots recognize plant signals derived from an appropriate host plant. These signaling processes have been reviewed recently by

Bécard *et al.* (2004), Lambais (2006) and Paszkowski (2006). The first plant signals involved in AM establishment appear to be strigolactones (Akiyama *et al.* 2005). These compounds induce metabolic activity and hyphal branching of AM fungi and, consequently, increase the chances of the fungus encountering the plant root (Akiyama *et al.* 2005; Besserer *et al.* 2006; Reinhardt 2007). After the contact of a hypha with the root surface, the fungus differentiates to form an appressorium, from which a penetration hypha develops into the root. During appresorium formation the underlying epidermal cell reacts to the presence of the fungus with a reorganization program to accommodate the cell to the penetration hypha, which indicates an active contribution of the plant to root colonization by the AM fungus (Genre *et al.* 2005). Once inside the root, the fungus may form intracellular coils in the sub-epidermal cell layers, followed by intense proliferation of the mycelium by intercellular and/or intracellular growth into the cortex. In the inner cortical layers, hyphal branches penetrate the cortical cell walls and differentiate within the cells to form arbuscules (Smith and Read 1997). Arbuscules constitute highly specialized intracellular structures, extremely branched and with very fine hyphal tips, well suited to the exchange of nutrients and signals between symbionts. Although the fungal hypha penetrates the cell wall to form the arbuscule, all fungal branches remain surrounded by a new plant membrane, the periarbuscular membrane, which is continuous with the plasma membrane of the host cell (Smith and Gianinazzi-Pearson 1988). The morphological changes that occur in both plant and fungus result in a large surface area of contact between the symbionts. The space between the periarbuscular membrane and the fungal plasma membrane, which includes both apoplastic material of the plant and a minimal fungal cell wall, constitutes one of the most important interfaces in the symbiosis, as most of the nutrient exchange between symbionts occurs there (Smith and Gianinazzi-Pearson 1988).

Arbuscules are, in general, short-lived structures, with a turnover rate of about one to two weeks (Alexander *et al.* 1988). At the end of their lifespan, arbuscules senesce and collapse, and are completely degraded by the plant cell, which continues to live and can be colonized by other hyphae and form new arbuscules. Colonization by the fungus is a progressive process in which arbuscules are continuously being formed and degraded in different parts of the root system. As colonization progresses, most AM fungi form vesicles, which are intra or intercellular lipid-filled storage organs (Smith and Read 1997).

Once the fungus has entered the root cortex and arbuscules have been formed, an external mycelium develops in the soil, which constitutes a bridge connecting the root with its surrounding soil microhabitats. The AM fungal mycelium can spread through the soil over considerably longer distances (usually several centimeters) than root hairs (Jakobsen *et al.* 1992a,b). The extraradical mycelium is profusely branched and provides a very efficient nutrient-absorbing system beyond the Pi-depletion zone surrounding the plant roots, thereby reducing the distance that Pi must diffuse through the soil prior to its interception. As will be explained later, the ability of the AM hyphae to grow beyond the root Pi-depletion zone and deliver the intercepted Pi to the plant is thought to be the reason why AM mycorrhizae increase Pi accumulation and plant growth in soils with low P availability (Smith and Read 1997).

The AM fungi contribute to P acquisition and supply to plants by linking the geo-chemical and biotic portions of the soil ecosystem, thereby affecting rates and patterns of P cycling in both agricultural and natural ecosystems (Jeffries and Barea 2001). However, the AM symbiosis not only influences nutrient cycling in soil-plant systems but also improves plant health through increased protection against environmental stresses, whether they be biotic (e.g. pathogen attack) or abiotic (e.g. drought, salinity, heavy metals, organic pollutants), and enhancing soil structure through the formation of the aggregates necessary for good soil tilth (Kapulnik and Douds 2000; Gianinazzi *et al.* 2002; van der Heijden and Sanders 2002; Jeffries *et al.* 2003; Smith *et al.* 2003; Barea *et al.* 2005; Rillig and Mummey 2006; Turnau *et al.* 2006).

In this Chapter, the analysis of AM effects on plant biology and ecology will focus on the impact of AM symbiosis on P acquisition. In this context, as many studies have shown, a general and universally accepted conclusion is that mycorrhizal plants acquire P more effectively than non-mycorrhizal ones. Indirect evidence of this comes from the fact that mycorrhizal plants are frequently not only larger but also contain more P than non-mycorrhizal plants of the same species growing in the same environment. Greater P acquisition by AM plants is attributed to a combination of factors. The impact of these factors for AM functioning and, in particular, the mechanisms whereby the extraradical mycelium explores the soil and acquires P for AM-plants are described in the next section.

FUNCTIONAL BIOLOGY OF PHOSPHATE UPTAKE BY AM-PLANTS

Phosphorus sources for AM plants

It is generally assumed that AM fungi, like plant roots, take up P as Pi from the soil solution, but that the AM mycelium acquires Pi more efficiently than plant roots (Smith and Read 1997). However, some studies have shown that AM fungi have the ability to use P sources that are not directly available to a plant (Powell and Daniel 1978; Bolan *et al.* 1984). For example, Tawaraya *et al.* (2006) found that AM hyphal exudates were able to solubilize Pi more effectively than root exudates, suggesting that fungal exudates could contribute to increased P accumulation by AM-plants by solubilizing Pi. Hence, although the contribution of AM fungi to Pi solubilization is thought to be low, a limited contribution cannot be excluded. This has been proposed to occur through different mechanisms, either direct, as H^+ excretion or production of hydroxyacids with chelating activity (Bago and Azcón-Aguilar 1997; Smith and Read 1997), or indirect, through qualitative or quantitative changes in the microbial populations of the soil surrounding the hyphae (hyphosphere) or mycorrhiza (mycorrhizosphere) as reviewed by Barea *et al.* (2005). These microbial interactions are particularly relevant when Pi-solubilizing microorganisms are involved, as it will be described later in this Chapter. Additionally, it has been shown that the extraradical mycelium of AM fungi can excrete enzymes, such as phosphatases, that release Pi from some organic sources (Joner and Johansen 2000; Joner *et al.* 2000; Koide and Kabir 2000). However, the importance of these fungal enzymes for P nutrition of AM plants appears to be limited to soils where organic P compounds constitute a large fraction of the total soil P. The presence of AM fungal structures in decomposing leaves was reported by Aristizabal *et al.* (2004), who suggested that there was no saprophytic AM development but a mechanical entry through vascular tissues. These authors suggested that AM fungi were well positioned to obtain and recycle P, and other nutrients, released by decomposer microorganisms, thereby avoiding losses by leaching or immobilization by soil components.

Soil exploration by AM plants

One of the favored explanations for increased P acquisition by mycorrhizal plants is the extension of the extraradical mycelium into soil to take up and translocate P from soil solution to the plant over distances up to several cm from the root (Jakobsen *et al.* 1992a). In addition, since the density of mycorrhizal hyphae in soils is very high, the area for P uptake is much greater in mycorrhizal plants. Despite the difficulties in measuring AM hyphal length in the soil, several estimates are available using different methodological approaches and host plants. An average value of 1 m of AM hyphae per centimeter of colonized root has been found (Smith and Read 1997), which represents a large surface area for Pi uptake. Additionally, due to the smaller size of hyphae ($<20\,\mu m$) than roots, hyphae can explore soil pores inaccessible to roots. Thus, the higher Pi uptake by mycorrhizal plants can be explained in term of increased hyphal exploitation of the soil and the competitive ability of the hyphae to absorb localized and dilute sources of Pi. In this context, several studies have shown positive correlations between AM fungal variables, such as hyphal length or hyphal density, with growth response variables of the colonized plants, such as shoot biomass, Pi uptake and P content (Jakobsen *et al.* 2001; Avio *et al.* 2006). However, this cannot be taken as a general conclusion since advanced hyphal development does not always correlate with a plant growth response (Smith *et al.* 2004).

In addition to the physical extension to the root system in AM plants, the possibility that AM fungi might absorb Pi from the soil solution at lower concentrations than non-mycorrhizal roots has also been proposed as a mechanism for the increased efficiency of Pi acquisition by AM plants. The affinity of mycorrhizal plants for Pi uptake has been determined from kinetic studies. These suggest that Pi transporters in mycorrhizal plants have lower K_m values or, in other terms, higher affinity for Pi uptake, than non-mycorrhizal plants (Cress *et al.* 1979; Cardoso *et al.* 2006). As will be described below, high affinity Pi transporters have been characterized in the extraradical hyphae of certain AM fungi.

Mycorrhizal plants have two pathways for Pi uptake: the "direct" pathway at the plant-soil interface, through the root epidermal and hair cells, and the "mycorrhizal" pathway via the fungal mycelium (Smith *et al.* 2003). The mycorrhizal pathway involves Pi uptake from the soil solution by the extraradical AM fungal hyphae, translocation through the AM fungal structures from the external to the internal mycelium and transfer to the plant across the symbiotic interfaces. For some plant species/AM fungi combinations, it has been found that AM colonization results in complete inactivation of the direct Pi uptake pathway and 100% of the P in plant tissues is provided by the AM fungus (Smith *et al.* 2004).

Molecular mechanisms of plant adaptation for improved P acquisition

Phosphate is taken up by root cells against a steep concentration gradient resulting from micromolar external concentrations and about 2,000-fold higher concentrations inside the cell (Schachtman *et al.* 1998). This uptake process is energy dependent and is mediated by the concerted activities of specialized membrane proteins (H^+-ATPases and H^+/Pi-cotransporters). The following paragraphs will present the known molecular components of the complex mycorrhizal Pi uptake pathway and present an overview of the mechanism of plant Pi acquisition and its regulation in a mycorrhizal root.

Transport proteins putatively involved in Pi uptake by AM fungi have been identified. In particular, genes encoding high-affinity Pi transporters that are preferentially expressed in the extraradical mycelium and that are, therefore, likely to be involved in Pi acquisition from the soil solution have been identified in *Glomus versiforme* (*GvPT*; Harrison and Vanbuuren 1995), *Glomus intraradices* (*GiPT*; Maldonado-Mendoza *et al.* 2001) and *Glomus mosseae* (*GmosPT*; Benedetto *et al.* 2005). Functional expression of *GvPT* in yeast showed that its gene product is a high-affinity Pi transporter that operates by proton-coupled symport (Harrison and Vanbuuren 1995). The H^+-ATPases responsible for the generation of the proton-motive force driving the uptake of Pi across the plasma membrane of the extraradical hyphae have been identified in *G. mosseae* (Ferrol *et al.* 2000a,b; Requena *et al.* 2003). Detailed expression analyses of *GiPT* and *GmosPT* have revealed that both genes are regulated in a manner typical of genes encoding high-affinity Pi transporters, with maximal expression at the micromolar Pi concentrations usually found in the soil solution, and that expression of *GiPT* is modulated by the overall Pi status of the mycorrhiza. Interestingly, a relatively high expression level of *GmosPT* was also observed in the intraradical fungal structures, suggesting that the fungus may exert control over the amount of Pi that is delivered to the plant (Maldonado-Mendoza *et al.* 2001; Benedetto *et al.* 2005).

Phosphate taken up by the fungus is first incorporated into the cytosolic Pi pool, which maintains a constant Pi concentration to sustain various fungal functions, such as energy generation and biosynthesis of phospholipids, nucleic acids and precursors of carbohydrate polymers. Excess cytosolic Pi is transported into the fungal vacuoles and condensed into polyphosphates, linear polymers of three to thousands of Pi residues connected by high-energy bonds, which are believed to play a central role in Pi supply to the plant (Ezawa *et al.* 2002). The mechanism of polyphosphate translocation from the extraradical to the intraradical hyphae has not been yet elucidated, but it may be driven by cytoplasmic streaming and/or a motile tubular vacuole system (Cooper and Tinker 1981; Olsson *et al.* 2002; Uetake *et al.* 2002). The observation that the length of polyphosphate polymers is shorter in mycorrhizal roots than in extraradical hyphae supports the hypothesis that polyphosphate is hydrolyzed in the intraradical fungal structures before Pi is supplied to the host plant (Ohtomo and Saito 2005). The detection of strong alkaline phosphatase activity in the intraradical fungal structures raised the idea that these enzymes are involved in Pi efflux from arbuscules (Tisserant *et al.* 1993; Ezawa *et al.* 1995, 1999; Kojima and Saito 2004). Expression analyses of the genes putatively encoding alkaline phosphatases in the AM fungi *G. intraradices* (*GintALP*) and *Gigaspora margarita* (*GmALP*) revealed that these enzymes were expressed more highly in intraradical than in extraradical hyphae (Aono *et al.* 2004). However, transcription of these genes was not regulated by Pi availability in the environment. Although the involvement of alkaline phosphatase in Pi efflux from arbuscules has been questioned (Ezawa *et al.* 1999), the enzyme was shown to be involved in Pi efflux from isolated intraradical hyphae of AM fungi (Kojima and Saito 2004). It is clear that further studies, such as determining the substrate specificity of the *ALP* gene products, are needed to uncover the mechanisms of polyphosphate break-down in the intraradical hyphae.

The mechanisms involved in the release of Pi from the fungus in colonized cells are currently unknown. Because Pi efflux across the fungal plasma membrane probably follows a concentration gradient, it could be facilitated by anion channels, carriers or pumps. As mentioned before, the expression of *GmosPT* in the intraradical fungal structures suggests that Pi efflux at the arbuscular interface occurs in competition with Pi uptake, and that the fungus may exert control over the amount of Pi that is delivered to the plant (Benedetto *et al.* 2005). The Pi released from the fungus is subsequently taken up by root cortical cells through high-affinity Pi transporters. Gene expression analyses of plant Pi transporters and H^+-ATPases in mycorrhizal roots of various plant species have provided evidence that the development of the symbiosis is accompanied by large rearrangements in plant Pi transport pathways and a coincident change in the expression of Pi transporters and H^+-ATPases within plant roots. An extensive review of mycorrhizal-regulated Pi transporters is provided by Javot *et al.* (2007). Phosphate transporters and H^+-ATPases operating at the root-soil interfaces are often down-regulated in mycorrhizal roots (Chiou *et al.* 2001; Ferrol *et al.* 2002). However, mycorrhiza-specific Pi transporters, which are expressed exclusively in mycorrhizal roots, and mycorrhiza up-regulated Pi transporters, which are strongly induced by the development of the symbiosis but have a basal expression in non-mycorrhizal roots, have been identified in several plant species. *In situ* hybridization and *in vivo* promoter analyses of genes encoding several mycorrhiza-induced Pi transporters have demonstrated their predominant, or exclusive, expression in cortical cells colonized by arbuscules, which suggests that the encoded Pi transporters are probably involved in taking up the Pi that is transferred by the fungus to the apoplast of the arbuscular interface (Rausch *et al.* 2001; Nagy *et al.* 2005). In particular, immunolocalization of the *Medicago truncatula* Pi transporter MtPT4 revealed its specific location in the plant periarbuscular membrane surrounding fine branches of developing and mature arbuscules (Harrison *et al.* 2002). This expression pattern correlates with previous studies showing the accumulation of plasma membrane H^+-ATPases around fine arbuscular branches (Gianinazzi-Pearson *et al.* 2000). Together, these expression patterns indicate that activation of the mycorrhizal uptake pathway is characterized by the induction of mycorrhiza-specific Pi transporters and partial down-regulation of the Pi transporters of the direct uptake pathway. These data also reflect the fine balance between fungal and root Pi uptake pathways. Since the down-regulated plant Pi transporters are also generally responsive to plant Pi status, it is not clear whether their down-regulation is a result of a direct regulation of their expression triggered in response to the symbiosis or an indirect regulation of their expression through the fungal-induced improvement of the plant Pi status.

Despite the large number of mycorrhiza-inducible phosphate transporters identified to date, the functional genomics of Pi transport at the arbuscular interface is not well understood. Functional analysis *in planta* of the mycorrhiza-specific phosphate transporter LePT4 from tomato using a *LePT4* knock-out mutant revealed considerable redundancy between Pi transporters at the arbuscular interface in this solanaceous species (Nagy *et al.* 2005). The *LePT4* knock-out mutant displayed no detectable phenotype. There was no alteration of fungal Pi transport, and the expression of other known Pi transporters, including the mycorrhiza-inducible (*LePT3*) and mycorrhiza-specific (*LePT5*) transporters, remained unchanged. The authors suggested that redundancy within mycorrhiza-inducible Pi transport pathways might ensure that symbiotic Pi transfer will be evolutionary robust and relatively insensitive to mutation.

Recently, experimental evidence was provided for the mycorrhizal Pi transport function of the mycorrhiza-specific Pi transporters *LjPT3* from *L. japonicus* and *MtPT4* from *M. truncatula*. Reducing expression of the *LjPT3* gene using RNAi technology resulted in reduced growth of transgenic plants when colonized by a mycorrhizal fungus, reduced allocation of radiotracer Pi to the shoot and reduced AM fungal colonization (Maeda *et al.* 2006). Mutants of *M. truncatula*, in which either the gene encoding the mycorrhiza-specific Pi transporter *MtPT4* was down-regulated using RNAi technology or harboring a loss-of-function MTP4 allele identified by target-induced local lesions in genomes (TILLING), have demonstrated that MtPT4 is not only essential for symbiotic Pi transport but is also required for the development of the symbiosis (Javot *et al.* 2007). Loss of MtPT4 function resulted in a block in symbiotic Pi transfer from the arbuscule to the cortical cell, premature death of the arbuscules, inability of the fungus to proliferate within the root and no further development of the symbiosis. Based on these data, the authors proposed that Pi uptake through MtPT4 serves, either directly or indirectly, as a signal to the plant cell of the presence of a beneficial fungal symbiont. This hypothesis would explain why symbiosis does not develop under extremely low Pi conditions and why colonization is dramatically stimulated by the application of small amounts of Pi.

ECOLOGICAL IMPACT OF AM SYMBIOSES ON PLANT COMMUNITY STRUCTURE AND PRODUCTIVITY

The impact of AM fungi on plant community composition and functioning has been the subject of many studies over the last decades and this research has led to the conclusion that the activity of AM fungi is a key mechanism for linking biodiversity and ecosystem functioning (Read 1998). Reciprocally, disturbance of natural ecosystems affects the diversity of AM fungal populations (Lovera and Cuenca 2007).

A pioneering experiment on the impact of AM fungal diversity on plant community diversity by Grime *et al.* (1987) showed that the presence of undefined mixtures of AM fungi increased the floristic diversity in a microcosm trial. The first demonstration that the diversity and identity of AM fungi, and not merely their presence, was a determinant of plant diversity and/or ecosystem productivity came from the field plots and mesocosms studies of van der Heijden *et al.* (1998b). This influential publication prompted a number studies aimed at ascertaining whether AM fungal diversity could determine the species composition and functioning of plant communities. Elucidating the mechanisms and ecological factors involved in the diversity/functioning interactions between plant and AM fungal populations have recently been the objectives of several research groups. Their studies will be discussed in the following paragraphs with a special emphasis on the role of P capture by the AM fungi and P supply to the plant in such ecological interactions.

Several mechanisms/factors have been proposed to be responsible for the ecological interactions between plant and fungal communities. Among these are (1) the functional specificity of the different plant-fungus combinations, particularly concerning P nutrition (van der Heijden *et al.* 1998a,b, 2006; Klironomos 2002), (2) the mycorrhizal dependency/responsiveness of the plant species involved, (3) the dominant *vs.* subordinate character of these species in the community (O'Connor *et al.* 2002; van der Heijden 2002; van der Heijden *et al.* 2006), (4) the "niche differentiation in P use" (Reynolds *et al.* 2006; Vogelsang *et al.* 2006) and (5) the so-called "sampling effect" (Vogelsang *et al.* 2006).

Despite the fact that specificity *sensu strictum* does not exist in AM associations, in which almost all plants in the community can be colonized simultaneously by several species of AM fungi, it is accepted that different AM fungal taxa induce more positive responses in some plant species than in others (Sanders 2002). It is also noteworthy that not all plants form AM associations, and that not all plants that form AM associations obtain nutritional benefits from AM fungi under all growth conditions (O'Connor *et al.* 2002). The differential effects on plant growth and development of specific plant-fungus associations, known as their "functional compatibility/specificity" (Gianinazzi-Pearson and Gianinazzi 1988), has a critical significance in AM ecology/functioning.

The results of van der Heijden *et al.* (1998b) were re-analyzed by Read (1998) and Hart and Klironomos (2002) in discussing the specificity relationships in the AM symbiosis. These review articles highlight the role of AM fungi in increasing productivity and/or diversity of plant communities and support the hypothesis that the differential effects of specific plant-fungus combinations (functional compatibility) explain the observations of van der Heijden *et al.* (1998b). In particular, the ability of specific fungi to supply plants with P was found to be the basis of the differential effects of functional compatibility (van der Heijden *et al.* 1998b). Hart and Klironomos (2002) also pointed out that the diverse AM fungal communities produced an intermingling, extensive extra-radical mycelium that allowed a more efficient exploitation of soil nutrients, which could account for the increase in plant tissue P concentrations and decrease in soil P observed by van der Heijden *et al.* (1998b). The establishment of such an external hyphal network gives way to a mycelial web around the roots, constituting a diverse inoculum source for the different plant species that benefit from the nutrient (P) flow through the soil-fungus-plant system (Read 1998; Jakobsen 2004).

However, more recent work investigating the influence of AM fungi on the diversity and/or productivity of plant communities has presented some conflicting results. Van der Heijden (2002) re-analyzed publications on this topic and concluded that the main factors influencing the ecological interactions between plants and AM-fungi were the degree of mycorrhizal dependency (MD) or mycorrhizal responsiveness (MR) of the plant species involved and their dominant or subordinate character in the plant community. The terms MD and MR are often used synonymously in an ecological context. However, Janos (2007) distinguishes between these two terms. According to this author, MR is represented by the difference in growth between plants with and without AM colonization, at any level of P availability, being also a measure of AM fungus effectiveness, and MD is defined by the lowest level of P availability at which plants can grow without AM colonization. All in all, MD is a measure of the extent to which a plant benefits from association with AM fungi, which is in turn is related to the ability of AM fungi to supply P to the plant (van der Heijden *et al.* 1998b). In this context, van der Heijden (2002) concluded that the differential responses of a plant species to different AM fungal species are higher when the target plant species has a higher MD. The presence of AM fungi promotes plant diversity when most of the subordinate plant species in the community have a high MD so that their growth is stimulated specifically (Grime *et al.* 1987; van der Heijden *et al.* 1998b). Conversely, AM fungi reduce the diversity of plant communities when most of the plant species have a negative or low MD (Hartnett and Wilson 1999). Moreover, when dominant plant species in the community have a high degree of mycorrhizal responsiveness, and thereby derive greater benefit from AM fungi, plant diversity is not promoted, irrespective of the level of MD of the subordinate species (O'Connor *et al.* 2002). By labeling the dominant but low mycorrhiza-dependent plant species with ^{14}C, Grime *et al.* (1987) also found a belowground transport of C via the AM mycelium to the subordinate highly mycorrhiza-dependent species. This C and P redistribution facilitates the establishment and growth of subordinate species to increase plant diversity in the community.

With regard to the impact of AM fungal diversity on community productivity, van der Heijden (2002) concluded that AM fungi enhance community productivity when the communities are dominated by mycorrhizal-dependent plant species, able to reap the benefit from P supply from AM fungi. When the community is dominated by plant species with a low MD, van der Heijden (2002) concluded that there was no increase in productivity, because P supply by AM fungi is a less influential process. A recent publication shows that mixtures of mycorrhiza-dependent and non-dependent plant species do not result in an increase in the productivity of a grassland community as a whole (van der Heijden *et al.* 2006).

In the context of the mechanisms underlying the relationship between AM fungal diversity and plant community diversity and/or productivity it is important to consider whether the benefits generated by individual fungal isolates are greater than those produced by mixed inoculum treatment ("sampling effect") or conversely, the benefits generated by a mixed inoculum, i.e. the diversity *per se*, are greater than those from any single inoculum. In studying this, Vogelsang *et al.* (2006) found that AM fungal identity benefited plant diversity and productivity more than the diversity *per se*. These results conflict with those of van der Heijden *et al.* (1998b), who correlated AM fungal diversity *per se* with plant diversity and productivity. However, these conflicting reports might be explained by the confounding effects of plant species dominance or mycorrhizal-dependency (van der Heijden *et al.* 2006). Since different plant-AM fungus combinations access different P sources, Vogelsang *et al.* (2006) also investigated whether competition for limited P resources could favor AM fungus-mediated P-niche partitioning. These authors manipulated the P sources in mesocosm experiments and found little support for such AM fungal-facilitated complementarity in P use by the community.

MYCORRHIZOSPHERE INTERACTIONS AND ECOSYSTEM P CYCLING

The AM fungi, as components of soil microbiota, are immersed in a framework of interactions with other soil microorganisms that are fundamental to rhizosphere functioning. Some of these interactions have beneficial consequences for plant health and productivity (Barea *et al.* 2002a,b). Particularly relevant to P cycling in ecosystems are interactions between AM fungi and phosphate-solubilizing-microorganisms (Barea *et al.* 2002c, 2007).

Some microorganisms are known to benefit mycorrhiza establishment (Gryndler 2000; Barea *et al.* 2002a). In particular, the "mycorrhiza-helper-bacteria" (MHB), as identified by Garbaye (1994), are known to stimulate mycelial growth and/or improve mycorrhizal formation of mycorrhizal fungi and have been the subject of many studies (Barea *et al.* 2005). Likewise, the establishment of an AM fungus in the root cortex can affect rhizosphere microbial populations. This effect is in part due to AM formation changing many key aspects of plant physiology, such as the mineral nutrient composition of plant tissues, hormonal balance, and the patterns of C allocation. As a consequence of these changes, the AM symbiotic status usually modifies root exudation and the chemical characteristics of root exudates (Tawaraya *et al.* 2006). In addition, the development of an external AM mycelium introduces physical modifications to the soil environment surrounding roots (Barea *et al.* 2005). The microbial populations change both quantitatively and qualitatively in the rhizosphere of AM plants, the so-called *mycorrhizosphere* (*sensu lato*) resulting in features that differ from those of a non-mycorrhizal plant (Gryndler 2000; Barea *et al.* 2002a; Johansson *et al.* 2004). The use of molecular techniques has demonstrated specificity in the bacterial populations associated with AM-roots (Offre *et al.* 2007), AM-mycelium (Toljander *et al.* 2005; Rillig and Mummey 2006; Rillig *et al.* 2006) and AM-spores (Roesti *et al.* 2005). In addition to their effects on natural bacterial populations, AM-associations also affect the establishment in the rhizosphere of bacterial inoculants based on Plant Growth Promoting Rhizobacteria (PGPR; Barea *et al.* 2005). It has been observed that AM inoculation improved the establishment of both inoculated and indigenous phosphate-solubilizing PGPR, which in turn benefited AM formation (Toro *et al.* 1997; Barea *et al.* 2002c).

There are many studies describing the interactions between AM fungi and specific bacterial groups that benefit plant growth and development. These have been reviewed recently and some of the underlying mechanisms discussed (Artursson *et al.* 2005; Barea *et al.* 2005). As stated previously, the interactions between AM fungi and phosphate-solubilizing-microorganisms in the mycorrhizosphere are particularly relevant to P acquisition by plants and merit more detailed consideration here.

Among the microbiological processes involved in P cycling are those responsible for increasing Pi availability in soils (Kucey *et al.* 1989; Richardson 2001). Two general types of processes have been described in this context: those promoting the solubilization of recalcitrant P-sources in soils and those improving the uptake of solubilized Pi by plants. The solubilization/mineralization of unavailable P compounds is carried out by diverse saprophytic bacteria and fungi acting by several mechanisms (Marschner 2008; George and Richardson 2008). As indicated before, the external mycelium of the AM fungi acts as a bridge between roots and the surrounding soil microhabitats, which give access to Pi from soil solution beyond the Pi-depletion zone surrounding the roots (Barea 1991).

Because of the importance of microbial solubilization processes in soil-plant systems, Pi-solubilizing bacteria (PSB) have been assayed as seed inoculants (Kucey *et al.* 1989). For bacterial Pi solubilization to be effective in soil, there are a number of ecological aspects to be considered. First of all, PSB inoculants must establish in root-associated soil habitats. For this reason, it has been recommended that PSB for inoculation are selected from the subset of PGPR populations since their ability to colonize the rhizosphere can be a pre-requisite for effectiveness (Glick *et al.* 1995). Unfortunately, because of the transient availability of the released compounds, and their possible re-fixation on their way to the root surface, solubilizing Pi using inoculated bacteria may have limited utility (Kucey *et al.* 1989). The Pi made available by PSB acting on sparingly-soluble P sources may not reach the root surface due to limited diffusion. However, it has been proposed that if the solubilized Pi were taken up by an AM mycelium, this synergistic microbial interaction could improve P supply to a plant. This has been investigated in a number of studies carried out in our laboratory (Barea *et al.* 2007), which involved the application of poorly reactive rock phosphate (RP) and the use of ^{32}P-tracer methodologies (Toro *et al.* 1997). Upon adding a small amount of ^{32}P to label the exchangeable soil P pool, the isotopic composition, or "specific activity" (SA = ^{32}P/^{31}P quotient), was determined in plant tissues (Zapata and Axmann 1995). These studies found that dual inoculation reduced the SA of the host plant, indicating that these plants acquired P from sources, either endogenous or from added RP, which were not directly available to non-inoculated or singly-inoculated plants.

To validate these results, a series of experiments were carried out under controlled conditions in the glasshouse and in the field (Barea *et al.* 2002c). These experiments and their conclusions can be summarized as follows. The main aim of the experiments was to investigate the effects of PSB × AM interactions on P capture, cycling and supply under field conditions. A co-lateral objective was to investigate the agronomic effectiveness of RP as a source of P for crop production as affected by the target microbial interactions. The possibility of using the sparingly soluble RP as a P source for sustainable agriculture is a topic of current interest (Truong and Zapata 2002). A neutral to alkaline, but low-Ca, soil was used. The experiments involved a factorial combination of four microbial and two chemical treatments. The microbial treatments were: (1) AM inoculation; (2) PSB inoculation; (3) AM plus PSB dual inoculation; and (4) non-inoculated controls, exposed to the naturally existing AM fungi and PSB. The two chemical treatments were: (1) non-amended control without P application, and (2) RP application. For the greenhouse experiment, the exchangeable soil P pool was labelled with ^{32}P. The ^{32}P activity in the plant material was measured, and tissue SA was determined (Zapata and Axmann 1995). Both RP addition and microbial inoculation improved biomass production and P accumulation in the test plants, with dual microbial inoculation being the most effective treatment. Independently of RP addition, AM-inoculated plants showed a lower SA than the non-inoculated controls, particularly when they were co-inoculated with PSB. This confirmed that AM-inoculated plants could acquire soil P from sources unavailable to non-inoculated plants. Possibly, the PSB were effective in releasing ^{31}P from sparingly soluble sources, either from the soil components or from the added RP. This release of Pi would constitute a part of the total ^{31}P pool from which the AM mycelium acquired P and transferred it to the plants. Such microbial activities could result in the lower SA in dually-inoculated plants. Results from the field trial corroborated the hypothesis that interactions between AM fungi and PSB can have a cooperative fundamental role in P-cycling, which is of considerable importance for both ecosystems in general and for the agronomic efficiency of RP in particular.

Apart from co-inoculation of AM and PSB, other biological approaches to improve the use of RP in non-acidic soils have been considered (Vassilev *et al.* 2002). These approaches are based on a fermentation process involving agrowastes (sugar beet), application of *Aspergillus niger*, a microorganism possessing a high acid-producing activity (citric acid), and RP. At the end of the solid state fermentation process, the mineralization of the organic matter and solubilization of RP was demonstrated. A series of microcosm experiments were then carried out aimed at evaluating the effectiveness of the fermentation products in a neutral, non-calcareous, P-deficient soil. The host plant (*Trifolium repens*) was inoculated with an AM fungus. It was shown that the fermentation products improved plant growth and P content and that AM inoculation further improved the effectiveness of the fermentation product. Thus, plants benefited from the P solubilized from RP by the microbial activities. This biotechnological approach has potential for sustainable agriculture. Recently, Vassilev *et al.* (2006) reported the effect of four agro-industrial wastes (sugar beet, olive cake, olive mill waster wasters and dry olive cake) as substrates for microbial solubilization of RP. Amendments resulting from all these fermented products improved plant growth and P acquisition, which were further enhanced by AM inoculation. In another experiment, the fermentation products from agro-wastes plus RP treated with *Yarowia lipolityca*, a dry-soil-adapted yeast with RP-solubilizing ability, were assayed (Vassilev *et al.* 2001). The products were applied to a degraded and P-limited soil for restoration purposes in combination with AM fungal inoculation (Medina *et al.* 2005). These biotechnological products improved plant growth, soil structure and soil biochemical characteristics.

CONCLUSIONS

The establishment of mycorrhizal symbiosis, one of the most ancient mutualisms on Earth, has been demonstrated to be a successful and widespread strategy to maximize the access of plant roots to Pi in the soil. The AM association is the most widespread mycorrhizal type, and influences the P nutrition of land plants considerably. The morphological intra- and extra-radical integration of both symbionts in AM associations allows the development of the extraradical fungal mycelia, which function as an absorptive structure that forages effectively P sources in the soil, and allows Pi to be transferred from fungus to plant through the arbuscules. Several of the morphological, physiological, biochemical and molecular adaptations of plants and fungi required for AM functioning have been elucidated.

Phosphate uptake is energy dependent and mediated by the concerted activities of specialized membrane-spanning proteins (Pi transporters and H^+-ATPases) in the plasma membranes of both plant and fungus. Proteins putatively involved in Pi uptake by plants and AM fungi have been identified. Plant genes encoding Pi transporters are specifically expressed in arbuscules. Fungal genes encoding high-affinity Pi transporters that are preferentially expressed in the extraradical mycelium, likely involved in Pi acquisition from the soil solution, have also been identified. It appears that an alkaline phosphatase activity in the intraradical fungal structures is involved in Pi efflux from the fungus to the plant.

Because it was suggested that AM fungal diversity determines the species composition, functioning and productivity of plant communities, the impact of several processes on these ecological relationships have been analyzed. Some of these relate to the ability of AM fungi to supply plants with P. The conclusion is that the most influential factors are (a) the functional specificity of the different plant-fungus combinations, (b) the degree of mycorrhizal dependency/responsiveness of the plant species in the community and (c) the dominant *vs.* subordinate character of the plant species responding to AM symbiosis. Diversity of AM fungi appears to promote plant diversity when most of the subordinate plant species in the community have a high mycorrhizal dependency, so that their growth is specifically stimulated by P acquisition. The productivity of plant communities is enhanced when they are dominated by mycorrhizal-dependent plant species, capable of obtaining a great benefit from the P supplied by AM fungi. Mycorrhizosphere interactions involving Pi-solubilizing saprophytic bacteria and fungi improve plant P nutrition and favor P-cycling within ecosystems.

REFERENCES

Akiyama K, Matsuzaki K, Hayashi H (2005) Plant sesquiterpenes induce hyphal branching in arbuscular mycorrhizal fungi. Nature 435: 824–827

Alexander T, Meier R, Toth R, Weber HC (1988) Dynamics of arbuscule development and degeneration in mycorrhizas of *Triticum aestivum* L. and *Avena sativa* L. with reference to *Zea mays* L. New Phytol 110: 363–370

Aono T, Maldonado-Mendoza IE, Dewbre GR, Harrison MJ, Saito M (2004) Expression of alkaline phosphatase genes in arbuscular mycorrhizas. New Phytol 162: 525–534

Aristizabal C, Rivera EL, Janos DP (2004) Arbuscular mycorrhizal fungi colonize decomposing leaves of *Myrica parvifolia*, *M. pubescens* and *Paepalanthus* sp. Mycorrhiza 14: 221–228

Artursson V, Finlay RD, Jansson JK (2005) Combined bromodeoxyuridine immunocapture and terminal-restriction fragment length polymorphism analysis highlights differences in the active soil bacterial metagenome due to *Glomus mosseae* inoculation or plant species. Environ Microbiol 7: 1952–1966

Avio L, Pellegrino E, Bonari E, Giovannetti M (2006) Functional diversity of arbuscular mycorrhizal fungal isolates in relation to extraradical mycelial networks. New Phytol 172: 347–357

Awramik SM (1981) The pre-Phanerozic biosphere: Three billion years of crises and opportunities. In: Nitecki MH (ed), Biotic Crises in Ecological and Evolutionary Time. Academic, New York, pp 83–102

Bago B, Azcón-Aguilar C (1997) Changes in the rhizospheric pH induced by arbuscular mycorrhiza formation in onion (*Allium cepa* L). Z Pflanz Bodenkunde 160: 333–339

Barea JM (1991) Vesicular-arbuscular mycorrhizae as modifiers of soil fertility. In: Stewart BA (ed), Advances in Soil Science. Springer, New York, pp 1–40

Barea JM, Azcón R, Azcón-Aguilar C (2002a) Mycorrhizosphere interactions to improve plant fitness and soil quality. Anton Van Leeuwenhoek 81: 343–351

Barea JM, Gryndler M, Lemanceau P, Schüepp H, Azcón R (2002b) The rhizosphere of mycorrhizal plants. In: Gianinazzi S, Schüepp H, Barea JM, Haselwandter K (eds), Mycorrhiza Technology in Agriculture: From Genes to Bioproducts. Birkhäuser, Basel, Switzerland, pp 1–18

Barea JM, Toro M, Orozco MO, Campos E, Azcón R (2002c) The application of isotopic (^{32}P and ^{15}N) dilution techniques to evaluate the interactive effect of phosphate-solubilizing rhizobacteria, mycorrhizal fungi and *Rhizobium* to improve the agronomic efficiency of rock phosphate for legume crops. Nutr Cycl Agroecosyst 63: 35–42

Barea JM, Pozo MJ, Azcón R, Azcón-Aguilar C (2005) Microbial co-operation in the rhizosphere. J Exp Bot 56: 1761–1778

Barea JM, Toro M, Azcón R (2007) The use of ^{32}P isotopic dilution techniques to evaluate the interactive effects of phosphate-solubilizing bacteria and mycorrhizal fungi at increasing plant P availability. Plant Soil 287: 1–8

Barghoorn ES (1974) Two billion years of prokaryotes and the emergence of eukaryotes. Taxon 23: 259

Bécard G, Kosuta S, Tamasloukht M, Séjalon-Delmas N, Roux C (2004) Partner communication in the arbuscular mycorrhizal interaction. Can J Bot 82: 1186–1197

Benedetto A, Magurno F, Bonfante P, Lanfranco L (2005) Expression profiles of a phosphate transporter gene (GmosPT) from the endomycorrhizal fungus *Glomus mosseae*. Mycorrhiza 15: 620–627

Besserer A, Puech-Pages V, Kiefer P, Gomez-Roldan V, Jauneau A, Roy S, Portais JC, Roux C, Becard G, Sejalon-Delmas N (2006) Strigolactones stimulate arbuscular mycorrhizal fungi by activating mitochondria. Plos Biol 4: 1239–1247

Bieleski RL (1973) Phosphate pools, phosphate transport, and phosphate availability. Ann Rev Plant Physiol 24: 225–252

Bolan NS, Robson AD, Barrow NJ, Aylmore LAG (1984) Specific activity of phosphorus in mycorrhizal and non-mycorrhizal plants in relation to the availability of phosphorus to plants. Soil Biol Biochem 16: 299–304

Brundrett MC (2002) Coevolution of roots and mycorrhizas of land plants. New Phytol 154: 275–304

Cardoso IM, Boddington CL, Janssen BH, Oenema O, Kuyper TW (2006) Differential access to phosphorus pools of an oxisol by mycorrhizal and nonmycorrhizal maize. Commun Soil Sci Plant Anal 37: 1537–1551

Chiou TJ, Liu H, Harrison MJ (2001) The spatial expression patterns of a phosphate transporter (MtPT1) from *Medicago truncatula* indicate a role in phosphate transport at the root/soil interface. Plant J 25: 281–293

Cooper KM, Tinker PB (1981) Translocation and transfer of nutrients in vesicular-arbuscular mycorrhizas. 4. Effect of environmental variables on movement of phosphorus. New Phytol 88: 327–339

Cress WA, Throneberry GO, Lindsey DL (1979) Kinetics of phosphorus absorption by mycorrhizal and non-mycorrhizal tomato roots. Plant Physiol 64: 484–487

Evans RD, Johansen JR (1999) Microbiotic crusts and ecosystem processes. Crit Rev Plant Sci 18: 183–225

Ezawa T, Saito M, Yoshida T (1995) Comparison of phosphatase localization in the intraradical hyphae of arbuscular mycorrhizal fungi, *Glomus* spp and *Gigaspora* spp. Plant Soil 176: 57–63

Ezawa T, Kuwahara S, Sakamoto K, Yoshida T, Saito M (1999) Specific inhibitor and substrate specificity of alkaline phosphatase expressed in the symbiotic phase of the arbuscular mycorrhizal fungus, *Glomus etunicatum*. Mycologia 91: 636–641

Ezawa T, Smith SE, Smith FA (2002) P metabolism and transport in AM fungi. Plant Soil 244: 221–230

Ferrol N, Barea JM, Azcón-Aguilar C (2000a) Molecular approaches to study plasma membrane H^+-ATPases in arbuscular mycorrhizas. Plant Soil 226: 219–225

Ferrol N, Barea JM, Azcón-Aguilar C (2000b) The plasma membrane H^+-ATPase gene family in the arbuscular mycorrhizal fungus *Glomus mosseae*. Curr Genet 37: 112–118

Ferrol N, Pozo MJ, Antelo M, Azcón-Aguilar C (2002) Arbuscular mycorrhizal symbiosis regulates plasma membrane H^+-ATPase gene expression in tomato plants. J Exp Bot 53: 1683–1687

Garbaye J (1994) Helper bacteria, a new dimension to the mycorrhizal symbiosis. New Phytol 128: 197–210

Gehrig H, Schussler A, Kluge M (1996) *Geosiphon pyriforme*, a fungus forming endocytobiosis with *Nostoc* (Cyanobacteria), is an ancestral member of the Glomales: evidence by SSU rRNA analysis. J Mol Evol 43: 71–81

Genre A, Chabaud M, Timmers T, Bonfante P, Barker DG (2005) Arbuscular mycorrhizal fungi elicit a novel intracellular apparatus in *Medicago truncatula* root epidermal cells before infection. Plant Cell 17: 3489–3499

George TS, Richardson AE (2008) Potential and limitations to improving crops for enhanced phosphorus utilization. In: White PJ, Hammond JP (eds), The Ecophysiology of Plant-Phosphorus Interactions. Springer, Dordrecht, The Netherlands, pp 247–270

Gianinazzi S, Schüepp H, Barea JM, Haselwandter K (2002) (eds) Mycorrhizal Technology in Agriculture: From Genes to Bioproducts. Birkhäuser, Basel, Switzerland
Gianinazzi-Pearson V, Gianinazzi S (1988) Morphological integration and functional compatibility between symbionts in vesicular-arbuscular endomycorrhizal associations. In: Scannerini S, Smith DC, Bonfante-Fasolo P, Gianinazzi-Pearson V (eds), Cell to Cell Signals in Plant, Animal and Microbial Symbiosis. NATO ASI, Series H, Cell Biology, Springer, Berlin, pp 73–84
Gianinazzi-Pearson V, Arnould C, Oufattole M, Arango M, Gianinazzi S (2000) Differential activation of H^+-ATPase genes by an arbuscular mycorrhizal fungus in root cells of transgenic tobacco. Planta 211: 609–613
Glick BR, Karaturovic DM, Newell PC (1995) A novel procedure for rapid isolation of plant-growth promoting pseudomonads. Can J Microbiol 41: 533–536
Grime JP, Mackey JML, Hillier SH, Read DJ (1987) Floristic diversity in a model system using experimental microcosms. Nature 328: 420–422
Gryndler M (2000) Interactions of arbuscular mycorrhizal fungi with other soil organisms. In: Kapulnik Y, Douds DDJ (eds), Arbuscular Mycorrhizas: Physiology and Function. Kluwer, Dordrecht, The Netherlands, pp 239–262
Harley JL, Smith SE (1983) Mycorrhizal Symbiosis. Academic, New York
Harrison MJ, Dewbre GR, Liu JY (2002) A phosphate transporter from *Medicago truncatula* involved in the acquisition of phosphate released by arbuscular mycorrhizal fungi. Plant Cell 14: 2413–2429
Harrison MJ, Vanbuuren ML (1995) A phosphate transporter from the mycorrhizal fungus *Glomus versiforme*. Nature 378: 626–629
Hart M, Klironomos JN (2002) Diversity of arbuscular mycorrhizal fungi and ecosystem functioning. In: van der Heijden MGA, Sanders IR (eds), Mycorrhizal Ecology. Springer, Berlin/Heidelberg/New York, p 242
Hartnett DC, Wilson GWT (1999) Mycorrhizae influence plant community structure and diversity in tallgrass prairie. Ecology 80: 1187–1195
Jakobsen I (2004) Hyphal fusion to plant species connections-giant mycelia and community nutrient flow. New Phytol 164: 4–7
Jakobsen I, Abbott LK, Robson AD (1992a) External hyphae of vesicular-arbuscular mycorrhizal fungi associated with *Trifolium subterraneum* L. 1. Spread of hyphae and phosphorus inflow into roots. New Phytol 120: 371–380
Jakobsen I, Abbott LK, Robson AD (1992b) External hyphae of vesicular arbuscular mycorrhizal fungi associated with *Trifolium subterraneum* L. 2. Hyphal transport of ^{32}P over defined distances. New Phytol 120: 509–516
Jakobsen I, Gazey C, Abbott IK (2001) Phosphate transport by communities of arbuscular mycorrhizal fungi in intact soil cores. New Phytol 149: 95–103
Janos DP (2007) Plant responsiveness to mycorrhizas differs from dependence upon mycorrhizas. Mycorrhiza 17: 75–91
Javot H, Pumplin N, Harrison MJ (2007) Phosphate in the arbuscular mycorrhizal symbiosis: transport properties and regulatory roles. Plant Cell Environ 30: 310–322
Jeffries P, Barea JM (2001) Arbuscular mycorrhiza - a key component of sustainable plant-soil ecosystems. In: Hock B (ed), The Mycota. Vol. IX. Fungal Associations. Springer, Berlin/Heidelberg/New York, pp 95–113
Jeffries P, Gianinazzi S, Perotto S, Turnau K, Barea JM (2003) The contribution of arbuscular mycorrhizal fungi in sustainable maintenance of plant health and soil fertility. Biol Fertil Soils 37: 1–16
Johansson JF, Paul LR, Finlay RD (2004) Microbial interactions in the mycorrhizosphere and their significance for sustainable agriculture. FEMS Microbiol Ecol 48: 1–13
Joner EJ, Johansen A (2000) Phosphatase activity of external hyphae of two arbuscular mycorrhizal fungi. Mycol Res 104: 81–86
Joner EJ, Ravnskov S, Jakobsen I (2000) Arbuscular mycorrhizal phosphate transport under monoxenic conditions using radio-labelled inorganic and organic phosphate. Biotechnol Lett 22: 1705–1708

Kapulnik Y, Douds DDJ (2000) (eds) Arbuscular Mycorrhizas: Physiology and Function. Kluwer, Dordrecht, The Netherlands

Kar RK, Saxena RK (1976) Algal and fungal microfossils from matanomadh formation (Paleocene), Kutch, India. Paleobotanist 43: 1–19

Kenrick P (2003) Fishing for the first plants. Nature 425: 248–249

Kenrick P, Crane PR (1997) The origin and early evolution of plants on land. Nature 389: 33–39

Klironomos JN (2002) Another form of bias in conservation research. Science 298: 749–749

Koide RT, Kabir Z (2000) Extraradical hyphae of the mycorrhizal fungus *Glomus intraradices* can hydrolyse organic phosphate. New Phytol 148: 511–517

Kojima T, Saito M (2004) Possible involvement of hyphal phosphatase in phosphate efflux from intraradical hyphae isolated from mycorrhizal roots colonized by *Gigaspora margarita*. Mycol Res 108: 610–615

Kucey RMN, Janzen HH, Leggett ME (1989) Microbially mediated increases in plant-available phosphorus. Adv Agron 42: 199–228

Lambais MR (2006) Unraveling the signaling and signal transduction mechanisms controlling arbuscular mycorrhiza development. Sci Agric 63: 405–413

Lovera M, Cuenca G (2007) Diversity of arbuscular mycorrhizal fungi (AMF) and mycorrhizal potential of the soil from a natural and a disturbed savannah from La Gran Sabana, Venezuela. Interciencia 32: 108–114

Maeda D, Ashida K, Iguchi K, Chechetka SA, Hijikata A, Okusako Y, Deguchi Y, Izui K, Hata S (2006) Knockdown of an arbuscular mycorrhiza-inducible phosphate transporter gene of *Lotus japonicus* suppresses mutualistic symbiosis. Plant Cell Physiol 47: 807–817

Maldonado-Mendoza IE, Dewbre GR, Harrison MJ (2001) A phosphate transporter gene from the extra-radical mycelium of an arbuscular mycorrhizal fungus *Glomus intraradices* is regulated in response to phosphate in the environment. Mol Plant-Microbe Interact 14: 1140–1148

Marschner H (1995) Mineral Nutrition of Higher Plants (2nd edition). Academic, London

Marschner P (2008) The effect of rhizosphere microorganisms on P uptake by plants. In: White PJ, Hammond JP (eds), The Ecophysiology of Plant-Phosphorus Interactions. Springer, Dordrecht, The Netherlands, pp 165–176

Medina A, Vassilev N, Barea JM, Azcón R (2005) Application of *Aspergillus niger*-treated agrowaste residue and *Glomus mosseae* for improving growth and nutrition of *Trifolium repens* in a Cd-contaminated soil. J Biotech 116: 369–378

Nagy R, Karandashov V, Chague W, Kalinkevich K, Tamasloukht M, Xu GH, Jakobsen I, Levy AA, Amrhein N, Bucher M (2005) The characterization of novel mycorrhiza-specific phosphate transporters from *Lycopersicon esculentum* and *Solanum tuberosum* uncovers functional redundancy in symbiotic phosphate transport in solanaceous species. Plant J 42: 236–250

O'Connor PJ, Smith SE, Smith EA (2002) Arbuscular mycorrhizas influence plant diversity and community structure in a semiarid herbland. New Phytol 154: 209–218

Offre P, Pivato B, Siblot S, Gamalero E, Corberand T, Lemanceau P, Mougel C (2007) Identification of bacterial groups preferentially associated with mycorrhizal roots of *Medicago truncatula*. Appl Environ Microbiol 73: 913–921

Ohtomo R, Saito M (2005) Polyphosphate dynamics in mycorrhizal roots during colonization of an arbuscular mycorrhizal fungus. New Phytol 167: 571–578

Olsson PA, van Aarle IM, Allaway WG, Ashford AE, Rouhier H (2002) Phosphorus effects on metabolic processes in monoxenic arbuscular mycorrhiza cultures. Plant Physiol 130: 1162–1171

Paszkowski U (2006) A journey through signaling in arbuscular mycorrhizal symbioses 2006. New Phytol 172: 35–46

Phipps CJ, Taylor TN (1996) Mixed arbuscular mycorrhizae from the Triassic of Antarctica. Mycologia 88: 707–714

Pirozynski KA, Malloch DW (1975) Origin of land plants: matter of mycotropism. Biosystems 6: 153–164

Powell CL, Daniel J (1978) Mycorrhizal fungi stimulate uptake of soluble and insoluble phosphate fertilizer from a phosphate-deficient soil. New Phytol 80: 351–358

Rausch C, Daram P, Brunner S, Jansa J, Laloi M, Leggewie G, Amrhein N, Bucher M (2001) A phosphate transporter expressed in arbuscule-containing cells in potato. Nature 414: 462–466

Read D (1998) Biodiversity - plants on the web. Nature 396: 22–23

Redecker D, Kodner R, Graham LE (2000a) Glomalean fungi from the Ordovician. Science 289: 1920–1921

Redecker D, Morton JB, Bruns TD (2000b) Ancestral lineages of arbuscular mycorrhizal fungi (Glomales). Mol Phylogenet Evol 14: 276–284

Reinhardt D (2007) Programming good relations - development of the arbuscular mycorrhizal symbiosis. Curr Opin Plant Biol 10: 98–105

Remy W, Taylor TN, Hass H, Kerp H (1994) Four hundred-million-year-old vesicular arbuscular mycorrhizae. Proc Natl Acad Sci USA 91: 11841–11843

Requena N, Breuninger M, Franken P, Ocon A (2003). Symbiotic status, phosphate, and sucrose regulate the expression of two plasma membrane H^+-ATPase genes from the mycorrhizal fungus *Glomus mosseae*. Plant Physiol 132: 1540–1549.

Reynolds HL, Vogelsang KM, Hartley AE, Bever JD, Schultz PA (2006) Variable responses of old-field perennials to arbuscular mycorrhizal fungi and phosphorus source. Oecologia 147: 348–358

Richardson AE (2001) Prospects for using soil microorganisms to improve the acquisition of phosphorus by plants. Aust J Plant Physiol 28: 897–906

Rillig MC, Mummey DL (2006) Mycorrhizas and soil structure. New Phytol 171: 41–53

Rillig MC, Mummey DL, Ramsey PW, Klironomos JN, Gannon JE (2006) Phylogeny of arbuscular mycorrhizal fungi predicts community composition of symbiosis-associated bacteria. FEMS Microbiol Ecol 57: 389–395

Roesti D, Ineichen K, Braissant O, Redecker D, Wiemken A, Aragno M (2005) Bacteria associated with spores of the arbuscular mycorrhizal fungi *Glomus geosporum* and *Glomus constrictum*. Appl Environ Microbiol 71: 6673–6679

Sanders IR (2002) Specificity in the arbuscular mycorrhizal symbiosis. In: van der Heijden MGA, Sanders IR (eds), Mycorrhizal Ecology. Springer, Berlin, pp 415–437

Schachtman DP, Reid RJ, Ayling SM (1998) Phosphorus uptake by plants: from soil to cell. Plant Physiol 116: 447–453

Schopf JW (1983) (ed) Earth's Earliest Biosphere. It's Origin and Evolution. Princeton University Press, Princeton, NJ

Schüβler A, Gehrig H, Schwarzott D, Walker C (2001a) Analysis of partial *Glomales* SSU rRNA gene sequences: implications for primer design and phylogeny. Mycol Res 105: 5–15

Schüβler A, Schwarzott D, Walker C (2001b) A new fungal phylum, the *Glomeromycota*, phylogeny and evolution. Mycol Res 105: 1413–1421

Simon L, Bousquet J, Levesque RC, Lalonde M (1993) Origin and diversification of endomycorrhizal fungi and coincidence with vascular land plants. Nature 363: 67–69

Smith FW (2002) The phosphate uptake mechanism. Plant Soil 245: 105–114

Smith FW, Mudge SR, Rae AL, Glassop D (2003) Phosphate transport in plants. Plant Soil 248: 71–83

Smith SE, Gianinazzi-Pearson V (1988) Physiological interactions between symbionts in vesicular-arbuscular mycorrhizal plants. Ann Rev Plant Physiol Plant Mol Biol 39: 221–244

Smith SE, Read DJ (1997) Mycorrhizal Symbiosis. Academic, San Diego, CA

Smith SE, Smith FA, Jakobsen I (2004) Functional diversity in arbuscular mycorrhizal (AM) symbioses: the contribution of the mycorrhizal P uptake pathway is not correlated with mycorrhizal responses in growth or total P uptake. New Phytol 162: 511–524

Stubblefield SP, Taylor TN, Trappe JM (1987) Fossil mycorrhizae: a case for symbiosis. Science 237: 59–60

Tawaraya K, Naito M, Wagatsuma T (2006) Solubilization of insoluble inorganic phosphate by hyphal exudates of arbuscular mycorrhizal fungi. J Plant Nutr 29: 657–665

Tisserant B, Gianinazzi-Pearson V, Gianinazzi S, Gollotte A (1993) In planta histochemical staining of fungal alkaline phosphatase activity for analysis of efficient arbuscular mycorrhizal infections. Mycol Res 97: 245–250

Toljander JF, Artursson V, Paul LR, Jansson JK, Finlay RD (2005) Attachment of different soil bacteria to arbuscular mycorrhizal fungal extraradical hyphae is determined by hyphal vitality and fungal species. FEMS Microbiol Lett 254: 34–40

Toro M, Azcón R, Barea JM (1997) Improvement of arbuscular mycorrhiza development by inoculation of soil with phosphate-solubilizing rhizobacteria to improve rock phosphate bioavailability (^{32}P) and nutrient cycling. Appl Environ Microbiol 63: 4408–4412

Truong BJ, Zapata F (2002) Standard characterization of phosphate rock samples from the FAO/IAEA a phosphate project. In: Assessment of Soil Phosphorus Status and Management of Phosphatic Fertilisers to Optimise Crop Production. IAEA, Vienna, pp 9–23

Turnau K, Jurkiewicz A, Lingua G, Barea JM, Gianinazzi-Pearson V (2006) Role of arbuscular mycorrhiza and associated microorganisms in phytoremediation of heavy metal-polluted sites. In: Prasad MNV, Sajwan KS, Naidu R (eds), Trace Elements in the Environment. Biogeochemistry, Biotechnology and Bioremediation. CRC/Taylor & Francis, Boca Raton, FL, pp 235–252

Uetake Y, Kojima T, Ezawa T, Saito M (2002) Extensive tubular vacuole system in an arbuscular mycorrhizal fungus, *Gigaspora margarita*. New Phytol 154: 761–768

van der Heijden MGA (2002) Arbuscular mycorrhizal fungi as a determinant of plant diversity: in search for underlying mechanisms and general principles. In: van der Heijden MGA, Sanders IR (eds), Mycorrhizal Ecology. Springer, Berlin, pp 243–265

van der Heijden MGA, Sanders IR (2002) (eds) Mycorrhizal Ecology. Springer, Berlin

van der Heijden MGA, Boller T, Wiemken A, Sanders IR (1998a) Different arbuscular mycorrhizal fungal species are potential determinants of plant community structure. Ecology 79: 2082–2091

van der Heijden MGA, Klironomos JN, Ursic M, Moutoglis P, Streitwolf-Engel R, Boller T, Wiemken A, Sanders IR (1998b) Mycorrhizal fungal diversity determines plant biodiversity, ecosystem variability and productivity. Nature 396: 69–72

van der Heijden MGA, Streitwolf-Engel R, Riedl R, Siegrist S, Neudecker A, Ineichen K, Boller T, Wiemken A, Sanders IR (2006) The mycorrhizal contribution to plant productivity, plant nutrition and soil structure in experimental grassland. New Phytol 172: 739–752

Vassilev N, Vassileva M, Azcón R, Medina A (2001) Application of free and Ca-alginate-entrapped *Glomus deserticola* and *Yarowia lipolytica* in a soil-plant system. J Biotechnol 91: 237–242

Vassilev N, Vassileva M, Azcón R, Barea JM (2002) The use of ^{32}P dilution techniques to evaluate the effect of mycorrhizal inoculation on plant uptake of P from products of fermentation mixtures including agrowastes *Aspergillus niger* and rock phosphate. In: Assessment of Soil Phosphorus Status and Management of Phosphatic Fertilisers to Optimise Crop Production. IAEA, Vienna, pp 47–53

Vassilev N, Medina A, Azcón R, Vassileva M (2006) Microbial solubilization of rock phosphate on media containing agro-industrial wastes and effect of the resulting products on plant growth and P uptake. Plant Soil 287: 77–84

Vogelsang KM, Reynolds HL, Bever JD (2006) Mycorrhizal fungal identity and richness determine the diversity and productivity of a tallgrass prairie system. New Phytol 172: 554–562

Wang B, Qiu YL (2006) Phylogenetic distribution and evolution of mycorrhizas in land plants. Mycorrhiza 16: 299–363

Zapata F, Axmann H (1995) ^{32}P isotopic techniques for evaluating the agronomic effectiveness of rock phosphate materials. Fert Res 41: 189–195

Chapter 8
THE ROLE OF RHIZOSPHERE MICROORGANISMS IN RELATION TO P UPTAKE BY PLANTS

Petra Marschner

INTRODUCTION

The rhizosphere is defined as the soil around the roots that is influenced by the root (Hiltner 1904). Due to the release of easily decomposable compounds by the roots (root exudates), the rhizosphere is characterized by high microbial density. Rhizosphere microorganisms strongly influence nutrient uptake by plants by either enhancing or decreasing nutrient availability.

Rhizosphere microbial communities are a subset of the soil microbial community, but are often quite distinct from those in the bulk soil (Foster 1986; Marilley and Aragno 1999; Gomes *et al.* 2001; Berg *et al.* 2002). Rhizosphere communities are influenced by soil and plant factors. Soils can have distinct microbial communities (Gelsomino *et al.* 1999; Carelli *et al.* 2000), as a result of the soil physical and chemical characteristics (e.g. soil texture, nutrient and organic matter content and pH) and environmental factors such as climate and vegetation. Plants contribute to these physical and chemical properties by depositing between 1% and 25% of their net photosynthetic production, which includes dead roots, sloughed-off cells and soluble compounds (Merbach *et al.* 1999). A large proportion of the root exudates such as sugars, organic acid anions or amino acids are easily degradable by microorganisms in the rhizosphere resulting in high microbial density and activity in the rhizosphere (Foster 1986; Kandeler *et al.* 2001).

The main driver of rhizosphere community composition is the plant species; different plant species growing in the same soil often have distinct rhizosphere communities (Ibekwe and Kennedy 1998; Marschner *et al.* 2001b); a given plant species may have a similar community structure when grown in different soils (Grayston *et al.* 1998; Miethling *et al.* 2000). The effects of plant species on rhizosphere community composition often become more pronounced during plant development (Gomes *et al.* 2003; Marschner *et al.* 2001b). It is generally accepted that plant species-specific rhizosphere communities are the result of differential rhizodeposition and, in particular, root exudate amount and composition.

In agreement with the view that a large proportion of root exudates are easily degradable, microbial species with high growth rates and relatively high nutrient requirements, such as Pseudomonas spp., are often found to dominate in the

P.J. White, J.P. Hammond (eds.) *The Ecophysiology of Plant-Phosphorus Interactions*,

rhizosphere (De Leij *et al.* 1993; Marilley and Aragno 1999). Although the rhizosphere is a habitat with large amounts of readily-available C sources, it is however, also the site of intense competition between microorganisms (Whipps and Lynch 1983). Indeed, as an increasing number of rhizosphere microbial species are identified, it becomes clear that they have a wide range of growth strategies (Gomes *et al.* 2001; Smalla *et al.* 2001; Mansfeld-Giese *et al.* 2002).

Root exudation, and consequently microbial density and community structure, vary along the root axis. Root exudates are primarily released in the zone of elongation behind the root tip (Hoffland *et al.* 1989; Römheld 1991; Marschner *et al.* 1997). In the older root parts, the main substrates for microbial growth are cellulose and other recalcitrant cell wall materials from sloughed-off root cortex tissues. The differences in the type and quantity of carbon available in different root zones influence microbial growth and result in distinct rhizosphere community structures (Yang and Crowley 2000; Marschner *et al.* 2001a). Certain components of root exudates have a selective influence on rhizosphere microorganisms by repelling some species and attracting others (Geurts and Franssen 1996). Examples for the latter are flavonoids released by legume roots that specifically attract Rhizobium. In contrast, microorganisms can influence the conditions in the rhizosphere by enhancing root exudation (Meharg and Killham 1995) or producing growth factors that influence root growth (Frankenberger and Poth 1987).

From this brief overview it is evident that microbial community composition and activity in the rhizosphere are temporarily and spatially highly variable and affected by plant species, soil and environmental factors. These not only make studying rhizosphere microorganisms challenging, but are also important considerations when the role of rhizosphere microorganisms in P uptake by plants is discussed.

Rhizosphere microorganisms may increase or decrease the availability of phosphate (Pi) to plants (Figure 8.1). Solubilization and mineralization of soil P or stimulation of root and root hair growth by rhizosphere microorganisms will lead to increased plant Pi uptake. Alternatively, plant Pi uptake may be decreased by competition for Pi or inhibition of root and root hair growth. The amount of P in the microbial biomass available for plant P uptake is variable and influenced by the carbon supply. Phosphorus uptake into the microbial biomass (net immobilization) may reduce plant P availability, whereas turnover of the biomass may release P.

ROLE OF RHIZOSPHERE MICROORGANISMS IN INCREASING PLANT AVAILABLE P

As a result of the low P availability in most soils, the capacity to mobilize P is widespread among soil microorganisms. Mechanisms by which P availability can be increased include (i) solubilization of poorly soluble inorganic P forms by releasing organic acid anions or protons to modify soil pH (Illmer *et al.* 1995; Illmer and Schinner 1995) and (ii) mineralization of organic P by release of phosphatases.

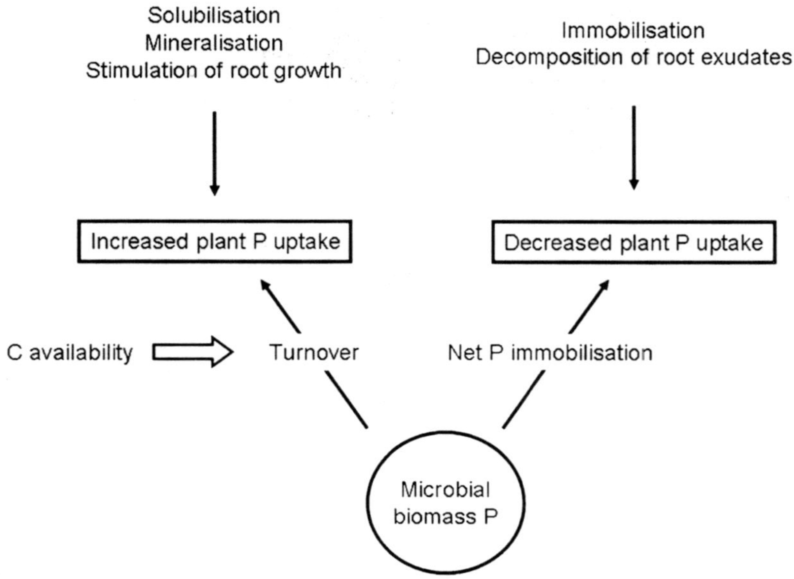

Fig. 8.1 Mechanisms by which rhizosphere microorganisms can increase or decrease plant P uptake

The effectiveness of these mechanisms will depend on a number of factors, namely the pH buffering capacity of the soil, decomposition rate and/or sorption of organic acid anions and phosphatases to soil particles.

P solubilization

A large number of microorganisms have been isolated that show high P solubilization *in vitro* (Table 8.1; Banik and Dey 1983; Whitelaw *et al.* 1999). However, these *in vitro* tests should be viewed with caution. On nutrient-rich media, a large number of soil microorganisms can solubilize poorly soluble Ca-phosphates (Louw 1969), because high growth rates are often associated with proton release and hence dissolution of Ca-phosphates. Proton release is not effective in mobilizing P from Fe or Al phosphates or Pi adsorbed to Fe or Al oxides. For these forms of P, mobilization requires the release of organic acid anions, which release Pi via ligand exchange or chelation of Fe or Al. Hence, the capacity to mobilize Pi from Fe- or Al-phosphates is less common, and Pi mobilization rates are often lower than for Ca-phosphates (Barthakur 1978; Banik and Dey 1983).

Under nutrient-poor conditions, and hence lower microbial growth rates, P solubilization is often strongly reduced. As mentioned above, the rhizosphere may not be as nutrient-rich as previously thought. Therefore isolates selected as strong P solubilizers *in vitro* may not be effective in the rhizosphere due to lack of carbon

Table 8.1 Examples of microbial genera that have been shown to be P solubilizers or phytase producers

Genus	Reference
P solubilizers	
Aspergillus, Penicillium, Trichoderma	Barthakur 1978
Bradyrhizobium, Rhizobium	Antoun *et al.* 1998
Enterobacter	Kim *et al.* 1997b
Gordonia	Hoberg *et al.* 2005
Panthoea	Deubel *et al.* 2000
Pseudomonas	Deubel *et al.* 2000; Hoberg *et al.* 2005
Rahella	Kim *et al.* 1997a
Phytase producers	
Aspergillus, Emmericella, Penicillum	Yadav and Tarafdar 2003
Peniophora	George *et al.* 2007
Pseudomonas	Richardson and Hadobas 1997
Telephora, Suillus (ectomycorrhizal fungi)	Colpaert *et al.* 1997

and other nutrients. Additionally, P mobilization may be transient because of the re-formation of poorly soluble P forms (Delvasto *et al.* 2006) and the uptake of Pi by microorganisms (Hoberg *et al.* 2005).

Due to the ease of isolating microorganisms with apparently high P solubilization capacity, many studies have been conducted to investigate the effect of inoculation with P solubilizers on plant growth and Pi uptake. In several pot and field experiments, inoculation with P-solubilizing microorganisms resulted in increased plant growth and Pi uptake (Gerretsen 1948; Kundu and Gaur 1980; Kumar and Narula 1999; see also Table 8.1). Compared to plants grown in sterile media, Pi uptake by oat plants inoculated with a soil microbial community increased by 120% and 320% when grown with Fe-phosphate and Ca-phosphate (rock-P), respectively (Gerretsen 1948). But there are also reports in which inoculation did not increase plant growth and Pi uptake (Azcon-Aguilar *et al.* 1986; Badr el-Din *et al.* 1986). It is likely that there are a large number of similar 'disappointing' results that have not been published. The poor effectiveness of inoculated strains to increase the plant available Pi, may be the result of poor growth and survival due to lack of nutrients and/or low competitiveness compared to the indigenous microflora. Successful inoculants must be 'rhizosphere-competent.' A number of traits have been shown to be important for rhizosphere competence, including motility, high growth rate, ability to synthesize amino acids and vitamin B1, ability to utilize organic acids, presence of certain cell surface proteins as well as rapid adjustment to changing conditions (Lugtenberg and Dekkers 1999).

Often, a combination of microorganisms with different characteristics, such as P solubilizers combined with N_2 fixers or with AM fungi, is superior to inoculation with the P solubilizers alone (Kundu and Gaur 1980; Toro *et al.* 1997; Sahin *et al.* 2004).

P mineralization

Up to 80% of soil P can be in organic form (Richardson 2001). Hence, the ability to access organic P can contribute substantially to increasing plant available Pi. Phosphatases hydrolyze organic P and release Pi. The activity of phosphatases decreases with increasing distance from the root surface (Tarafdar and Jungk 1987; Kandeler *et al.* 2001). Phosphomonoesterases and phosphodiesterases can be released by plant roots, rhizosphere microorganisms and also mycorrhizal fungi (Joner and Johansen 2000; Koide and Kabir 2000). Therefore, it is difficult to quantify the contribution of rhizosphere microorganisms to phosphatase activity in the rhizosphere.

Phytate, which is considered to be the dominant form of organic P in soils (Turner *et al.* 2003), is a poor P source for plants grown under sterile conditions, because plant roots have low extracellular phytase activity (Hayes *et al.* 2000; Richardson *et al.* 2001). However, microorganisms can excrete phytase (Table 8.1; Richardson and Hadobas 1997; George *et al.* 2007). Indeed, compared with plants grown under sterile conditions, inoculation with a soil suspension strongly increased plant Pi uptake from phytate, suggesting that microorganisms play an important role in mobilizing P from phytate (Richardson *et al.* 2001). However, the effectiveness of phytase in the soil is unclear because (i) phytate is adsorbed to Fe or Al oxides, which strongly reduces its availability, and (ii) phytase is rapidly adsorbed to soil particles, leading to decreased activity (George *et al.* 2005, 2007). Moreover, phytate may not be the dominant form of organic P in soils (Smernik and Dougherty 2007). It appears that some of the compounds in NMR spectra may have been falsely identified as phytate. In a range of different Australian pasture soils, phytate comprised <5% of organic P and <3% of total P (Smernik and Dougherty 2007).

Indirect stimulation of plant P uptake

Rhizosphere microorganisms can also enhance plant Pi uptake indirectly by releasing plant growth regulators or stimulating mycorrhizal colonization. Release of plant growth regulators such as IAA by rhizosphere microorganisms can enhance root hair formation (Schmidt *et al.* 1988). Mycorrhiza helper bacteria (MHB), which can be readily isolated from AM and ectomycorrhiza, have been shown to stimulate mycorrhizal colonization even in non-sterile soil and when present at low numbers (Garbaye 1994; Frey-Klett *et al.* 1997). In bulk soil it has been shown that organic acids produced during microbial decomposition of plant residues can increase P availability by competition for sorption sites and complexation of Al and Fe (Iyamuremye *et al.* 1996). Decomposition of root exudates or sloughed-off root cells could therefore also increase P availability in the rhizosphere.

ROLE OF RHIZOSPHERE MICROORGANISMS IN DECREASING PLANT AVAILABLE P

Rhizosphere microorganisms can also reduce the availability of Pi to plants by taking up P into the microbial biomass, decomposing Pi-mobilizing root exudates or by inhibiting root growth.

P uptake into the microbial biomass (Immobilization)

In the rhizosphere, plant and microbial solubilization and mineralization processes occur simultaneously. For bulk soil it has been shown that mineralization/solubilization dominate at low soil C/P ratios, whereas immobilization (uptake of P by the microbial biomass) exceeds mineralization/solubilization at high C/P ratios (He *et al.* 1997). Root exudates consist predominantly of sugars, hence are C-rich. Therefore, it can be assumed that microbial immobilization of P dominates in the rhizosphere. Thus, plants and microorganisms compete for P. McLaughlin *et al.* (1988) investigated the distribution of P in the plant-soil system after addition of isotopically labeled residues or inorganic P fertilizer. Of the fertilizer P, 18% was taken up by the plants and 29% by the soil microbial biomass. The distribution of residue P showed a similar bias to the microbial biomass, with 65% being taken up by the microbial biomass and 16% being taken up by plants. This indicates that the microbial biomass is more competitive at acquiring P than plants (McLaughlin and Alston 1986). However, as explained below, microbial P demand is a function of C availability. Thus, microbial competition for P is mediated by the concentration of easily available C compounds.

Decomposition of root exudates

Organic acid anions released by plant roots, which could potentially mobilize P, are rapidly decomposed in the soil (Van Hees *et al.* 2002). Together with organic acid anion sorption (Ström *et al.* 2001), this explains the discrepancy between the often high exudation rates of P deficient plants, such as white lupin in solution culture, and the relatively low organic acid anion concentrations measured in the rhizosphere. Therefore, the importance of organic acid anions for Pi mobilization has been questioned by Jones (1998), who argued that the organic acid anion concentrations measured in the rhizosphere of most plants would not be high enough to mobilize sufficient amounts of Pi. Hence, by decomposing organic acid anions, rhizosphere microorganisms can reduce the availability of Pi for uptake by plants. However, organic acid anion exudation may not be as ineffective as the above discussion suggests. The main sites of organic acid anion exudation are the root tips and, in some plant species, specialized root structures such as cluster roots. Microbial density at the root tip is lower than in the older root zones. Additionally, the low pH and release of phenolic compounds in cluster roots may inhibit microorganisms. Hence, in certain root zones, decomposition of organic acid anions may be lower than in the root system in general.

Finally, rhizosphere microorganisms and, particularly deleterious or pathogenic microorganisms, may reduce plant Pi uptake by inhibiting root growth or damaging roots. Similarly, inhibition of mycorrhizal colonization could reduce Pi uptake.

THE ROLE OF THE MICROBIAL BIOMASS IN P UPTAKE BY PLANTS

In the presence of a readily available carbon source, P is rapidly immobilized in the microbial biomass. However, upon C depletion microbial growth rates decrease and part of the microbial biomass may die off, releasing P. The rapid turnover of biomass P in response to addition of glucose to soil is shown in Figure 8.2. On day two, increasing the amount of glucose-C added to the soil resulted in increasing P in the microbial biomass and decreasing plant available P (resin P). Over time, microbial biomass P decreased due to depletion of C and, as a result of net P release from the microbial biomass, plant available P increased. This clearly shows (i) the importance of C supply for P immobilization in the microbial biomass, (ii) the rapid turnover of microbial biomass once C is depleted, and (iii) the direct and negative relationship between microbial P and plant available P. It should be noted that this study was carried out in a soil with low P fixation capacity. The relationship between microbial P and plant available P may be less clear in soils with high P fixation capacity, when P released from the biomass is fixed rather than becoming available. Interestingly, P addition did not increase microbial biomass P in the presence of glucose in this study, indicating that the microbial biomass in this soil was C and not P limited.

Hence, an active microbial biomass with a high turnover rate can rapidly take up P, but may also represent a slow and constant source of available P through decomposition of dead microbial cells (Seeling and Zasoski 1993; Oberson *et al.* 2001). Seeling and Zasoski (1993) suggested that P uptake by the microbial biomass could be beneficial for plants, because it would decrease P fixation by maintaining low inorganic P concentrations in the soil solution.

The importance of the microbial biomass for plant P uptake appears to vary among plant families. In an acidic soil with low P availability, microbial biomass P in the rhizosphere was positively correlated with P uptake of three Poaceae, but not with P uptake of three brassicas (Table 8.2; Marschner *et al.* 2007), although the concentration of microbial biomass P in the rhizosphere was similar in the two plant families. In the brassicas, plant P uptake was correlated with root length and P availability in the rhizosphere, suggesting that P uptake was mainly governed by plant-inherent properties.

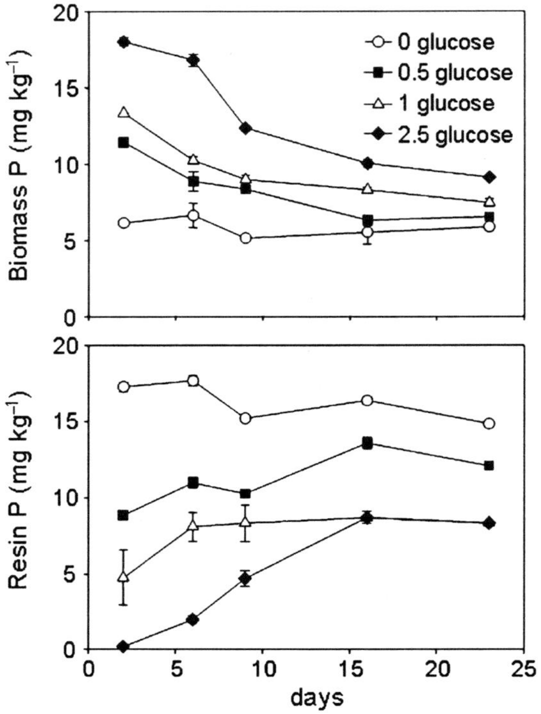

Fig. 8.2 Available P (resin P) and microbial P in soil after addition of C as glucose at 0, 0.5, 1 and 2.5 g C kg^{-1} soil

Table 8.2 Correlation coefficients between shoot P uptake (grams per plant) and rhizosphere properties (resin P, microbial P and acid phosphatase activity) or root length in Poaceae and brassicas at low and high P (across all growth stages) and at different growth stages (six-leaf, tillering/flowering and maturity; across P levels). (Data for Poaceae and brassicas are taken from Marschner *et al.* 2006 and Marschner *et al.* 2007, respectively. With permission from Elsevier.)

		Microbial P (mg kg^{-1})	Available P (mg kg^{-1})	Acid phosphatase activity (n Kat g^{-1})	Total root length (m)
Data set	Plant family	Correlation coefficient with plant P uptake			
Low P[a]	Poaceae	0.34	–	0.53	0.50
	brassicas	–	−0.57	0.53	0.77
High P[b]	Poaceae	0.68	–	–	0.61
	brassicas	–	−0.55	–	0.80
Six-leaf	Poaceae	0.97	0.77	0.81	0.85
	brassicas	0.45	0.88	0.46	0.79
Tillering/ flowering	Poaceae	0.78	0.96	0.73	0.67
	brassicas	–	0.75	0.48	0.47
Maturity	Poaceae	0.85	0.70	–	0.96
	brassicas	–	0.89	–	0.48

Only values with $P \leq 0.05$ are shown
[a]Low P: Poaceae 0; brassicas 25 mg P kg soil^{-1} added as $FePO_4$
[b]High P: Poaceae 120; brassicas 100 mg P kg soil^{-1} added as $FePO_4$

It can not be excluded that rhizosphere microorganisms contributed to P mobilization in the rhizosphere of the brassicas. However, differences in microbial community structure were not consistently related to differences in growth of Poaceae and brassicas (Marschner *et al.* 2006, 2007; Wang *et al.* 2007), indicating that presence or absence of certain microbial species or groups (e.g. P solubilizers) is not important for growth and P uptake.

CONCLUSIONS

We can conclude that rhizosphere microorganisms play an important role in the availability of Pi for plants. However, this is variable and affected by plant genotype, soil and environmental conditions. Moreover, the ratio of P mobilization to P immobilization and hence plant Pi availability are likely to vary along the root axis. The high rate of root exudation close to the root tip will lead to high growth rates of rhizosphere microorganisms accompanied by strong P mobilization. The mobilized P may be available to the plant, however exponentially growing microbial cells have a high P demand (low C/P ratio; Vrede *et al.* 2002). Thus, most of the mobilized P is likely to be taken up by the microorganisms. In the older root zones, the lower amount of carbon compounds results in lower microbial growth rates and hence P demand, as well as death of microorganisms (Vrede *et al.* 2002). Hence, microbial biomass P is likely to become available to the plant. Due to the widespread capacity among microorganisms to solubilize and/or mineralize poorly available P, the composition of the rhizosphere microbial community is probably less important than its activity.

To further elucidate the role of rhizosphere microorganisms it will be important to (i) quantify microbial biomass P turnover and the contribution of microbial biomass P to plant Pi uptake, e.g. by labeling the biomass with ^{32}P; (ii) measure the expression of genes involved in phosphatase and organic acid anion release by plant roots and rhizosphere microorganisms; and (iii) differentiate between plant and microbial phosphatases in the rhizosphere, for example using the techniques of proteomics.

REFERENCES

Antoun H, Beauchamp CJ, Goussard N, Chabot R, Lalande R (1998) Potential of Rhizobium and Bradyrhizobium species as plant growth promoting rhizobacteria on non-legumes: effect on radishes (*Raphanus sativus* L.). Plant Soil 204: 57–67

Azcon-Aguilar C, Gianinazzi-Pearson V, Fardeau JC, Gianinazzi S (1986) Effect of vesicular-arbuscular mycorrhizal fungi and phosphate-solubilizing bacteria on growth and nutrition of soybean in a neutral-calcareous soil amended with ^{32}P-^{45}Ca-tricalcium phosphate. Plant Soil 96: 3–15

Badr el-Din SMS, Khalafallah MA, Moawad H (1986) Response of soybean to dual inoculation with *Rhizobium japonicum* and phosphate dissolving bacteria. Z Pflanz Bodenkunde 149: 130–135

Banik S, Dey BK (1983) Phosphate-solubilizing potentiality of the microorganisms capable of utilizing al phosphate as a sole phosphorus source. Zbl Mikrobiol 138: 17–23
Barthakur HP (1978) Solubilization of relatively insoluble phosphate by some fungi isolated from the rhizosphere of rice. Indian J Agric Sci 48: 762–766
Berg G, Roskot N, Steidle A, Eberl L, Zock A, Smalla K (2002) Plant-dependent genotypic and phenotypic diversity of antagonistic rhizobacteria isolated from different *Verticillium* host plants. Appl Environ Microbiol 68: 3328–3338
Carelli M, Gnocchi S, Fancelli S, Mengoni A, Paffetti D, Scotti C, Bazzicalupo M (2000) Genetic diversity and dynamics of *Sinorhizobium meliloti* populations nodulating different alfalfa cultivars in Italian soils. Appl Environ Microbiol 66: 4785–4789
Colpaert JV, Van Laere A, Van Tichelen KK, Van Assche JA (1997) The use of inositol hexaphosphate as a phosphorus source by mycorrhizal and non-mycorrhizal Scots pine (*Pinus sylvestris*). Funct Ecol 11: 407–415
De Leij FAAM, Whipps JM, Lynch JM (1993) The use of colony development for the characterization of bacterial communities in soil and on roots. Microbiol Ecol 27: 81–97
Delvasto P, Valverde A, Ballester A, Igual JM, Munoz JA, Gonzalez F, Blazques ML, Garcia C (2006) Characterization of brushite as a re-crystallization product formed during bacterial solubilization of hydroxyapatite in batch cultures. Soil Biol Biochem 38: 2645–2654
Deubel A, Gransee A, Merbach W (2000) Transformation of organic rhizosdeposits by rhizosphere bacteria and its influence on the availability of tertiary calcium phosphate. J Plant Nutr Soil Sci 163: 387–392
Foster RC (1986) The ultrastructure of the rhizoplane and rhizosphere. Ann Rev Phytopathol 24: 211–234
Frankenberger WT, Poth M (1987) Biosynthesis of indole-3-acetic acid by the pine ectomycorrhizal fungus *Pisolithus tinctorius*. Appl Environ Microbiol 53: 2908–2913
Frey-Klett P, Pierrat JC, Garbaye J (1997) Location and survival of mycorrhiza helper *Pseudomonas fluorescens* during establishment of ectomycorrhizal symbiosis between *Laccaria bicolor* and Douglas fir. Appl Environ Microbiol 63: 139–144
Garbaye J (1994) Helper bacteria: a new dimension to the mycorrhizal symbiosis. New Phytol 128: 197–210
Gelsomino A, Keijzer WAC, Cacco G, Van Elsas JD (1999) Assessment of bacterial community structure in soil by polymerase chain reaction and denaturing gradient gel electrophoresis. J Microbiol Meth 38: 1–15
George TS, Richardson AE, Simpson RJ (2005) Behaviour of plant-derived extracellular phytase upon addition to soil. Soil Biol Biochem 37: 977–988
George TS, Simpson RJ, Gregory PJ, Richardson AE (2007) Differential interaction of *Aspergillus niger* and *Peniophora lycii* phytases with soil particles affects the hydrolysis of inositol phosphates. Soil Biol Biochem 39: 793–803
Gerretsen FC (1948) The influence of microorganisms on the phosphate intake by the plant. Plant Soil 1: 51–81
Geurts R, Franssen H (1996) Signal transduction in *Rhizobium*-induced nodule formation. Plant Physiol 112: 447–453
Gomes NCM, Heuer H, Schoenfeld J, Costa R, Mendoca-Hagler L, Smalla K (2001) Bacterial diversity of the rhizosphere of maize (*Zea mays*) grown in tropical soil studied by temperature gradient gel electrophoresis. Plant Soil 232: 167–180
Gomes NCM, Fagbola O, Costa R, Rumjanek NG, Buchner A, Mendona-Hagler L, Smalla K (2003) Dynamics of fungal communities in bulk and maize rhizosphere soil in the Tropics. Appl Environ Microbiol 69: 3758–3766
Grayston SJ, Wang S, Campbell CD, Edwards AC (1998) Selective influence of plant species on microbial diversity in the rhizosphere. Soil Biol Biochem 30: 369–378
Hayes JE, Simpson RJ, Richardson AE (2000) The growth and phosphorus utilisation of plants in sterile media when supplied with inositol hexaphosphate, glucose-1-phosphate of inorganic phosphate. Plant Soil 220: 165–174
He ZL, O'Donnell AG, Syers JK (1997) Seasonal responses in microbial biomass carbon, phosphorus and sulphur in soils under pasture. Biol Fert Soils 24: 421–428

Hiltner L (1904) Über neuere Erfahrungen und Probleme der Bodenbakteriologie unter besonderer Berücksichtigung der Gründüngung und Brache. Arb Deut Landwirt Ges 98: 59–78
Hoberg E, Marschner P, Lieberei R (2005) Organic acid exudation and pH changes by *Gordonia* sp. and *Pseudomonas fluorescens* grown with P adsorbed to goethite. Microbiol Res 160: 177–187
Hoffland E, Findenegg GR, Nelemans JA (1989) Solubilization of rock phosphate by rape. II Local root exudation of organic acids as a response to P-starvation. Plant Soil 113: 161–165
Ibekwe AM, Kennedy AC (1998) Fatty acid methyl ester (FAME) profiles as a tool to investigate community structure if two agricultural soils. Plant Soil 206: 151–161
Illmer P, Schinner F (1995) Solubilization of inorganic calcium phosphates - solubilization mechanisms. Soil Biol Biochem 27: 257–263
Illmer P, Barbato A, Schinner F (1995) Solubilization of hardly-soluble $AlPO_4$ with P-solubilizing microorganisms. Soil Biol Biochem 27: 265–270
Iyamuremye F, Dick RP, Baham J (1996) Organic amendments and phosphorus dynamics. I. Phosphorus chemistry and sorption. Soil Sci 161: 426–435
Joner EJ, Johansen A (2000) Phosphatase activity of external hyphae of two arbuscular mycorrhizal fungi. Mycol Res 104: 81–86
Jones DL (1998) Organic acids in the rhizosphere - a critical review. Plant Soil 205: 25–44
Kandeler E, Marschner P, Tscherko D, Gahoonia TS, Nielsen NE (2001) Microbial community composition and functional diversity in the rhizosphere of maize. Plant Soil 238: 301–312
Kim KY, Jordan D, Krishan HB (1997a) *Rahella aquatilis*, a bacterium isolated from soybean rhizosphere, can solubilize hydroxyapatite. FEMS Microbiol Lett 153: 273–277
Kim KY, McDonald GA, Jordan D (1997b) Solubilization of hydroxyapatite by *Enterobacter agglomerans* and cloned *Escherichia coli* in culture medium. Biol Fert Soils 24: 347–352
Koide RT, Kabir Z (2000) Extraradical hyphae of the mycorrhizal fungus *Glomus intraradices* can hydrolyse organic phosphate. New Phytol 148: 511–517
Kumar V, Narula N (1999) Solubilization of inorganic phosphates and growth emergence of wheat as affected by *Azotobacter chroococcum* mutants. Biol Fert Soils 28: 301–305
Kundu BS, Gaur AC (1980) Establishment of nitrogen-fixing and phosphate-solubilizing bacteria in the rhizosphere and their effect on yield and nurtrient uptake of wheat crop. Plant Soil 57: 223–230
Louw HA (1969) A study of the phosphate-dissolving bacteria in the root region of wheat and lupin. Phytophylactica 2: 21–26
Lugtenberg BJJ, Dekkers LC (1999) What makes Pseudomonas bacteria rhizosphere competent? Environ Microbiol 1: 9–13
Mansfeld-Giese K, Larsen J, Bodker L (2002) Bacterial populations associated with mycelium of the arbuscular mycorrhizal fungus *Glomus intraradices*. FEMS Microbiol Ecol 41: 133–140
Marilley L, Aragno M (1999) Phylogenetic diversity of bacterial communities differing in degreee of proximity of *Lolium perenne* and *Trifolium repens* roots. Appl Soil Ecol 13: 127–136
Marschner P, Crowley DE, Higashi RM (1997) Root exudation and physiological status of a root-colonizing fluorescent pseudomonad in mycorrhizal and non-mycorrhizal pepper (*Capsicum annuum* L.). Plant Soil 189: 11–20
Marschner P, Crowley DE, Lieberei R (2001a) Arbuscular mycorrhizal infection changes the bacterial 16S rDNA community composition in the rhizosphere of maize. Mycorrhiza 11: 297–302
Marschner P, Yang CH, Lieberei R, Crowley DE (2001b) Soil and plant specific effects on bacterial community composition in the rhizosphere. Soil Biol Biochem 33: 1437–1445
Marschner P, Solaiman Z, Rengel Z (2006) Rhizosphere properties of *Poaceae* genotypes under P-limiting conditions. Plant Soil 283: 11–24
Marschner P, Solaiman Z, Rengel Z (2007) Brassica genotypes differ in growth, phosphorus uptake and rhizosphere properties under P-limiting conditions. Soil Biol Biochem 39: 87–98
McLaughlin MJ, Alston AM (1986) The relative contribution of plant residues and fertiliser to the phosphorus nutrition of wheat in a pasture/cereal system. Aust J Soil Res 24: 517–526
McLaughlin MJ, Alston AM, Martin JK (1988) Phosphorus cycling in wheat-pasture rotations. II. The role of the microbial biomass in phosphorus cycling. Aust J Soil Res 26: 333–342
Meharg AA, Killham K (1995) Loss of exudates from the roots of perennial ryegrass inoculated with a range of microorganisms. Plant Soil 170: 345–349

Merbach W, Mirus E, Knof G, Remus R, Ruppel S, Russow R, Gransee A, Schulze J (1999) Release of carbon and nitrogen compounds by plant roots and their possible ecological importance. J Plant Nutr Soil Sci 162: 373–383

Miethling R, Wieland G, Backhaus H, Tebbe CC (2000) Variation of microbial rhizosphere communities in response to crop species, soil origin, and inoculation with *Sinorhizobium meliloti* L33. Microbiol Ecol 40: 43–56

Oberson A, Friesen DK, Rao IM, Buehler S, Frossard E (2001) Phosphorus transformations in an oxisol under contrasting land-use systems: the role of the soil microbial biomass. Plant Soil 237: 197–201

Richardson AE (2001) Prospects for using soil microorganisms to improve the acquisition of phosphorus by plants. Aust J Plant Physiol 28: 897–906

Richardson AE, Hadobas PA (1997) Soil isolates of *Pseudomonas* spp. that utilize inositol phosphates. Can J Microbiol 43: 509–516

Richardson AE, Hadobas PA, Hayes JE, O'Hara CP, Simpson RJ (2001) Utilisation of phosphorus by pasture plants supplied with myo-inositol hexaphosphate is enhanced by the presence of soil micro-organisms. Plant Soil 229: 47–56

Römheld V (1991) The role of phytosiderophores in acquisition of iron and other micronutrients in graminaceous species: an ecological approach. Plant Soil 130: 127–134

Sahin F, Cakmakci R, Kantar F (2004) Sugar beet and barley yields in relation to inoculation with N_2 fixing and phosphate solubilizing bacteria. Plant Soil 265: 123–129

Schmidt W, Martin P, Omay SH, Bangerth F (1988) Influence of *Azospirillum brasiliense* on nodulation of legumes. In: Klingmüller W (ed), Azospirillum IV: Genetics, Physiology, Ecology. Springer, Berlin/Heidelberg/New York, pp 92–100

Seeling B, Zasoski RJ (1993) Microbial effects in maintaining organic and inorganic solution phosphorus concentrations in a grassland topsoil. Plant Soil 148: 277–284

Smalla K, Wieland G, Buchner A, Zock A, Parzy J, Kaiser S, Roskot N, Heuer H, Berg G (2001) Bulk and rhizosphere soil bacterial communities studied by denaturing gradient gel electrophoresis: plant-dependent enrichment and seasonal shifts revealed. Appl Environ Microbiol 67: 4742–4751

Smernik RJ, Dougherty WJ (2007) Identification of phytate in phosphorus-31 nuclear magnetic resonance spectra: the need for spiking. Soil Sci Soc Am J 71: 1045–1050

Ström L, Owen AG, Godbold DL, Jones DL (2001) Organic acid behaviour in a calcareous soil: sorption reactions and biodegradation rates. Soil Biol Biochem 33: 2125–2133

Tarafdar JC, Jungk A (1987) Phosphatase activity in the rhizosphere and its relation to the depletion of soil organic phosphorus. Biol Fert Soils 3: 199–204

Toro M, Azcon R, Barea JM (1997) Improvement of arbuscular mycorrhiza development by inoculation of soil with phosphate-solubilizing rhizobacteria to improve rock phosphate bioavailability (^{32}P) and nutrient cycling. Appl Environ Microbiol 63: 4408–4412

Turner BL, Mahieu N, Condron LM (2003) The phosphorus composition of temperate pasture soils determined by NaOH-EDTA extraction and solution ^{31}P NMR spectroscopy. Org Geochem 34: 1199–1210

Van Hees PAW, Jones DL, Godbold DL (2002) Biodegradation of low molecular weight organic acids in coniferous forest podzolic soils. Soil Biol Biochem 34: 1261–1272

Vrede K, Heldal M, Norland S, Bartbak G (2002) Elemental composition (C, N, P) and cell volume of exponentially growing and nutrient-limited bacterioplankton. Appl Environ Microbiol 68: 2965–2971

Wang D, Marschner P, Solaiman Z, Rengel Z (2007) Belowground interactions between intercropped wheat and *Brassicas* in acidic and alkaline soils. Soil Biol Biochem 39: 961–971

Whipps JM, Lynch JM (1983) Substrate flow and utilization in the rhizosphere of cereals. New Phytol 95: 605–623

Whitelaw MA, Harden TJ, Helyar KR (1999) Phosphate solubilization in solution culture by the soil fungus *Penicillium radicum*. Soil Biol Biochem 31: 655–665

Yadav RS, Tarafdar JC (2003) Phytase and phosphatase producing fungi in arid and semi-arid soils and their efficiency in hydrolyzing different organic P compounds. Soil Biol Biochem 35: 1–7

Yang CH, Crowley DE (2000) Rhizosphere microbial community structure in relation to root location and plant iron nutritional status. Appl Environ Microbiol 66: 345–351

Chapter 9
SOIL AND FERTILIZER PHOSPHORUS IN RELATION TO CROP NUTRITION

Ernest A. Kirkby and A. Edward (Johnny) Johnston

INTRODUCTION

Phosphorus (P) plays a pivotal role in the nutrition of all plants as an essential element participating in a wide array of physiological and biochemical processes occurring in all living organisms (Vance *et al.* 2003). Historically, of all the nutrients required by plants, P was frequently the one that most limited growth; until P deficiency was corrected many crops did not respond to nitrogen (N), and this is still the case for many soils worldwide. Most crops grown for human food, animal feed, fiber and now for biofuels contain between 0.2% and 0.5% P in their dry matter when sufficient P is available in the soil (Sanchez 2007). In intensive agriculture much of this P can be applied in inorganic P fertilizers and organic manures. Inorganic P fertilizers were first available some 160 years ago after JB Lawes, of Rothamsted (UK), patented a commercially successful method of producing superphosphate, containing water-soluble monocalcium phosphate, from phosphate rock (PR). From the mid 19th century, superphosphate quickly proved to be effective in providing plant-available P on almost all soil types in the UK (Johnston 1994) and has since been used worldwide for this purpose. With the opportunity to use inorganic P fertilizers and organic manures to minimize the risk of soil P deficiency limiting crop growth, there exists the possibility of increasing crop yields to improve food security for an increasing world population.

However, the global distribution of PR, the consumption of inorganic P fertilizers and the distribution of P-deficient soils are all poorly matched. Currently, over 70% of the PR reserves that are economically exploitable are located in three countries: the USA, Morocco and Western Sahara, and South Africa (Heffer *et al.* 2006). The production of PR reached 171 million tonnes (Mt) in 2005, the main producers being USA (24%), China (20%) and Morocco (19%). About 85% of this production was processed for use in agriculture (80% as fertilizer and 5% as animal feed supplement), whilst 15% was used for industrial products, such as detergents (12%; Heffer *et al.* 2006). In 2005/2006, the annual consumption of P fertilizer was estimated to be 36.8 Mt P_2O_5 of which 74% was consumed in four regions: East Asia, South Asia, Latin America and North America. Almost 32% was consumed in China and Vietnam, which the International Fertilizer Industry Association's regional classification includes in East Asia (Heffer *et al.* 2006).

P.J. White, J.P. Hammond (eds.) *The Ecophysiology of Plant-Phosphorus Interactions*,

Phosphate fertilizers are essential for maintaining the production of staple foods. On a global scale, P is often regarded as the main mineral nutrient restricting plant growth in acid, neutral and calcareous soils. About 67% of the total farmland used worldwide contains too little readily plant-available P (Batjes 1997) and it is thought that some 30–40% of all soils growing arable crops are deficient in P (Runge-Metzger 1995; von Uexküll and Mutert 1995).

Phosphorus is a non-renewable resource and, although estimates of global reserves of PR and how they will be used vary, it is not disputed that P reserves are limited (Heffer *et al.* 2006). The US Geological Survey's Commodity Summary Report estimated that the world reserve base of PR was 46,000 Mt in 2005 (Heffer *et al.* 2006). This reserve base is sufficient to supply more than 300 years' consumption at the current rate but the "reserves" (that part of the reserve base that can be economically exploited with current technology with the present relative costs of production and value of product) is sufficient for only 100 years (Heffer *et al.* 2006). However, as yet unknown deposits may be discovered in the future, to benefit the continued existence of mankind. Vance *et al.* (2003) have drawn attention to the four- to five-fold increase in the use of P fertilizer between 1960 and 2000 and a projected further annual increase of 20 Mt by 2030. This projected increase in the use of P fertilizer will be required to meet the food needs of an increasing world population. This will be mainly in the tropics and subtropics, where the majority of the population of the world live already and where farmers are resource poor and the infrastructure to purchase and distribute fertilizers is only weakly developed.

Vance *et al.* (2003) distinguish clearly between problems associated with the use of P fertilizers in intensive agriculture and the lack of P inputs in extensive agriculture. In both of these very different systems there is a need to manage P in crop nutrition better to achieve sustainable yields and food security. The latter goal requires the achievement of two objectives. The first is a better understanding of the behavior of P in soils. The second is developing "P-efficient" cultivars of crop plants that can use soil and fertilizer P more effectively. Success in both these objectives will benefit both the use of P on soils where P supply is limited and crop yields are small, and the recovery of soil P reserves, which have accumulated in soil over many years from past applications of fertilizers and manures. Plant roots take up P as inorganic phosphate (Pi) from the soil solution and factors that influence the concentration of Pi in the soil solution and its rate of replenishment will affect yield. Thus, when P is strongly bonded to soil constituents Pi concentration may be less than required for optimal yield. The concentration of Pi in the soil solution can be increased by the addition of P as a water-soluble P fertilizer, or as a water-insoluble P compound which in an acid environment of either the rhizosphere or the bulk soil allows the release of Pi. The root system must be of an adequate size and any factors, biological, chemical or physical, and interactions between them that adversely affect root growth, size and function will decrease above ground growth and final yield. It also has to be clearly recognized that any improvement in P uptake from the soils reserves that is not replaced depletes ("mines") these reserves and cannot go on indefinitely without jeopardizing the ability of the soil to produce crops.

It is still widely believed that P fertilizers are used inefficiently in crop production. In part, this is because the recovery of P applied to a crop as ^{32}P-labelled fertilizer, is usually only 10–15%, rarely 25%, of that applied (Syers *et al.* 2007). It has generally been assumed that one reason for this poor recovery is because the P has become "fixed" by soil components and is unavailable for crop use. However, current thinking is that P is used efficiently, with recoveries of between 50% and 90%, if an appropriate time scale and method of estimation is used (Syers *et al.* 2007). The belief that fertilizer P is used inefficiently has led to the frequently held view that the transfer of P from soil to surface water bodies, with the consequent adverse effect of eutrophication on the biological balance in the water body, is due to the excessive use of P fertilizers. This is not totally correct, as is discussed later in this chapter.

In intensive agriculture, large yields of modern cultivars of arable crops annually remove 20–35 kg P ha^{-1}. For example, using the P content, kg P t^{-1} in fresh harvested material (MAFF 2000), the amount of P removed in 10 t ha^{-1} cereal grain is 34 kg, in 4 t ha^{-1} oilseed rape seed it is 24 kg, in 70 t ha^{-1} potatoes it is 30 kg and in 70 t ha^{-1} sugar beet roots it is 25 kg. In the past and often today, a varying, but frequently substantial amount of this P is completely lost from the soil when a crop is consumed directly or indirectly by humans because their excreta, containing most of the ingested P, is collected in a sewage treatment works. Traditionally, water-soluble phosphates have been discharged in the effluent to an adjacent river, where the bioavailable P is potentially able to cause environmental problems through eutrophication before being lost irreversibly to the sea (Neeteson *et al.* 2006). This one-way movement of P means that to maintain crop production new P has to be introduced into the system by application of fertilizer P obtained from the finite reserve of PR. Recently introduced changes of sewage treatment should be beneficial in allowing P recycling in agriculture because larger sewage treatment works are now required to use a tertiary treatment to remove P from the effluent. To further close the cycle of P use will require smaller treatment works to recover P from the effluent water. The solid sewage sludge remaining after treatment has traditionally been returned to the land and although the organic P it contains mineralizes slowly it does eventually increase the readily plant-available P in the soil (Johnston 1975). Fears about the return to soil of heavy metals in sewage sludge have largely been assuaged by the strict controls on their concentration in sludge.

This chapter is divided into seven sections. Following the Introduction we give an up-to-date account of the interactions between soil and fertilizer P in relation to the availability of P to plants. This summarizes the findings of part of a comprehensive review on the efficiency of soil and fertilizer P use commissioned by FAO and four other institutions (Syers *et al.* 2007). The acquisition of P by the roots of crop plants is then considered in relation to its availability in soil. In the following two sections we discuss first crop nutrition and the efficient use of P where soil P is adequate, and then the acquisition of P by plants where P supply is limited and to the adaptive mechanisms induced in plants by P deficiency and their possible exploitation. In both these sections, which discuss crop production in very different agro-environments, we present some possible ways to increase the efficiency of use of both soil and fertilizer P. In the final two sections we deal briefly with environmental and ecological aspects related to the use of P in crop production.

INTERACTIONS BETWEEN SOIL AND FERTILIZER P AND THEIR CONSEQUENCES FOR P AVAILABILITY TO CROPS

No attempt is made here to present a comprehensive account of the chemistry of soil P, nor give in detail the changing concepts of the behavior of soil and fertilizer P, as these have been reviewed recently by Syers *et al.* (2007). Instead, we consider recent concepts about the behavior of soil and fertilizer P in relation to plant nutrition and crop production.

Early views on the behavior of soil and fertilizer P

Until recently there was a long-held and persistent view that when a water-soluble P compound, like monocalcium phosphate, was added to soil any residue of the P remaining after P was taken up by the crop was to a large extent retained ("fixed") in the soil in forms that were unavailable to subsequent crops. This view can be traced back to a paper by Way (1850) and since that time soil scientists have been intrigued by, experimented on, and discussed at great length, the fate of P added to soils. Way (1850) percolated solutions of water-soluble potassium (K) and ammonium (NH_4) phosphates through columns of soil and found that the Pi was retained in the soil whereas other anions, often chloride or sulfate, appeared in the leachate. To explain the observed retention of Pi, Way (1850) proposed that the accompanying K and NH_4 cations had been retained by exchange with the divalent calcium (Ca) cation and that the Pi had combined with the free Ca cation to produce a water-insoluble phosphate of lime. He conjectured, that this calcium phosphate (not specified in his paper) would be soluble in the dilute carbonic acid solution, occurring in soil, so that Pi would be released into the soil solution for uptake by roots. The amount of Pi solubilized by this process, however, is probably very small judged by the results from the early years of the Rothamsted experiments, which showed that adding superphosphate fertilizer gave large increases in crop yields (Johnston 1994).

Using data from P balance studies (P applied *minus* P removed in harvested crops) for two long-term experiments at Rothamsted, Dyer (1894, 1902) concluded that much of the positive P balance in these experiments was retained or "fixed" in the surface soil. Also at this time, field experiments assessing the value to crops of the P retained in soil showed that it had very little or no effect on subsequent crop yields. In other words, there was no residual effect of the applied P fertilizer. This was taken as further evidence that P residues were fixed in soil in forms that were not available for uptake by plant roots. Based perhaps in part on these observations,

Russell (1912) supported the view that the retention of P applied to soil as water-soluble monocalcium phosphate was due to the precipitation of a water-insoluble calcium phosphate.

Beginning in the 1920s, there were many studies on the reactions and fate of fertilizer P added to soil that produced conflicting results and conclusions (see reviews by Pierre and Norman 1953; Larsen 1967; Khasawneh *et al.* 1980). From the 1950s, especially in the USA, many laboratory experiments and modeling exercises were performed on the reactions that occur when water-soluble phosphates are added to soil. Laboratory experiments showed the formation of "discrete-phase", water-insoluble minerals like variscite (aluminium phosphate) and strengite (iron phosphate) under acid conditions and a range of calcium phosphates under near-neutral and alkaline conditions when monocalcium phosphate was added to soil. It was widely assumed that these water-insoluble minerals were formed when water-soluble, fertilizer P reacted with soil components, and this would explain why the plant availability of fertilizer P was so small in many soils (Kurtz 1953; Hemwall 1957; Jackson 1963; Huffman 1962, 1968; Larsen 1967; Sample *et al.* 1980). In attempts to show that such compounds were produced, soil P was sequentially extracted with reagents thought to extract distinct phase, inorganic P compounds. The latter included iron, aluminium and calcium phosphates, the type and amounts formed depending on the acidity of the soil (Dean 1938; Chang and Jackson 1958). Barrow (1983a) suggested that the formation of these insoluble compounds was debatable because the conditions used in many of these laboratory experiments were far removed from the conditions in the heterogeneous environment of the soil.

In the late 1950s and 1960s, thermodynamic models, particularly solubility isotherms, were combined with laboratory data to produce elegant descriptions for the formation of pure crystalline compounds in equilibrium with Pi in solution (Lindsay and Moreno 1960; Larsen 1967; Lindsay 1979). However, these models often ignored the conditions that exist in soil and Barber (1995) suggested that P compounds in soil are likely to be impure and of unknown solubility and, therefore, not likely to be the reaction products following addition of water-soluble P to soil.

Evidence for the need to review ideas about the behavior of P in soil

Two earlier papers that questioned the assumption of P fixation appear to have received little attention. Coleman (1942) noted that the failure of a crop to respond to fertilizer P was not necessarily due to the rapid fixation of P by soil. Rather, it could be because there was already sufficient plant-available P in the soil and that large amounts of soil P "formerly considered fixed" are available to plants. Kurtz (1953) also argued that the results of experiments that implied that P was held in soil by simple precipitation reactions sometimes led to rather questionable conclusions. He noted that, "Contrary to the apparent belief of two decades ago, more recent evidence indicates that the reactions of phosphorus with soils are not entirely irreversible and that for most soils the term "fixation" is an exaggeration". Major changes in the concepts of the behavior of P in soil came in the late 1960s and 1970s with the work of Posner and Barrow (1982) and Barrow (1983b) on the adsorption/absorption and desorption of Pi in soils. The slow reaction between Pi and soil was attributed to the diffusive penetration of adsorbed phosphate ions into soil components as shown by Evans and Syers (1971). This explained the decrease in extractability, isotopic exchangeability and plant availability of P added to soil over time (Barrow 1980).

In a comprehensive review of the literature, Sample *et al.* (1980) concluded that both sorption and precipitation reactions were likely to occur simultaneously following the addition of fertilizer P to soil. These authors also considered that initial P reaction products and the initially adsorbed Pi may be metastable with important changes over time such that there would be lower concentrations of Pi in the soil solution.

Barrow (1983b) used data from Leamer (1963) to support his suggestion that absorbed Pi could be released over time (i.e. that adsorption/absorption was largely reversible), but that this could be a slow process. Data from the Exhaustion Land experiment at Rothamsted, using "Olsen P" (P extracted from the soil with 0.5 M $NaHCO_3$, pH 8.5) as an estimate of plant available P, provides evidence for the reversibility of adsorption/absorption of P in soil (Johnston and Poulton 1977). From 1856 to 1901, P was added first as fertilizer to wheat (1856–1875) and then as both fertilizer and farmyard manure (FYM) to potatoes (1876–1901). The P balance was positive with both P treatments, but only about 15% of this P was extracted by the Olsen method, the remainder being in forms that were no longer immediately plant available. After 1901 no more P was applied, and from 1901 to 1958 the P balance was negative. However, only about 37% of this negative balance could be accounted for by a decline in Olsen P. This implies that some P that was not measured as plant available in 1901 had become available over time. Both sets of data indicate a degree of reversibility of sorbed P. Further evidence for reversible transfer of P between fractions of soil P is provided by experiments in North America (Barber 1979; Halvorson and Black 1985; McCollum 1991).

Additional evidence for the release of plant-available P that was not initially measured as Olsen P comes for an experiment on a sandy clay loam soil at Saxmundham in Suffolk, England. From 1965–1968, differential P fertilization established eight soils with Olsen P ranging from 3–67 mg kg^{-1}. For the next 16 years these soils were cropped with potatoes, sugar beet, spring barley, winter wheat and field beans grown in rotation and without addition of P but with adequate N and K. The cumulative P offtake in the crops was determined and, in alternate years, the top 23 cm soil was sampled to determine Olsen P. The decline in Olsen P ranged from 43% on the most P enriched soil to less than 10% on least P enriched soil. However, the decrease in Olsen P did not account for all the P removed in the harvested crops. Thus, P measured by the Olsen method was being replenished by P from P reserves not measured by the Olsen method in 1968. Of equal interest was the observation that the individual decay curves for Olsen P from each of the eight

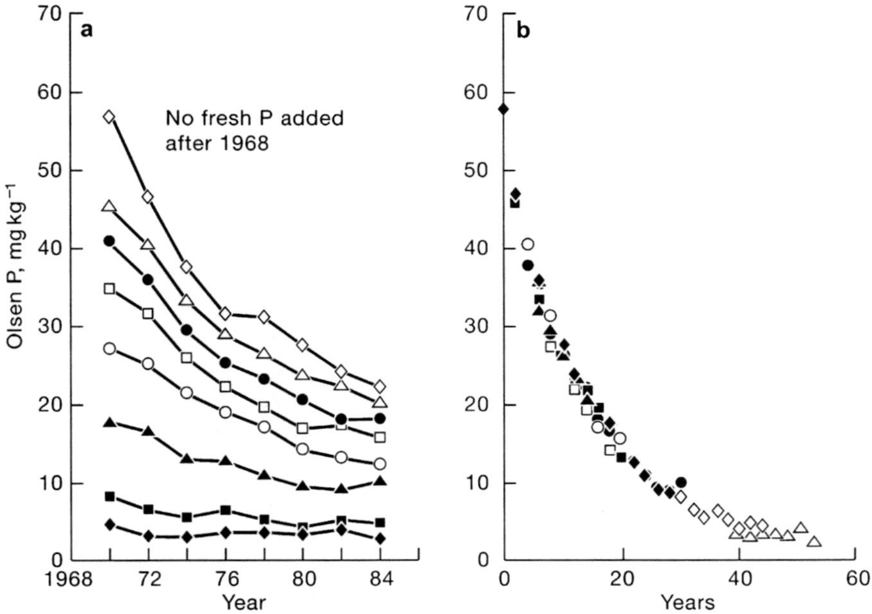

Fig. 9.1 Decline in Olsen P during 16 years in eight soils having different Olsen P (**a**). Development of a coincident decline curve by making horizontal shifts (**b**). Data are from an experiment at Saxmundham, UK. (Adapted from Johnston and Syers 2006.)

soils (Figure 9.1a) could be brought into coincidence by horizontal shifts to produce a unified decay curve for a 50-year period (Figure 9.1b; Johnston and Poulton 1992). This is critically important to the present discussion for it indicates that there are different pools of soil P in equilibrium with one another. As P in one pool is being depleted it is replenished (buffered) by P from other pools and the faster the replenishment the greater the buffer capacity of the soil. A unified decay curve from a large initial value of Olsen P supports the contention that there are no specific, well-defined and discrete fractions of soil P, as was previously widely believed. If there were discrete fractions of soil P it is probable that decay curves, such as those presented in Figure 9.1, would be a series of steps rather that a smooth curve.

Current concepts of the behavior of P in soil

The view that water-soluble P added to soil in fertilizers and not used by the crop to which it was applied became mostly fixed in soil in forms unavailable to future crops was largely supported by work prior to the 1950s. Subsequently, this view began to be challenged for a number of reasons. Field experiments showed that plant-available P residues could accumulate, at least in some soils, and that these residues increased crop yields. It was also realized that the results from many of the laboratory experiments were unlikely to relate to what happened in field soils because the conditions used for the laboratory experiments were inappropriate. This suggested that new concepts about the behavior of P in soil needed to be formulated.

These new concepts, accepted at least among many involved in soil and fertilizer P research, relate to P equilibria in soils which explain reasonably well changes in the extractability of soil P, and the decrease in plant availability of added P with time. These equilibria primarily involve adsorption and absorption reactions that may be largely reversible with time. For P, which in the short- and long-term will be plant available, the current concept is that this P is held by soil components with a continuum of bonding energies. Building on this concept, pools of soil P related to the accessibility and extractability, and thus the plant-availability of the P, can be categorized and conceptualized and expressed diagrammatically (Figure 9.2). Phosphorus in the soil solution, the first pool, is immediately available for uptake by plant roots and is present in solution in ionic forms. The second pool represents Pi that is only weakly bonded to the surfaces of soil components. This Pi is readily available because it is in equilibrium with Pi in the soil solution and is readily transferred to the soil solution as plant roots take up Pi. Readily available Pi was often described as labile P in papers published in the 1950s. The P in the third pool is less readily available for plant uptake, but it can become available over time. This P is more strongly bonded to soil components, or is present within the matrices of soil components as absorbed P (i.e. P adsorbed on internal surfaces). The P in the fourth pool is only very slowly available, often over periods of many years. It has a low or very low extractability. It is P that is very strongly bonded to soil components, or is P that has been precipitated as slightly-soluble P compounds, or it is part of the soil mineral complex, or it is unavailable due to its position within the soil matrix.

The most important feature of this conceptual model is the reversible transfer of P between the first three pools and it is this that clearly confronts the concept of irreversible fixation of P in soil. However, when a water-soluble P source is added

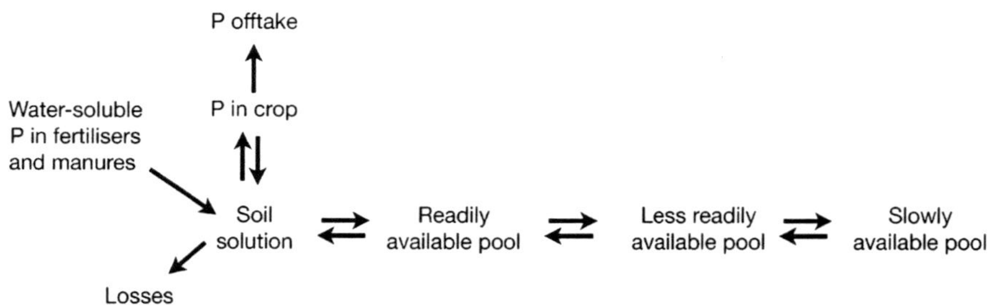

Fig. 9.2 A simple schematic representation of phosphorus pools in the plant-soil system. Soil analysis to estimate P in the readily available pool includes that in the soil solution. (Adapted from Syers *et al*. 2007.)

to soil only a small fraction remains in the soil solution, the remaining P becomes distributed between the readily- and less readily-available pools. For example in the long-term experiments at Rothamsted, Woburn and Saxmundham, where P has been applied as fertilizer or FYM for over 40 years, only about 13% of the increase in soil P is Olsen P (Johnston *et al.* 2001).

It is P in the soil solution and the readily available pool that is measured by routine soil analysis, and many different methods are used for this purpose. These include equilibration with resin and the methods of Olsen, Bray, Mehlich, Morgan and CAL (Kamprath and Watson 1980). Although these reagents all extract weakly bonded P, the amount of P extracted varies from reagent to reagent, which suggests that the readily available pool is not a finite quantity or a specific form of P. However, provided that the method characterizes the soil well, such that there is a strong relationship between the amount of P extracted and the response of a crop to an application of P fertilizer, then this fraction of P can be thought of as being well defined.

There is no routine method of soil analysis employed to estimate the amount of P in the less-readily available pool, and there is little information on the relation between the amounts of P in the readily- and less-readily-available pools of P. Recently sequential extraction of soil P has been used to follow the changes in different P fractions as the P balance has changed over time (Beck and Sanchez 1996; Blake *et al.* 2003). In these experiments, the quantity of P in all soil P fractions extracted increased as P accumulated in the soil and decreased as soil P reserves were depleted. However, no clear relationships have yet been shown between the amounts of P extracted by each reagent, or the rates of transfer of P between P pools. Both require further investigation, realizing that sequential extraction is time consuming and does not lend itself to routine soil analyses.

This concept of soil P existing in various pools related to the availability of Pi for plant uptake applies to all soils and may be considered in relation to crop production in two greatly contrasting agro-environments. One is where intensive agricultural systems operate, mainly in some developed countries. Here, large amounts of plant-available P have accumulated in soil from past additions of P in fertilizers and manures, and crop genotypes have been selected on the basis of a large yield potential. The requirement in this situation is to use soil and fertilizer P most efficiently. The other, contrasting, agro-environment is where the soil contains only very small amounts of plant-available P, which limits crop production, and where the availability of P fertilizer is often restricted by its cost or lack of infrastructure to transport it to where it is required. Estimates suggest that globally perhaps 67% of total farmland contains too little plant-available P (Batjes 1997) and much of this land is in less developed countries. Native plants that grow in many of these soils have adapted to the low P conditions in various ways. These adaptations and what can be learnt from them of benefit to the growth of crops is discussed later in this chapter, and in other chapters of this book (Lynch and Brown 2008; Raven 2008; White and Hammond 2008). For both these very different agro-environments it is possible to consider how soil and fertilizer P can best be made more accessible and available for plant uptake.

P ACQUISITION BY ROOTS OF CROP PLANTS

Plants acquire P by the roots taking up orthophosphate anions (Pi), mainly $H_2PO_4^-$ and to a lesser extent HPO_4^{2-}, from the soil solution. The size and efficacy of the root system and the presence of these anions at the soil-root interface therefore control the ability of the plant to take up sufficient P to achieve its yield potential when all other factors affecting growth are optimum. Plant and soil factors that directly or indirectly affect to availability and uptake of P by crops are summarized in Table 9.1.

Transport of Pi from the bulk soil to the rhizosphere

The concentration of Pi in the soil solution is very small and even in the most fertile soils rarely exceeds 8 μM (Barber 1995). In soils that are highly weathered, sandy or alkaline, Pi concentrations are commonly less than 1 μM (Reisenauer 1966). Thus, the amount of P in the soil solution that is immediately available for crop uptake as Pi is very small and, without frequent replenishment, is inadequate to meet a crop's large demand for P for optimal growth. This is illustrated by the data in Table 9.2. Based on the assumption that the concentration of Pi in the soil solution is 1 mg L^{-1} (approximately 3 μM), and that a crop producing a large yield requires a

Table 9.1 Plant and soil properties that control the availability of P uptake by crops either directly or indirectly. (Adapted from Horst *et al.* 2001.)

Plant properties	Soil properties
Root uptake efficiency	**Chemical**
Uptake capacity	pH
Uptake affinity	Organic ligands
Minimal concentration for uptake	Cation exchange capacity
Root morphology	CO_2
Root length density	Redox potential
Root fineness	Buffer capacity
Root hairs	Toxicity (e.g. Al)
Infection with mycorrhizal fungi	Deficiency (e.g. K, Ca)
Rhizosphere modification	**Physical**
H^+/OH^- balance	Temperature
Excretion of organic acid anions	Moisture
Increase in reduction capacity	Bulk density
Phosphatase production	Mechanical impedance
	Biological
	Microbial activity
	Soil fauna
	Pathogenic fungi

Table 9.2 Illustration, using two different approaches (A and B), of the inability of the P in the soil solution to supply the P requirement of a crop based on a total requirement of 30 kg P ha^{-1}. (After Kirkby and Römheld 2006.)

A	**P requirement:** 30 kg P ha^{-1}	
	(10 t DM with 0.3% P; 100 day growth period)	**300 g P ha^{-1} day^{-1}**
	P concentration in soil solution: 0.1 mg P L^{-1} (~3 μM)	↓↓ *f* = 5
	(soil - bulk density 1.3 g cm^{-3}; depth 25 cm, water content 18%)	↓↓
	= 0.6 × 10^6 L → 0.1 mg × (0.6 × 10^6 L)	↓ **60 g P ha^{-1}**
	Spatial availability	↓ *f* = 25
	(only 20% of topsoil exploited by roots)	**12 g P ha^{-1}**
B	**Water requirement:** 500 L kg^{-1} shoot dry weight produced	
	(transpiration coefficient: 300–600)	
	P solubility: 0.1 mg P L^{-1} → 0.1 mg × 500	**50 mg P kg^{-1} DM**
		↓ *f* = 60
	P requirement: 0.3% P in leaf DM	**3 g P kg^{-1} DM**

total P uptake of about 30 kg P ha^{-1} over a growth period of about 100 days, even allowing for complete spatial accessibility of Pi to the roots, the Pi in soil solution must be replenished five times each day. However, because not more than 20% of the topsoil is explored by roots during a growing season, the Pi in the soil solution has to be replenished at least 25 times daily to meet plant demand (Marschner 1995). A very similar conclusion is reached using another theoretical calculation. This calculation is based on the water use by a crop, the Pi concentration in the soil solution and the actual P concentration in the plant. The data clearly demonstrate that the amount of Pi transported by convective flow of water to plant roots (mass flow), is far too small to meet plant demand. Lambers *et al.* (1998) consider that mass flow accounts for less than 5% of the P demand of a plant and the amount of P intercepted by plant roots as they push their way through the soil is only half this amount.

The principle method by which Pi is replenished in the soil solution is by diffusion (Jungk 1994). Uptake of Pi by plant roots from the surrounding solution, which is not usually a growth limiting step, is energy demanding and takes place very rapidly against a steep concentration gradient between the soil solution and the cytoplasm of the plant cell (Schachtman *et al.* 1998; Amtmann *et al.* 2006). In this process, two different but equally important effects are induced. The removal of Pi from the root surface creates a gradient in the concentration of Pi between the root surface and the bulk soil. This gradient is the driving force for the diffusion of Pi in the soil solution and results in the transport of Pi from the bulk soil to the root surface. This, in turn, disturbs the equilibrium between the Pi concentration in the soil solution and that in the readily plant-available pool, resulting in the release of Pi from the latter to maintain the Pi concentration in the soil solution. If delivery of Pi to the root by this process of diffusion is too slow to meet plant demand for normal growth, zones of Pi depletion around roots are created, indicative of deprivation of Pi supply (Mengel and Kirkby 2001).

Maintaining Pi concentrations in the soil solution

Maintaining the concentration of Pi in the soil solution is dependent on the Pi buffer capacity of the soil. This can vary greatly between soils depending to a large extent on the amount of P in the readily plant-available pool of soil P. The latter is likely to be larger in heavy textured, clayey soils than in light textured sandy soils. Schofield (1955) introduced the concept of an intensity factor, which represents the concentration of Pi in the soil solution, and a capacity factor, indicative of the amount of more strongly held P in the soil. The approach accords with current thinking of P being held by soil components with a continuum of bonding energies (Figure 9.2). Generally the main component of the capacity factor as envisaged by Schofield is the pool of low energy sorbed P together with the readily-mineralizable organic P. This organically bound P may be important in relation to the P nutrition of crops in some soils but it is not taken into account in most routine methods of soil analysis used to estimate P availability. In part this is because the mineralization of organic P is highly dependent on ambient conditions in the field, such as soil temperature and moisture, which cannot be measured by extracting dry soils in the laboratory. In addition, the capacity factor is also dependent on the volume of soil occupied by roots. However, as discussed above, soil analysis using an appropriate method can provide an assessment in most agricultural soils of P that is potentially available to plants under conditions favorable for root growth and activity.

P acquisition over the growth period

Young plants have a very large demand for nutrients but only small root systems. Consequently, in the early stages of growth, the rate of nutrient uptake per unit length of root is extremely high. For example, the P uptake rate per unit length of root by 20-day-old maize plants is about ten times that of 30-day-old plants (Mengel and Barber 1974). Further evidence for the importance of maintaining a sufficient level of readily plant-available P in the soil appropriate to the growth stage to optimize both yield and efficient use of P is illustrated by the data presented in Figure 9.3. This shows the daily rate of P uptake throughout the growth of spring barley crop grown on two soils, one with 100 mg kg^{-1} and the other with only 5 mg kg^{-1} Olsen P. The larger Olsen P value is more than sufficient for all arable crops, and the maximum daily P uptake rate was 0.6 kg ha^{-1} compared with only 0.2 kg ha^{-1} when the soil was deficient in available P. This difference was reflected in the grain yields at harvest which were 6.4 and 2.9 t ha^{-1}, respectively (Leigh and Johnston 1986).

Modeling P acquisition by plants

Mechanistic mathematical models to estimate Pi transfer from the soil into the plant have been developed based on the assumption that Pi transport from the soil to the root is equal to Pi uptake by the plant (Claassen *et al.* 1986; Barber 1995). Transport

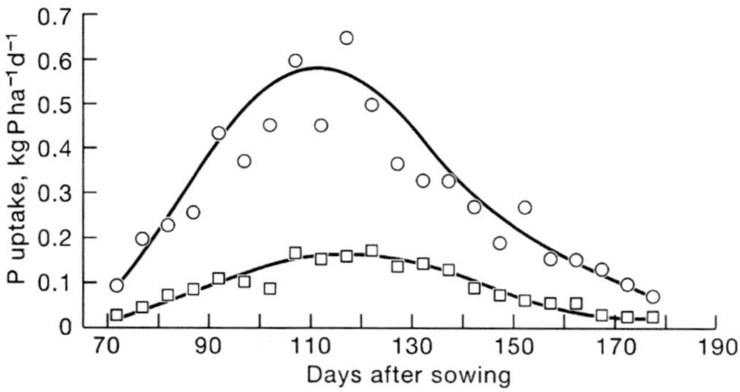

Fig. 9.3 Daily phosphorus uptake rate of spring barley grown on a soil with adequate P reserves (○) and too little readily available soil P (□). (Adapted from Leigh and Johnston 1986.)

from soil to roots is assumed to proceed by mass flow and diffusion, and Pi uptake into the root follows Michaelis-Menten kinetics. The higher the rate of Pi uptake, the faster is the decrease in Pi concentration outside the root and this, in turn, results in a steeper Pi concentration gradient between the root surface and the bulk soil. Thus the diffusive flux of Pi from the soil towards the root increases. At the same time, as the Pi concentration at the root surface is depleted, Pi uptake into the root slows. The model also takes into account various root parameters because of the low mobility of Pi in soil and the need for roots to grow into regions where Pi is located to acquire Pi (see Jungk 1994).

The Barber-Claassen model has shown good agreement between actual and predicted P uptake by plants, particularly for cereals where soil Pi was adequate and the supply of Pi via diffusion was sufficient to meet the demands of the crop. However, where soil P was less than adequate, and the rate of Pi supply via diffusion was likely to have been limiting, the crop acquired more P than was predicted by the model (Jungk and Claassen 1989). This discrepancy has been interpreted by those developing the model as an indication that under conditions of low Pi supply, roots develop properties that enhance Pi acquisition and these properties are not taken into consideration by the model that has a physico-chemical basis only. Such properties include the acquisition of relatively large amounts of P by VA mycorrhiza that effectively increase the surface area of roots (Jungk and Claassen 1989) and the possible release of organic anions from the roots into the rhizosphere, thereby increasing the availability of Pi in the rhizosphere (Ryan *et al.* 2001). Such processes will be discussed later in this chapter.

EFFICIENCY OF P USE WHERE SOIL P SUPPLY IS ADEQUATE

In developed countries, and in particular those with a long history of productive agriculture, plant nutrients, as organic manures or fertilizers, have been added to the soil over many years and, since the 1950s, their amounts have increased dramatically.

In the case of P, the balance between that applied to the soil and that removed by the harvested crops became positive, with more being P applied than is removed. Consequently the level of plant-available P in soils has increased, so that the need to achieve economically optimum crop yields in environmentally benign ways needs to be reconsidered. In this section, the principles involved in achieving this are considered. Various measurements of P-use efficiency are discussed and, using data from long-term experiments with known P fertilizer inputs, evidence is provided of much greater efficiency of P fertilizer use than is commonly recognized. This is followed by a brief description of P modeling aimed at efficient P fertilizer programming. Finally, some of the more agronomic aspects which are or could be available to improve the efficient use of soil and fertilizer P are mentioned.

Critical levels of soil P for crop production

There is an important outcome to the conceptual model that soil P exists in a number of pools of different extractability and, therefore, of differing availability to plants (Figure 9.2). If the P in the readily extractable pool supplies the bulk of Pi for plant uptake then it is only necessary to accumulate a certain amount of P in this pool to achieve optimum crop yield. This is consistent with the concept of a critical P value for a particular crop in a given situation. This point is well illustrated by the asymptotic relationship between yield and Olsen P for sugar beet, barley and winter wheat crops grown at three sites in Southeast England (Figure 9.4; Johnston 2005). As Olsen P was increased from very low levels, yield increased rapidly at first and then more slowly to reach an asymptote. The Olsen P associated with yield reaching the asymptote can be considered as the critical value. Below this value there is a loss of yield, representing a financial loss to the farmer. Above the critical value there is no increase in crop yield with further P additions. Adding more P is not only a financial loss and a squandering of a precious resource but may lead to the risk of adverse environmental effects (Johnston and Dawson 2005). Importantly, even though the yields of each of the three crops differed appreciably between years due to weather and N supply, the critical value for Olsen P was very similar for all years for each crop (Figure 9.4). Ideally, the critical value should be determined for each soil type, crop and farming system. In the UK, however, experience suggests that, in the absence of detailed information, a general value can be used as a first approximation. For soils growing arable crops, Olsen P values in the range of 15–25 mg kg^{-1} are satisfactory (MAFF 2000). Other examples of critical values for farming systems and different crops taken from the international literature are given by Syers *et al.* (2007). Applying the concept of a critical P value for crops and farming systems leads to greater agronomic efficiency of P use. Maintenance of the critical level is dependent on replacing the amount of P removed from the field in the harvested crop. This amount can be calculated from the yield and the P content per unit of yield removed from the field. The P concentrations of specific crops and cultivars can vary considerably (Sanchez 2007), indicating that locally determined

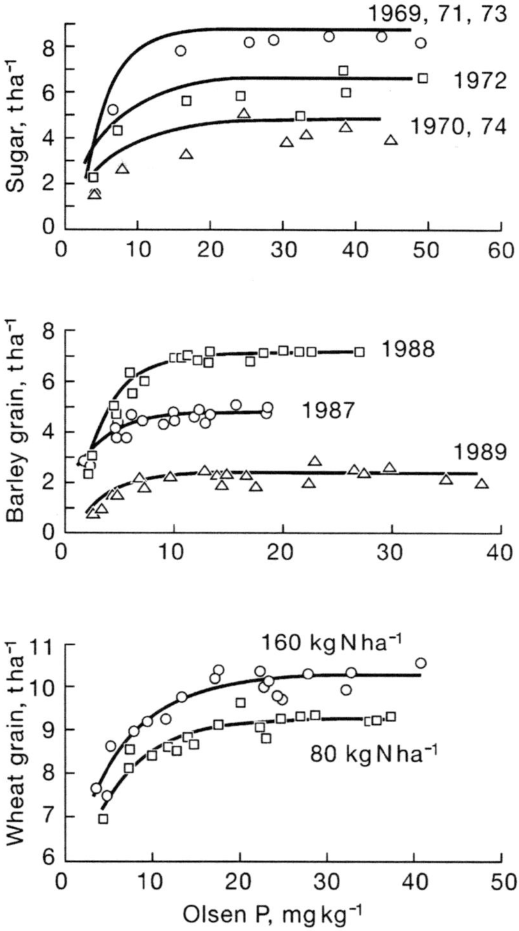

Fig. 9.4 Response of sugar beet, barley and winter wheat to increasing Olsen P in several years on three sites in southeast England. (Data from Johnston 2005.)

values should be used if available. This approach of maintenance P applications is now becoming widely adopted. Consistency of soil sampling and analytical techniques across years is essential, and soil sampling needs to be done on a frequent and regular basis to ensure that the soil is maintained at or about the critical level.

Measuring the efficient use of P fertilizer

The efficient use of fertilizers in crop production has always been important. In recent decades, however, it has become an issue of even greater importance especially for N and P fertilizers. This is because the presumption that they are not used efficiently has been linked with the loss of N and P from the crop-soil system and the risk that such losses pose to the wider environment. The new concepts about the behavior of soil and fertilizer P have led to a reappraisal of measuring the efficiency with which P is used in agriculture.

Frequently a large percentage recovery of an added nutrient is taken to imply an efficient use of that nutrient by a crop (Cassman *et al.* 1998). Several methods have been used for calculating the recovery of applied nutrients, including direct methods that employ uncommon isotopes to label the added nutrient or indirect methods that calculate differences between nutrient inputs and their concentrations in crops and soils. Direct methods generally use a fertilizer containing a tracer element, such as ^{15}N or ^{32}P, to estimate the uptake of a nutrient originating from the fertilizer applied. Nowadays, this is rarely used for P because of its expense, and the relatively short half-life of ^{32}P means that reliable results can only be obtained for crops grown for short periods. The difference method allows an apparent recovery of a nutrient to be determined, but includes a contribution from nutrient already in the soil in addition to the fertilizer applied. The difference method is widely used in agronomic studies (see Crowther *et al.* 1951 for earlier reviews). It is calculated as the nutrient content of a crop given the nutrient minus the nutrient content of a crop not given the nutrient divided by the amount of nutrient applied, and expressed as a percentage. Calculated in this way, the percentage recovery of P and, by inference, the efficiency with which the P has been used by a crop can vary greatly between experiments. This variability can be due to a combination of many factors. These include: (a) the yields of the crops with the two treatments, since these can be affected differently by weather, availability of other nutrients and soil physical conditions; (b) the amount of readily plant-available P in the soil to which no P is applied, because this greatly influences P uptake by the crop grown on this soil; (c) the extent to which the freshly applied P is mixed with the volume of soil where roots are taking up nutrients.

The balance method proposed and used by Syers *et al.* (2007) is another and better method of estimating the efficiency with which P fertilizers are used. It has been appreciated since the beginning of the 20th century that most or all of the P residues from fertilizers and manures accumulated in the topsoil. Since then this observation has been confirmed for many soils with the exception of very sandy soils in high rainfall areas. The direct method (^{32}P) for estimating percentage recovery gives values no larger than 20–25% and underestimates the P content of the crop. The remainder of the P in the crop comes from the reserve of plant-available P in the soil, mainly the topsoil. If soil fertility in terms of available P is not to be depleted then this P must be replaced. Hence added P that is taken up by the crop together with that which has gone to replace the P taken up from the soil reserves may be considered as part of the efficient use of added P fertilizer. The balance method simply calculates percentage P recovery of the added P. This method invariably gives higher percentage P recoveries and hence P use efficiencies than those calculated by the difference method. However, the actual value will still vary depending on the factors noted in the preceding paragraph.

Some examples comparing the difference and balance methods are given in Tables 9.3 and 9.4. Table 9.3 uses data from the Broadbalk winter wheat experiment at Rothamsted, in which the amounts of P and N added each year have remained unchanged since 1852. The yield potential of the cultivars grown has increased appreciably and the data for three separate periods shows how yield affects the estimates of P recovery. There was little increase in yield without N and percentage P recovery changed little when estimated by either of the two methods. Where N was applied yield increased and so did P uptake and hence the estimate of percent P recovery, which was always greater by the balance method. The effect of the existing level of plant-available soil P on the estimate of percentage P recovery is shown in Table 9.4. In this experiment, a four-course rotation of sugar beet, barley, potatoes and barley was grown on two soils with 4 and 33 mg kg^{-1} Olsen P, respectively. Three amounts of P, 27.5, 55.0 and 82.5 kg P ha^{-1} were tested on the sugar beet and potatoes. Thus in each four-year period the total P applied was 55, 110 and 165 kg P ha^{-1}. The rotations were phased, with one starting in 1969 and the other in 1970. The data in Table 9.4 are the mean of both rotations. All the crops were analyzed to prepare a P balance. Percent recovery was calculated by both the difference and balance methods and, for appropriate comparisons was always larger by the balance method. For both methods the percentage recovery always declined with increasing P application. For the soil with 33 mg kg^{-1} Olsen P, percentage P recovery was small when calculated by the difference method and for the balance method one value exceeded 140%. Any value over 100% implies that the amount of P removed in the harvested crop was larger than the amount of P applied. If P removed exceeds P added over a number of years especially when the soils are at or about the critical value, then this will decrease the amount of plant-available P and jeopardize soil fertility.

Table 9.3 Comparison of the difference and balance methods for estimating recovery of added phosphorus illustrating the effect of nitrogen (by increasing yield) and growing cultivars with a larger yield potential. Data from the Broadbalk Experiment, Rothamsted. (Adapted from Johnston and Syers 2006.)

		Annual treatment[b]				
		Without N		With N		
Period & cultivar	P in crop[a] & P recovery	None	PK	None	P	PK
1852–1871	P in crop (kg ha^{-1})	4.9	6.6	6.5	9.2	11.3
Red	Difference method (%)	–	5	–	8	14
Rostock	Balance method (%)	–	20	–	28	34
1970–1975	P in crop (kg ha^{-1})	6.2	6.8	9.0	13.1	17.3
Capelle	Difference method (%)	–	2	–	12	24
Desprez	Balance method (%)	–	21	–	40	52
1985–2000	P in crop (kg ha^{-1})	3.7	4.7	5.7	12.7	17.4
Brimstone	Difference method (%)	–	3	–	20	33
	Balance method (%)	–	13	–	36	50

[a] Total P in grain plus straw
[b] Annual treatment (kg ha^{-1}): N = 96, P = 33 (35 since 1974), K = 90

Table 9.4 Comparison of two methods of estimating phosphorus recovery when phosphorus was applied at three amounts to a sandy clay loam soil at two levels of Olsen P at Saxmundham, UK. Data are means of two four-year rotations (1969–1972 and 1970–1973) of sugar beet, barley, potatoes and barley. (Data from Johnston and Syers 2006.)

	Soil Olsen P (mg kg^{-1})					
	4	33	4	33	4	33
P applied (kg ha^{-1})[a]	P uptake in four years (kg P ha^{-1})		Recovery (%) by difference method		Recovery (%) by balance method	
0	23.3	75.2				
55	46.9	77.0	43	3	85	140
110	57.2	79.4	31	4	52	72
165	63.8	82.2	24	4	39	50

[a]P applied at 27.5, 55 and 82.5 kg P ha^{-1} to potatoes and sugar beet only

Large P recoveries are found for a wide variety of crops grown on different soils (Syers *et al.* 2007). This is even the case for soils of high P adsorbing capacity that were initially acid and deficient in P as in the Cerrado (Savannah) semi-arid region of Brazil. Once soil acidity in the topsoil was corrected to 50% base saturation resulting in the absence of free Al ions in the soil solution, crops responded well to additions of water-soluble P fertilizers. In two experiments, the residual P from a single application of P fertilizer increased yields for the next 13 and 17 years. Recovery of the added P, as measured by the balance method, over this time period ranged from 35–62% with all-arable cropping and from 52–69% when a pasture grass was included in the rotation. Percent recovery was larger with the smaller amounts of P applied. Similarly Heinemann (1996, quoted by Singh and Lal 2005), reported that as much as 68% of a 150 kg P ha^{-1} application of diammonium phosphate was recovered by a maize-bean rotation over five years on an Oxisol in western Kenya.

It can be concluded that the long and widely held belief that P in fertilizers is irreversibly fixed in the soil is no longer tenable. Applied P is used most efficiently if the amount of P in the readily available pool is sufficient to ensure optimum yields. This critical value may vary for different crops and soil types, but it can be maintained by replacing the P removed in the harvested crop. Soil sampling and analysis should be carried out on a regular and frequent basis to ensure that sufficient P is applied to maintain the critical value. Percent recovery or use of the added P in the year of application can then be calculated as the P removed in the harvested crop relative to the P applied – the balance method. Percent recovery calculated in this way frequently exceeds 50% indicating that P is used much more efficiently than previously thought.

Modeling P fertilization

The need for using P fertilizers efficiently has also driven the development of models that predict the effects of P fertilization on crop yield, crop P content and long-term changes of soil P, enabling the consequences of different P fertilization strategies to be assessed *in silico*. Greenwood *et al.* (2001) have proposed a mechanistic model based on the maintenance of three active pools of soil P with an input for the net P addition, close in concept to that described earlier (Figure 9.2). The equations have been calibrated for some crops from the results of fertilizer experiments and underpin a dynamic version of the model that permits annual inputs and calculates the time course of the various P pools. The model includes equations for the dependence of crop P uptake on the content of P already in the plant and of the effects of the distribution of P in soil and its diffusive transport to the roots. Inputs to the model are Olsen-P, a measure of soil P buffer capacity, initial soil water content and the water content at field capacity, the dates of planting and harvesting, and daily pan evaporation, temperature and rainfall. The model was calibrated using data from one set of field experiments and tested against the results of independent experiments. With the aid of the model PHOSMOD, which runs interactively at www.qpais.co.uk/phosmod/phos.htm, it is possible to simulate plant growth and phosphate uptake from day to day. Recently the model has been successfully assessed using data from a set of field fertilizer trials with spring barley (*Hordeum vulgare* L.) in Norway (Kristoffersen *et al.* 2006). Plant growth and P concentrations were recorded at three stages of growth and the model was able to reproduce the observed beneficial responses to placed P and starter fertilizer P, and to predict difference in soil types in the response to applied fertilizers. The model has also been integrated with similar models predicting crop responses to N and K fertilization, thereby enabling strategies for the synergistic application of P, N and K fertilizers (Zhang *et al.* 2007). The model appears to be a promising tool for predicting effects of P fertilizer strategies and may well play a role in planning future fertilization programs.

Agronomic strategies to improve the efficiency of soil and fertilizer P use

Using P inefficiently is to waste a valuable and dwindling resource and there are a number of options for improving P-use efficiency. For soils well supplied with P the question of when and how much P to add has already been discussed. Another question is whether agronomic practices can be altered or modified to improve the accessibility of soil and fertilizer P for root uptake. The methods to achieve this mentioned here and in the next section of this chapter are not exclusive but are discussed separately for convenience. Where appropriate they can be used for increasing crop production on soils well supplied with P and those where yield is restricted by lack of P.

Maintaining a well structured soil with adequate soil moisture and aeration is of paramount importance for the efficient use of soil and added P. Many soils have approximately equal volumes of mineral matter and voids or pores. The latter contain air and water and the relationship between these two is vitally important because roots need oxygen to respire and water for both growth and the transport of nutrients to the roots by diffusion.

Some soil properties that affect the accessibility and availability of P to crops are to a significant extent inherent characteristics of the soil, related to its chemical and physical properties, such as the distribution of P between the various soil P pools and the retention of P in soil in forms that are not immediately available to plants. These could be difficult to change or modify. Many soil properties, however, can be modified comparatively easily to ensure that there is no restriction to root growth and/or the availability of P. Whether such modifications are adopted or not will be determined largely by costs and benefits.

Root growth is restricted in compacted surface soil and by the presence of dense subsurface layers, such as plough pans. Compacted soil tends to have few large pores, restricting water and air movement, and too many pores with diameters too small for root tips to enter. This restricts spatial access to Pi within the soil mass and also slows the rate of Pi diffusion to the roots. Such compaction can be avoided by timely soil cultivation, by minimum-tillage and by maintaining a permanent crop or crop-residue cover on the soil surface. Also, traffic over the soil surface and poaching by livestock when the soil is wet, as occurs on heavy-textured soils in humid temperate regions, should be avoided.

Minimum or zero tillage allows the maintenance of a vegetative cover on the soil and is also an approach to minimizing soil erosion. The impact of no-till systems on the availability of soil and fertilizer P has been studied extensively in the USA and the Canadian Prairies. Initial fears that because P was not cultivated into the soil profile its availability to crops would decline, with adverse effect on yield, have been overestimated (P. Fixen (2007), personal communication). Any adverse effects of P stratification may be countered by better root exploitation of the surface soil to take up P due to improved physical structure of the surface soil under reduced tillage.

In acid soils aluminium (Al) toxicity can limit root growth. Largely eliminating free Al ions in the soil solution by liming acid soils to increase base saturation to 50% is usually sufficient. The effects of modifying the pH of acid soils (usually by lime application) on P retention and extractability has been studied widely but the results have been inconsistent (Haynes 1982; Sumner and Farina 1986; Mansell *et al.* 1984; Holford *et al.* 1994; Curtin and Syers 2001).

Surface soil structure and structural stability also affect both soil moisture content and root growth. Soil structure is particularly important in the case of P because, in contrast to N, Pi moves to only a limited extent in most soils (Barber 1995). Structural stability is related to particle-size distribution and to the calcium carbonate and organic matter (SOM) contents of the soil. In an experiment at Rothamsted, spring barley, potatoes and sugar beet were grown in the same field on soils with two levels of soil organic matter (SOM). The same yield could be obtained on both soils, but much larger concentrations of Olsen P were required for maximal yield on the soil with less SOM (Johnston and Poulton 2005). In a subsequent glasshouse experiment it was observed that the effect of differences in SOM on crop (ryegrass) growth was eliminated after passing these soils through a 2 mm sieve. The authors concluded that the extra SOM improved soil structure in the field and, thereby, the ability of the roots of the three crops to acquire sufficient P for maximum growth at a lower Olsen P.

Soil water content controls the acquisition of P both by favoring root growth so that a greater volume of soil can be exploited and also by allowing transport of Pi to the root surface by diffusion. The predominant effect appears to be that on root growth *per se*. Mackay and Barber (1985) showed that when the volumetric water content declined from 27% (medium) to 22% (low), P uptake decreased by 50–56% as a consequence of a decrease in root growth but only by 27–32% due to a decrease in transport of P by diffusion to the root surface.

Soil lost by water or wind erosion also removes nutrients and, where soils contain little P, such losses can be serious. Erosion is more severe on sloping land and in high rainfall areas. Controlling such losses is important from a water-quality aspect because the transfer of even small amounts of P, insignificant from an agronomic standpoint, from soil to water can cause the adverse effects of eutrophication of surface waters. Well-tested technologies are available to minimize such losses. These include terracing the land, planting trees periodically along the contour and cultivating and planting annual crops along the contour. Additionally, maintaining a soil cover of actively-growing vegetation or plant residues can minimize soil erosion and this can often be achieved with minimum and zero tillage practices. Simple procedures to minimize losses of P in subsurface runoff include not applying P fertilizers and organic manures to cracking soils while the fissures remain open, and not applying them to soils that are dry and hard or saturated with water.

Applying P as fertilizer, organic manure or biosolids (i.e. sewage sludge) to a soil where there is sufficient readily plant-available P, such that there is no increase in yield or benefit to crop quality, is an inefficient use of P. The intensification of animal production, especially for dairy cows, pigs, and poultry, has resulted in large numbers of animals producing more manure than can be used effectively on the land associated with the production unit. There are many logistical problems in storing and applying animal manures and biosolids, mainly because of their bulk, and farmers should be encouraged to follow codes of good agricultural practice in using them.

Much of the P in such manures is Pi (often dicalcium phosphate) added to the animal feed. Opportunities exist to improve the efficiency of P use in these systems. The digestibility of organic P in animal feeds can be improved by adding the enzyme phytase, so that less Pi has to be added, and consequently less Pi will be present in manures (Steén 2006). Improved management of grazed pasture could improve recycling of P by a more uniform distribution of animal excreta (Gillingham *et al.* 1980). Similarly, where manure is collected from housed animals, the manure could be spread more uniformly on arable land and grassland.

Organic manures and biosolids should be considered as sources of P and other nutrients rather than as wastes to be disposed of and making more effective use of them might allow a decrease in P fertilizer use. Some of the environmental issues associated with the application of manure may be controlled by limiting the amount of P that can be applied to land in the manure. As with fertilizers, organic manures and biosolids should be used to maintain the critical level of plant-available P in soil.

When P is required, both the type of fertilizer, the amount applied and the timing of the application is important for improving the efficiency with which the P is used. The amount of P applied at any one time should match P uptake in the harvested crop when the soil is at or about the critical soil P level. Generally water-soluble P fertilizers are appropriate for most soils with pH above about 6. Phosphate rock (PR) could be used as a source of P for crop production but its effectiveness depends on its reactivity in the soil, i.e. the speed at which soil acidity releases the P and the fineness of grinding which increases the surface area. It can be very effective on acid soils (Johnston and Syers 1998 and references therein). There is a tradition in some countries of composting PR with organic manure. It can also be mixed with elemental sulphur, which on oxidation mobilizes P with the production of "biosuperphosphate" (Stockdale *et al.* 2006). There is abundant evidence that PR is not effective on neutral and calcareous soils. There may be a place for slow-release P fertilizers, for example PR on acid soils, and at least one new slow-release product containing water-soluble P is currently being marketed (see http://deltafarmpress.com/news/nutrisphere-bz-pg/). The main use for slow-release P fertilizers would be in situations where P is at risk of loss by leaching, for example on coarse-textured soils in high-rainfall areas.

A more efficient use of P fertilizers within fields can be achieved using the tools of precision agriculture. Plant-available P varies within fields for a number of reasons. For example, where P has been applied uniformly for many years but yields have varied within a field, a larger P offtake with the larger yield results in a smaller amount of plant-available P remaining in the soil. Providing that the level of available P in the high-yielding areas is maintained at the critical level, smaller amounts of P can be applied to those areas where yield is consistently less. Using variable P application rates requires relating yield maps to soil analysis data and using computer-controlled fertilizer spreaders guided by GPS to apply the appropriate amount of fertilizer to different areas within the field.

Temporal and spatial factors have a considerable influence on the acquisition of P by plants. Consequently the timing and placement of P fertilizers is important and the recovery of P is usually increased when the fertilizer is placed near the seed, for example by band placement and seed priming. Such applications can be extremely effective in providing a sufficiently high Pi concentration near the emerging root system to supply a young plant until the root system is capable of accessing a sufficiently large volume of soil with a lower Pi concentration. Benefits from placing water-soluble P fertilizers depend on the speed of the adsorption and absorption reactions that occur in soil soon after the application of the fertilizer. Generally, placing P fertilizer in a band leaves more of the P readily available after the initial P-soil reactions take place. Placing fertilizer with the seed or in a band close to it significantly improves the recovery of fertilizer P, especially in the year of application (Sewell and Ozanne 1970; Barrow 1980; McKenzie and Roberts 1990; Stone 1998). Singh *et al.* (2005) showed that placing 10–40 kg P ha^{-1} at 10–15 cm depth in the semi-arid cropping region of northern Australia increased winter wheat yield by up to 43% because the P remained in more moist soil as the surface soil layer dried out. Bundy *et al.* (2005) reported that a small amount of starter fertilizer is a well established practice for grain crops, especially maize, in the USA. Dramatic benefits of seed-placed NPK starter fertilizer on the early development of sorghum and grain yields were reported by Lamont and Whitney (1991).

Changing agricultural practices

There is increasing interest in growing crops for the production of biofuels. There is a long-established industry producing ethanol from sugar cane, especially in Brazil. As fossil fuel reserves decline there are suggestions for growing crops other than sugar cane to process for bioethanol and to grow appropriate crops for processing for biodiesel. This could result in large areas of land currently growing food crops being used for growing such crops. Already a number of bioethanol plants based on using maize are in production in the USA. Such changes will impact significantly, probably within the next few years, on nutrient use and management. In a scenario of possible change in the USA, Fixen (2007) looked at four possible ethanol sources and the effects on nutrient use. The sources were maize replacing soybean and other crops, refineries using maize stover and fourthly biomass crops like switchgrass or Miscanthus. Such a change could result in an annual increase in P use of some 130,000t, equal to 6.5% of current use in the USA. As none of the P in the crop will be in the biofuel it will be essential to look at ways to ensure that the P in the residue is returned to the land.

P ACQUISITION BY PLANTS WHERE P SUPPLY IS LIMITED

Phosphorus deficiency is a worldwide problem affecting crop production. It is estimated that crop growth on 5.7 billion hectares of land (equivalent to about 67% of the total farmland used worldwide) is limited by P availability (Batjes 1997). Nearly all the acid soils of tropical Africa, America, Asia and Australia are P deficient (Fairhurst *et al.* 1999). On these highly weathered soils of the humid and semi humid tropics, P deficiency relates to the inherently low levels of available P in soils characterized by a high capacity for P sorption. Cultivation of these acid soils is also restricted in particular by high Al concentrations in the soil solution, which are toxic to plants. Additionally the soils can suffer from lack of major nutrients including N, Ca, Mg and Mo, all of which limit crop growth.

In stark contrast to soils of the developed countries, where P has accumulated in soils, there is a negative P balance in the tropical soils of the poorer countries of the world. Here only small amounts of P fertilizer, if any, are applied to replace the P removed by harvested crops and that lost by soil erosion (Singh and Lal 2005). These soils are also in those areas of the world with the largest populations, where crop production is limited to the extensive agricultural systems with both small inputs and yields. Recommendations made to farmers have therefore to be easily adaptable in practice. Fertilizers, because of their relatively high cost to farmers, must be used very efficiently. On these low input, low yielding systems it is often more practical to make the plant "fit the soil" than the soil "fit the plant". For this reason there is a need to understand how plants are able to adapt to extreme environments not usually encountered in the temperate regions.

Phosphorus deficiency occurs on calcareous soils as well as on acid soils. These soils too are very widespread including areas of semi-arid and arid regions and cover more than 30% of the earth's surface (Chen and Barak 1982). The soils of most of north China and the Great Plains of the west USA are all calcareous. On these soils, which contain free $CaCO_3$ with a pH range from 7.5 to 8.5, crop growth can be limited by the low availability of Zn, Fe and Mn as well as P. Recommendations for the alleviation of P deficiency on these calcareous soils can thus be very different from those for acid P-deficient soils.

Plants grow and thrive in soils where P supply is limited by invoking a diverse array of adaptive mechanisms – morphological, physiological, biochemical and molecular which either conserve the use of P within the plant or enhance its acquisition from the soil (Vance *et al.* 2003). There is considerable research activity investigating these responses because knowledge gained about these adaptive mechanisms allows their exploitation to improve P acquisition by plants. This can be achieved by developing P-efficient crops both by targeted breeding as well as by gene transfer and manipulation. Additionally an understanding of how plants adapt to a lack of P provides the basis for management strategies and for efficient P fertilization both on adequately supplied soils as well as on low P soils. In this section some of these responses are considered and how they are being, and can be, exploited to improve utilization of soil and fertilizer P in agronomic practice is illustrated using a few selected examples.

Response of plants to a lack of P

A typical physiological response of plants to lack of P in the shoot is the diversion of a greater proportion of photosynthates to the root, thereby increasing the root: shoot biomass ratio (Hermans *et al.* 2006; Kirkby and Römheld 2006). This leads to an increase in P uptake that is achieved in two ways. First, the increased root surface area gives a greater contact between soil and root for Pi uptake and, second, metabolic changes in the root mobilize Pi in the rhizosphere, which can then be taken up by the plant. For recent excellent detailed accounts of these responses readers are referred to Vance *et al.* (2003), Amtmann *et al.* (2006) and Lambers *et al.* (2006).

Increased root growth in P-deficient soil is accompanied by marked changes in morphology. Roots are longer and more slender, thus increasing specific root length and surface area as they continuously explore the soil searching for nutrients. This strategy of increasing root exploration of a greater soil volume is highly effective in the acquisition of Pi, which is relatively immobile and unevenly distributed both in the soil and in the soil profile. In most soils, the topsoil contains more P than the subsoil and it is in the topsoil that profusely branched lateral roots with long root hairs proliferate especially in P rich patches (Drew 1975). An increase in root hair abundance and length under P deficiency also greatly increases the ability of roots to exploit the rhizosphere for P acquisition (Föhse and Jungk 1983; Bates and Lynch 1996).

Root hairs are produced more or less abundantly by most plant species particularly among the agronomically important crops of the Graminaceae, Chenopodeaceae and Brassicaceae (Jungk 2001). For plants lacking mycorrhizae, root hairs are the primary site for nutrient uptake. In many cases they may contribute up to 70% of the total root surface, thus increasing the surface area of the root cylinder by nearly 27 times (Jungk 2001). Also, under P deficiency the plasma membrane of root hairs is enriched in high-affinity Pi transporters (Liu *et al.* 1998). Of the possible ways of increasing effective root surface area, a change in root hair morphology is considered least costly metabolically (Hetrick 1991).

The importance of root hair length for P acquisition from the rhizosphere is illustrated well by an experiment contrasting Pi concentrations at varying distances from the roots of barley cultivars grown in soil (Figure 9.5; Gahoonia and Nielsen 1997). The cultivar Salka with long root hairs (1.00 ± 0.26 mm) exploited the rhizosphere twice as effectively as the cultivar Zita with short root hairs (0.63 ± 0.24 mm) by reducing the distance of Pi diffusion to the root. Accordingly, Salka also absorbed more P than Zita when grown in a low-P soil and also produced more shoot biomass (Gahoonia *et al.* 1999). More recent field experiments by these workers have shown that barley cultivars with long root hairs maintained stable

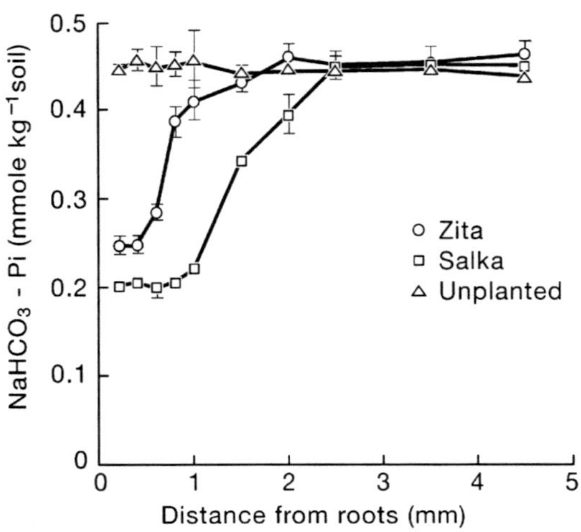

Fig. 9.5 Depletion of inorganic P (extracted with 0.5 M $NaHCO_3$) at varying distances from the roots of two barley cultivars, Salka (with long root hairs) and Zita (with short root hairs). (Figure redrawn from Gahoonia and Nielsen 1997.)

economic grain yields both in P-limited and P-sufficient soils. In contrast, cultivars with short root hairs produced small grain yields in P-limited soils and produced large grain yields only in P-sufficient soils (Gahoonia and Nielsen 2004).

A highly successful line of research that has exploited root adaptation to limited P supply has been the development of P-efficient genotypes of common bean (*Phaseolus vulgaris*) varying in root architecture (Lynch 1995; Lynch and Brown 2008). Several influential root traits underpinning genetic differences in P acquisition efficiency have been identified including topsoil foraging, via shallower basal roots and greater adventitious rooting, greater rhizosphere exploitation, via increased length and density of root hairs, and decreased root metabolic costs, via the formation of root cortical aerenchyma and root etiolation (Bates and Lynch 2001; Lynch and Brown 2008). Genotypic variation in basal root shallowness accounted for a 600% difference in P uptake from soil and was significantly correlated with yield when crops were grown in low P soils in field experiments. Genotypic variation in root hair length accounted for a 250% difference in P uptake in the field and genotypic variations in root metabolic costs for a 200% difference in root biomass and root exploration. Lynch and colleagues have also studied the effects of these traits on agro-ecosystem processes including drought resistance, nutrient cycling and soil erosion in field trials in Central America. Phosphorus efficient genotypes that yield between 200% and 300% more than existing cultivars are being used as parents in bean breeding programs in Latin America and Africa (Lynch 2007). In collaboration with Chinese colleagues these researchers are applying their knowledge to breeding P-efficient soybean lines, which are yielded 15–50% more than existing genotypes in low-P soils in south China (Yan *et al.* 2006).

Increasing root growth at the expense of the shoot is a hallmark response of plants to P deficiency. Just how important root growth is for P acquisition for plants grown under water stress is illustrated by the work of Mackay and Barber (1985). These workers observed that water shortage decreased both the growth of maize roots and Pi diffusion in the soil. However, they calculated that over half the decrease in P uptake under water shortage could be attributed to a reduction in root growth, which restricted exploitation of the soil volume, whereas the lower rate of Pi diffusion in the soil accounted for less than a third. In this experiment there was an increase in root hair growth due to the lower moisture level and although this increased root surface area, this only partially offset the lower rate of root growth, which was the primary cause of the reduced P uptake.

This example illustrates the principle that any factor which restricts root growth depresses P uptake primarily by limiting the volume of soil explored by the roots. Roots frequently encounter adverse abiotic factors, such as low or high temperatures, salinity, drought, as well as adverse biotic conditions including microbial pathogens, nematodes, viruses and plant parasites. All the factors that adversely affect root growth, and particularly the spatial distribution of roots in soil, will depress P uptake. Understanding that a restriction in root growth limits P acquisition underpins many agronomic measures to improve the efficient use of soils and fertilizer P.

Mycorrhizal associations

The roots of most crop plants are infected by arbuscular mycorrhizal (AM) fungi which enhance the ability of plant roots to explore a much greater volume of soil for nutrients and water (Smith and Read 1997). This symbiotic association between the host plant and fungus, based on the reciprocal transfer of P from the fungus to the plant and carbon from the plant to the fungus can be considered as another plant adaptation to P deficiency. Estimates suggest that the extent of the fungal mycelium may range from 10 to 100 m per root or per gram of soil under field conditions in P-deficient soil (McGonigle and Miller 1999). The occurrence of mycorrhizae in agro-ecosystems is almost universal but crop species differ in their dependence. Interestingly, some crops from families that are non-mycorrhizal, such as the Brassicaceae (e.g. oilseed rape) and Chenopodiaceae (e.g. sugar beet) appear to compensate by producing abundant root hairs. Evidence indicates that root hairs and external hyphae act similarly by shortening the distance for the diffusion of Pi to the root surface (Schweiger *et al.* 1995).

Attempts to exploit this adaptation to improve crop growth in the field through inoculation of selective AM populations have frequently been unsuccessful, not least because of competitive indigenous soil fungi. However, at CIAT, Colombia, when 50 kg P ha^{-1} as PR was added to an acid, severely P-deficient Oxisol, the yield of a number of crops was increased by inoculation with isolates of a native AM fungi (Howeler *et al.* 1987). Yields of cassava (*Manihot esculentum*) roots increased, on average, by 20–25% with AM inoculation both at the experimental station and in farmers' fields. Inoculation of various pasture legumes and grasses with AM fungi also in combination with applications of PR improved early plant growth and establishment. A pot experiment in which cassava, bean and maize were grown on a tropical soil with high P sorption to which triplesuperphosphate (TSP) fertilizer was added, suggested that the mycorrhizal dependency of growth depended greatly on root morphology (Table 9.5). Even when a large amount of TSP was supplied, growth of cassava was still extremely dependent on AM fungi

Table 9.5 Effect of mycorrhizal inoculation and phosphorus (P) application on the shoot dry weight (g pot^{-1}) of several plant species grown in a sterilized, strongly P adsorbing tropical soil supplied with three levels of P (0, 100, 500 kg P ha^{-1}) applied as a soluble phosphate fertilizer (TSP). (Data from Howeler *et al.* 1987.)

		Dry weight		
Plant species	Inoculation	P0	P100	P500
Cassava	–	0.34	0.72	0.54
(*Manihot esculentum*)	+	4.33	14.21	16.36
Bean	–	1.11	3.44	8.29
(*Phaseolus vulgaris*)	+	3.08	18.79	25.01
Maize	–	1.19	8.74	59.39
(*Zea mays*)	+	4.84	34.75	56.67

because of its coarse and poorly branched root system. Bean, with both short roots and short root hairs, was less dependent, and maize, which has a much more highly branched root system, was almost independent of AM fungi when large amounts of TSP were supplied (Howeler *et al.* 1987). However, not all grasses behave in this way. Mycorrhizal fungi can also have other effects in improving crop growth and these include increasing tolerance to root pathogens, water stress and enhancing N fixation by rhizobia (Marschner 1995). Harnessing mycorrhizal fungi to improve the utilization of soil and fertilizer P where the supply is limited and thus sustain soil fertility is therefore a worthwhile but very difficult and daunting challenge.

Biochemical adaptations

Mechanisms that increase P acquisition efficiency of roots include enhancing the capacity of the root to take up Pi by increasing the abundance of high-affinity P uptake systems in the plasma membranes of epidermal cells and the release of various exudates, mainly organic acid anions, protons and enzymes, into the rhizosphere. These exudates can increase rhizosphere Pi concentration by mobilizing inorganic and organic forms of P in the soil.

Under conditions of P deficiency, genes encoding high-affinity Pi transporters are up-regulated (Liu *et al.* 1998; Sas *et al.* 2001; Vance *et al.* 2003) providing the plant with a greater capacity for Pi uptake. For plants growing in soil, however, this increased uptake capacity may be of little relevance because Pi diffusion from the bulk soil to the root surface is usually the rate limiting step in P acquisition (Silberbush and Barber 1983; Raghothama and Karthikeyan 2005). This may explain why over-expression of a gene encoding a high-affinity Pi transporter in transgenic barley had no effect on Pi uptake by plants growing in soil (Rae *et al.* 2004). Nevertheless, in China, Yan *et al.* (2006) have reported the development, by physiological and genetic approaches, of a P-efficient wheat cultivar, Xiaoyan 52, in which a Pi transporter is highly expressed under both normal and P deprived conditions. In this case, however, other adaptations including organic anion release from the roots might also have favored P acquisition.

A number of crop species, (including oilseed rape, chickpea, alfalfa, white lupin and other legumes), respond to P deficiency by exuding carboxylates, such as citrate, malate and oxalate, from the roots. These carboxylates released into the rhizosphere mobilize Pi by chelating with metal cations that bind Pi and by displacing Pi from the soil matrix by ligand exchange (Gerke *et al.* 2000). The released Pi is then taken up by high-affinity Pi transporters. Potentially, tricarboxylates and dicarboxylates are the most able to mobilize Pi because of the high stability constants of their corresponding Fe, Al and Ca-complexes (Ryan *et al.* 2001). Citrate, which is commonly found in root exudates, is highly effective in chelating not only Ca in calcareous soils (Dinkelaker *et al.* 1989) but also Fe and Al in acid soils (Gardner *et al.* 1982) and can mobilize Pi from a range of sparingly soluble inorganic P complexes. Restricting the release of organic anions to the apical zone of the root, as in oilseed rape (*Brassica napus*) is a highly efficient means of extracting soil P (Hoffland 1992). A similarly located release of malate and its chelation of Al also provide a mechanism enabling Al tolerant wheat cultivars to grow in acid soils (Delhaize *et al.* 1993).

It is still a matter of debate whether the rate of carboxylate exudation from plant roots is adequate to provide sufficient Pi to meet plant requirements in all but a few species (Hinsinger *et al.* 2003). Neumann and Römheld (2006) suggest that large concentrations of carboxylates (millimolar) are required to release significant amounts of Pi from soil complexes. Additionally, carboxylates released from roots may be utilized in other ways, for example as substrates for soil microorganisms. Much is still to be learned of the reactions and processes in the rhizosphere that are governed by the release of organic anions leading to improved P nutrition (Rengel and Marschner 2005).

One of the adaptive responses of plants to P deficiency is to release protons from the roots into the rhizosphere to mobilize Pi. This is a strategy that can be exploited in crop nutrition. By supplying crops with NH_4-N the same effect can be achieved because of a higher cation:anion uptake ratio by the plants (Kirkby 1981; Mengel and Kirkby 2001). Soil acidification by plant roots can be used to benefit on high pH, calcareous soils for the mobilization and uptake of Pi from sparingly-soluble, native CaP complexes in soil. On neutral and more acid soils Pi can be mobilized from applied PR. This is illustrated by an experiment with common bean plants supplied with NH_4-N or NO_3-N growing in a soil with pH 6.6 and supplied with PR (Table 9.6; Thomson *et al.* 1993). Depressing the rhizosphere pH not only mobilized and increased the uptake of P but also the uptake of Fe, Mn, Zn and Cu, all nutrients which cause crop deficiencies on neutral and calcareous soils.

It is much more difficult to increase the pH of acid soils by managing the rhizosphere of plants. Graminaceous monocots, including pasture grasses and cereals, increase rhizosphere pH when supplied with NO_3-N because of a much higher anion:cation uptake ratio. Their effectiveness under field conditions is very dependent on the pH of the bulk soil. In general, a tendency to increasing acidification is the common trend in cropping systems (Lambers *et al.* 2006). Nodulated legumes offer an enormous potential for the supply of biologically fixed N_2 to low input agricultural systems (Vance 2001), but on neutral and acid soils they suffer the serious disadvantage of depressing pH, which leads to the formation of monomeric Al^{3+} in the soil and Al toxicity. Nitrogen fixation and proton release are closely

Table 9.6 Effect of the form of N fertilizer applied to a sandy loam soil (pH 6.8) on rhizosphere pH and nutrient uptake by bean (*Phaseolus vulgaris* L.) plants. Nitrogen was applied as $Ca(NO_3)_2$ or $(NH_4)_2SO_4$ plus a nitrification inhibitor. (Data from Thomson *et al.* 1993.)

		Uptake (μg m^{-1} root length)				
Form of N	Rhizosphere pH	P	Fe	Mn	Zn	Cu
$Ca(NO_3)_2$	6.6	815	68	23	11	2.7
$(NH_4)_2SO_4$	4.5	1,818	184	37	21	3.7

related (Raven and Smith 1976). Soil amelioration by liming can be important but does not always address sub-soil acidity. Moreover, liming materials are not always available and even when they are used, as for example on tropical acid soils they may promote Pi adsorption by the formation of Al-P complexes (Haynes 1984). On soils where Pi is strongly adsorbed, band placement of P fertilizers is recommended (Werner and Scherer 1995) because this decreases the immediate adsorption of Pi by soil particles. The innovative use of PR where soils are acid, combined with the application of organic manure can be effective in increasing P uptake (Stockdale *et al.* 2006). Pigeon pea (*Cajanus cajan*) is one crop species that has an outstanding capacity for P acquisition in acid but not alkaline soils. The roots of this crop release piscidic acid which chelates strongly with Fe^{3+} and thus mobilizes sparingly-soluble Fe phosphates. It is therefore used very successfully as an intercrop with cereals on acidic soils (Alfisols) in the Indian subcontinent (Ae *et al.* 1990).

The development of specialized root structures known as cluster roots (proteoid roots or bottlebrush roots), found on a number of species that grow in P-deficient soils is another adaptation to P deficiency. The most well researched plant species of this type is white lupin (*Lupinus albus*). This is the sole cluster root forming species of agricultural importance and is grown for grain and forage. This adaptation combines a high C investment in the root structure with a highly specialized root physiology (Le Bayon *et al.* 2006). Cluster roots consist of short lateral roots covered by a dense mat of root hairs that make up to 60% of the root biomass and are developed in patches of soil where P is most available, even in fertile soils. The roots periodically release large amounts of citrate and protons together with acid phosphatases. This localized excretion restricts microbial degradation of the exudates and simultaneously allows an intensive extraction of a limited volume of soil. As much as 23% of the dry weight of the plant can be released as citrate (Dinkelaker *et al.* 1989). It is thought that carboxylates released through anion channels in the root plasma membranes are able to mobilize both Pi by chelation and ligand exchange as described above. Organic P is hydrolyzed by acid phosphatases once it has been mobilized by the carboxylates. Inorganic P is taken up by high-affinity Pi transporters (Sas *et al.* 2001).

Interest has focused recently on other native species of Proteaceae and Cyperaceae that have cluster roots and grow on some of the most P-impoverished acid soils in the world in Western Australia and South Africa (Lambers *et al.* 2006). These plants are characterized by a very small P requirement and tissue P concentrations of only 0.02–0.04% P in dry wt, a concentration about ten times lower than most crop plants. Interestingly, Proteaceae are particularly sensitive to P toxicity at external P concentrations that are harmless to other plants, mainly because they have little capacity to down-regulate P uptake in response to increased P supply (Shane *et al.* 2004). In contrast to white lupin, these species, which grow on highly weathered acid soils, do not release protons together with carboxylates into the rhizosphere (Roelofs *et al.* 2001). This is presumably an adaptation to prevent acidification of the rhizosphere and Al toxicity, while still enabling the release of Pi from Fe and Al complexes. The concentration of root clusters as mats in surface soil horizons aids the scavenging of P from organic matter. This feature is comparable to topsoil foraging for efficient P acquisition by the lateral roots of agriculturally successful common bean genotypes (Lynch 2007). The potential use of these cluster-rooted species in the development of new crops with common root adaptations to enhance P acquisition has been discussed by Lambers *et al.* (2006).

In most agricultural soils about 20–80% of P is present as organic P in soil organic matter. In the rhizosphere there is a considerable turnover of organic P in which microorganisms act both to mobilize P as well as to incorporate it into the organic fraction. Hydrolysis of organic P is also mediated by the release from the root of phosphatases, which are increased in the rhizosphere in response to P deficiency in the plant (Vance *et al.* 2003). Plant roots release very little, if any phytase, the enzyme that breaks down phytate (inositol hexaphosphate), which is the predominant form of organic P in most soils (George and Richardson 2008; Marschner 2008). Plant access to this form of P is therefore mainly dependent on soil microorganisms. The complex interactions between enzymes released into the rhizosphere and rhizosphere microorganisms involved in the turnover of organic P are still not well understood. Knowing how these interactions operate, however, is especially important for intercropping farming systems on soils with little plant-available P, where plants depend considerably on the cycling of organic P, as discussed below (Horst *et al.* 2001).

P-Efficient crops and cropping systems

Crop species and crop cultivars that can make the most efficient use of soil P by mobilizing it from less plant-available pools and from fertilizer additions can play a key role in sustainable cropping systems. Mention has already been made of such P-efficient crops, like white lupin, brassicas, buckwheat and pigeon pea, which release organic exudates or protons into the rhizosphere. Phosphate mobilized by these crops can be used by neighboring plants of other species or crops following in the rotation. In a classic pot experiment Horst and Waschkies (1987) demonstrated that when white lupin and wheat were grown together on a P-deficient soil supplied with PR, the white lupin despite its much lower root length density than wheat was able by the release of citric acid to mobilize sparingly-soluble P sources not only to meet its own demand but also to provide additional P for the wheat.

Horst *et al.* (2001) consider the most promising agronomic approach to P sustainability on low P soils is the integration of P-mobilizing plant species as intercrops or as crops within a crop rotation. These authors have shown, in field experiments in the Northern Guinea Savannah of Nigeria, the beneficial effect of P mobilization by a number of leguminous grain and cover crops in addition to white lupin. The growth and grain yields of cereals were increased when grown in rotation with legumes. This beneficial effect of P-mobilizing plant species may result from a direct shift in the equilibrium between the less available to the more readily plant-available pool of soil P (Lambers *et al.* 2006). The results of Horst *et al.* (2001), however, suggest that the effects relate to the recycling of the mobilized

P via crop residues, which increase organic matter and enhance biological activity. This suggestion accords with enhanced colonization of the rhizosphere with P-mobilizing microorganisms on the roots of crops grown subsequently in the rotation. Organic materials including crop residues, manures and composts are used traditionally on low-P tropical soils and have also been shown to reduce P adsorption through the presence of organic anions. Of the different plant species used as green manure, the deep rooted woody species *Tithonia diversifolia* appears to be especially valuable. This species not only mines the deeper layers of the soil for P but applying the foliage to the soil also increases soil P availability (Singh and Lal 2005). Much is still to be learnt about rhizosphere interactions between released exudates and microorganisms and their effects on P utilization by plants. What is clear though is that applications of fertilizer P are required to sustain these cropping systems and particularly so on strongly P adsorbing acid soils (Horst *et al.* 2001; Singh and Lal 2005).

The Brassicaceae are extremely effective in acquiring P from the soil (Greenwood *et al.* 2005, 2006a) which appears to relate not only to their ability to mobilize Pi via organic acid exudation (Hoffland 1992) but also by a very extensive root system able to scavenge P from the soil (Dechassa *et al.* 2003). The benefit of including brassicas in the crop rotation on soils in the UK has been described by Greenwood *et al.* (2006b). In glasshouse and field experiments, they studied the responses of 12 genotypes of *Brassica oleraceae* to various levels of P fertilizer. All genotypes grew well on a sandy loam soil containing 20 mg kg^{-1} Olsen P, a value close to the critical level discussed earlier. The authors propose that high-yielding rotations of brassicas and cereals, which can also yield well on low P soils, may be sustained indefinitely on some soils at about this level of available P. Applications of only slightly more P (about 12 kg P $year^{-1}$) would be required in addition to that removed by the crop to maintain the level of Olsen P. In these experiments, the efficiency of fertilizer P use was about 62% in the year of application.

P-Efficient germplasm

There is considerable genetic variation both between and within crop species for many of the root traits related to plant adaptation to P deficiency (White *et al.* 2005). There is, therefore, interest worldwide to develop improved crop germplasm adapted to low-P soils to allow more efficient uptake and utilization of soil and fertilizer P. Considerable success has already been achieved in modifying root architecture in bean, maize and soybean cultivars so that the plants grow better under P-deficient conditions, as discussed earlier in this chapter and by Lynch and Brown (2008).

Phosphorus deficiency does not occur in isolation from other growth limiting factors. Crop improvement programs have therefore developed cultivars with multiple stress resistance to give larger yields and greater yield stability in different environments, for example, in the maize and sorghum improvement programs of CYMMT in Brazil. The aim of these programs has been to pyramid the genes that control plant adaptation to acid soils. As well as tolerance to P stress and improved efficiency in P uptake, these include tolerance of Al and of water deficiency, greater efficiency in ammonium utilization and a larger root system.

There are few reports on variation between crop genotypes in root release of organic acids in relation to P acquisition (Gahoonia and Nielsen 2004). Yet this would seem a promising approach as organic acid release enhances P mobilization irrespective of the pH of the bulk soil and, as shown by Delhaize *et al.* (1993), provides a mechanism to detoxify Al. There is some evidence for the feasibility of this line of research from the finding of more organic acid secretion from P-efficient than from P-inefficient genotypes of wheat in China and the utilization of these P-efficient crops (Yan *et al.* 2006). For rice too it is recognized that developing cultivars that grow on P-deficient soils may represent a more sustainable solution for continued production than sole reliance on P fertilizer applications. In assessing genotypic variation for tolerance to P deficiency, Wissuwa and Ae (2001) observed great variation among rice genotypes in their tolerance to P deficiency. In general, traditional varieties were superior to modern ones. By appropriate crossing of a modern variety Nipponbar with the P deficiency-tolerant landrace Kasalath, an improved line was developed, which surpassed Nipponbar in P uptake by 170% and in grain yield by 250%. By combining the large P uptake capacity of Kasalath with the high harvest index of Nipponbar, it was possible to triple the grain yields of Nipponbar when grown under P deficiency. More recently, Wissuwa (2003) has modeled for rice the response and tolerance to P deficiency and separated the effects of greater root growth and higher external root efficiency (i.e. the rate of P uptake per unit of root surface area). This has shown large genotypic differences associated with greater P uptake can be caused by relatively small changes in tolerance mechanisms. These small changes are therefore likely to be difficult to detect because they are overshadowed by secondary root growth effects.

Phosphorus efficient species and cultivars can be of benefit in improving the use of native soil P and residues of P applied as fertilizer on both low P soils and on soils adequately supplied with P. On the latter soils, crops such as white lupin and brassicas that can provide greater access to the less readily-available pool of P, may play an important role in enhancing the rate of P acquisition from such soils. The tremendous potential of root cluster species in agriculture and their innovative use in the development of new crops with similar root adaptations has been discussed by Lambers *et al.* (2006).

Genetic modification of plants to increase exudation of chelating and other compounds into the rhizosphere to increase P acquisition presents another means of developing P efficient plants. So far, however, progress has been rather slow. There are contradictory reports concerning the effects of increased exudation of carboxylates and P uptake by transgenic plants (Rengel and Marschner 2005). In transgenic plants engineered to increase phytase production and exudation to access organic P in soil organic matter, there is often no correlation between exudation and P uptake (George and Richardson 2008). Much therefore is still to be learnt before these approaches can give consistent and beneficial results which might be used in crop production.

SOME ENVIRONMENTAL ASPECTS TO THE USE OF P FERTILIZERS

There are two main aspects of environmental concern in the use of P fertilizers for crop production. The first relates to elements which occur as impurities in PR and can carry through into the manufactured product, be applied to soil and enter the food chain. Some of these elements can be detrimental to human and animal health and those causing most concern are uranium (U) and cadmium (Cd). The second relates to the transfer of P from the soil into the aqueous environment and the consequent risk of eutrophication. There is an enormous literature on both these topics. Consequently, only a brief account is presented here.

Radionuclides

Radioactive elements, like U and radium (Ra), are normal trace constituents of the earth's crust (Scholten and Timmermans 1996) and occur in PR, particularly sedimentary PR, at higher than average concentrations. The treatment of PR with sulfuric acid to produce phosphoric acid in the wet oxidation process is the starting point in the production of nearly all P fertilizers. Varying proportions of the radionuclides can carry through into the processed fertilizers or remain in the co-product phosphogypsum (Rutherford *et al.* 1994; Leikam and Achorn 2005). Uranium and other radionuclides are found to a lesser degree in fertilizer products than in phosphogypsum, and radioactivity levels of most P fertilizers are only slightly greater than those in the soil (Falk and Wymer 2006).

From the older literature there is good evidence that U uptake by plants is very small. In long-term experiments at Rothamsted where known quantities of single superphosphate fertilizer have been applied to a silty clay loam soil annually for over 120 years, most of the U applied in the fertilizer (1,300 g U ha^{-1}) could be accounted for in the top 23 cm of soils growing arable crops and in the organic surface layer under permanent grassland. Similarly, most of the 330 g U ha^{-1} added in superphosphate to permanent grassland in New Zealand was found in the surface layer (Rothbaum *et al.* 1979). More recently it has been shown that U is taken up by plants less readily than other toxic heavy metals and there is general agreement that soil-plant transfer is of no risk for contamination in the food chain (Kratz and Schnug 2006). However, The possibility of ground and surface water contamination through soil erosion, surface run-off and leaching, which could impair the quality of drinking water, is receiving increasing attention. The accumulation of U in soil arises not only from the use of P fertilizers, but also from the factories where they are manufactured, fossil fuel power plants and U mines (De Kok and Schnug 2007).

Phosphogypsum (PG) consists mainly of calcium sulfate but also contains small quantities of phosphate, fluorine and radium derived from the phosphate rock. In the wet process for manufacturing phosphoric acid, 5 t of PG are created for every tonne of phosphoric acid produced. Because of its small Ra content, the use and movement of PG is restricted in the USA and several other countries. Consequently, many hundreds of millions of tonnes of PG have been stored on land in stacks near the site of production and as much as 160–200 Mt are likely to be produced every year for the foreseeable future. The long-term management, use and disposal of this material are one of the most challenging problems facing phosphoric acid producers. Hilton (2006) has reviewed the various possible uses for PG including agriculture, road construction and the building industry. While these latter uses have been controversial, largely because of the radioactivity, stacking has environmental problems also (Hilton 2006). However, the risk from the radioactivity in PG is being reassessed and it may become possible to use PG as described by Hilton (2006) rather than stack it.

Cadmium

A wide range of elements other than those that are radioactive occur naturally in PR (Van Kauwenbergh 1997). Again, these can either be transferred to the finished fertilizer product or be retained in the PG. Cadmium is the element of principle concern because of its toxicity to man and animals. The world's major reserves of PR are sedimentary rocks which contain very variable amounts of Cd, ranging from less than 10 mg Cd kg^{-1} to more than 50 mg Cd kg^{-1} PR. Fertilizers produced from igneous rocks, which contain little Cd, contain very low Cd concentrations. Unfortunately supplies of these rocks are very limited.

The Scientific Committee on Problems of the Environment (SCOPE) organized an Environmental Cadmium Project holding two workshops. The first was in Brussels in 2000 (*http://www.icsu-scope.org/projects/cluster3/cadmium/Cd%20Brussels%20report.pdf*) and the second in Ghent in 2003 (*http://www.icsu-scope.org/cdmeeting/2003meeting/cdindex.htn*). Both meetings identified the importance of Cd as a toxic element and recognized that setting standards to protect soils, plants, animals and humans from exposure to Cd had become a high-profile policy issue. The project focused on two issues: (1) human and ecological health risks from cadmium and (2) impact of risk information on allowable levels of cadmium in P fertilizer. Any incorrect assessment of risk from Cd and its subsequent implementation through regulations on Cd applications could adversely affect the use of P, essential to food production worldwide, either as fertilizers and/or the recycling P through biosolids. The concern was that regulations introduced and designed to protect certain populations may have unforeseen effects on other populations involving nutrition, public health and social and economic factors. It was concluded that human exposure to Cd is largely driven by Cd in staple food crops, which originates from Cd in the soil. Of several potential sources of soil Cd that in P fertilizers received particular attention and it was agreed that there is no conclusive evidence of any adverse impact of Cd in P fertilizers on human health. This was the case even in Australia where there has been a long history of application of fertilizers with high Cd concentrations and a strict policy on their use.

Johnston and Jones (1995) discussed some agronomic aspects of Cd and P fertilizer use. One aspect on which there are divergent views is whether soil acidity affects the uptake of Cd by plants. Data from the Park Grass Experiment at Rothamsted suggests that it does. The Cd concentrations in the herbage harvested on these permanent grass plots during the past 150 years depends on soil pH and the addition of Cd in superphosphate and atmospheric deposition (Figure 9.6). On soils where the pH has been maintained at or slightly above 6.5, the concentration of Cd in the herbage is only a little larger where superphosphate has been applied than where none has been given since 1856. Conversely, where the soils have become acid, the concentration of Cd in the herbage is much larger where Cd has been added in superphosphate than where none has been applied. Interestingly, the Cd concentration in herbage grown on the acid plots receiving superphosphate has declined recently, possibly because the Cd taken up has been "diluted" by the increasing yield of dry matter on these plots (Nicholson *et al.* 1994). The discrepancy between these results and others that suggest that soil pH has no effect on herbage Cd content, may be due to experimental technique. Additions of lime that seek to change soil pH quickly in laboratory or field experiments may change the laboratory measurement of soil pH without affecting the chemistry of Cd in the soil and thus the uptake of Cd by the crop. Wu *et al.* (1989) showed that a difference in rhizosphere pH as a result of applying different sources of N affected the Cd content of *Lolium perenne* herbage. There was 4.2 mg Cd kg^{-1} dry matter at a rhizosphere pH of 6.8 and 12.2 mg Cd kg^{-1} dry matter at a rhizosphere pH of 5.5.

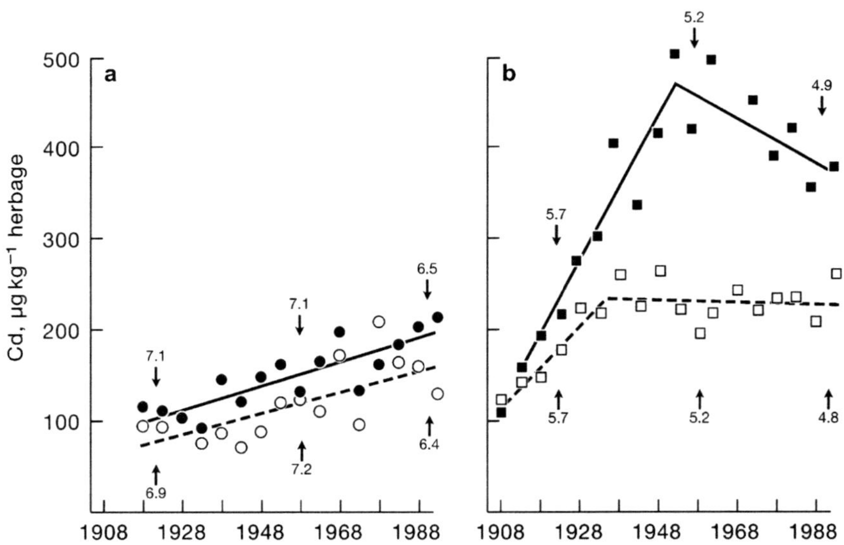

Fig. 9.6 Cadmium concentrations in herbage from the Park Grass Experiment at Rothamsted. Herbage from plots with neutral soils with (•) or without (○) superphosphate fertilizer applications (**a**). Herbage from plots with acid soils with (■) or without (□) superphosphate fertilizer applications (**b**). (Adapted from Nicholson *et al.* 1994.)

Transfer of P from soil to water

The increasing P concentration in some surface fresh water bodies and the adverse effects of this enrichment on the biological balance in the water body, has been attributed to the increasing use of P fertilizer in agriculture (Johnston and Dawson 2005). Current evidence suggests that P transferred from soils to water comes from three main sources. These are: (i) Surface runoff, especially in hilly terrain, when intense and prolonged rainfall follows the application of large amounts of slurry to either arable land or permanent grassland. (ii) Eroded soil that is excessively enriched with P as a result of the over, and often unnecessary, addition of P fertilizer and organic manures. (iii) Drainage water leaving the soil profile, which can be important on very coarse-textured, sandy soils where large amounts of P fertilizer or animal slurry have been applied. If the subsoil contains large amounts of clay the P can be retained there (Mattingly 1970). There is a relationship between Olsen P and very weakly bound soil P soluble in 0.01 M $CaCl_2$, which could move down the soil profile in drainage water (Heckrath *et al.* 1995; McDowell *et al.* 2001). The amount of P transferred from soil to water by these pathways can be minimized by the adoption of Codes of Good Practice as published in various countries.

Another major source of P entering water courses is that from sewage treatment works without a tertiary treatment facility that discharge the effluent to an adjoining river system. This could only be controlled if such a facility was installed and the cost effectiveness of doing this would have to be carefully assessed. However, recent evidence from 54 UK river catchments indicates that point sources of P from even small settlements, probably via sewage treatment works, provide a greater risk of river eutrophication than diffuse sources from agricultural land even where this has a large positive P balance (Jarvie *et al.* 2006). White and Hammond (2006) recently reassessed the data for the total P (TP) load on the waters of England, Wales and Scotland. Of the TP load, estimated to be between 41.6 and 51.1 kt year^{-1}, the contribution from agricultural land could be 22–28%. The larger TP loads, and lesser contribution from agriculture, assumed an annual discharge of 0.61 kg TP per capita from sewage treatment works.

SOME ECOLOGICAL ASPECTS TO THE USE OF P FERTILIZERS

There is considerable evidence that intensification of agriculture and the accompanying increase in soil fertility on sites previously of conservation value can lead to a loss of biological diversity because of the domination of invasive, vigorous plant species at the expense of existing, often slower growing species (Marrs 1993). In terms of fertilizer application the effects can depend in particular on changes in soil pH as well as increasing supplies of N, P and K. Interest, however, has focused mainly on P because the soils in many natural ecosystems contain much lower concentrations of this nutrient than are found in well-managed agricultural soils.

Withers *et al.* (2005) presented data from Critchley *et al.* (2002) for species richness measured in relation to plant-available soil P (Olsen P) taken from a survey of 571 grassland sites representing a wide rage of temperate grasslands in Environmentally Sensitive Areas in England, where fertilizer input would be low. Species rich grassland and high-value semi-natural grassland with large numbers of species well able to tolerate stress were found to occur mainly on soils with between 4–15 mg L^{-1} Olsen P. Grasslands with soil Olsen P less than 4 mg L^{-1} or greater than 15 mg L^{-1} were species poor and contained fewer stress-tolerant species. However, although, the maximum species abundance declines exponentially from 45–50 species m^{-2} to 5–10 species m^{-2} as Olsen P increases from 2 to 70 mg L^{-1}, species abundance ranges from less than 5 species m^{-2} to more than 40 species m^{-2} on soils with 4–12 mg P L^{-1}, and there are many more sites with species abundances between 5 and 10 species m^{-2} on soils with less than 10 mg P L^{-1} than on soils with more than 20 mg P L^{-1}. Thus, many sites with little plant-available P are species poor and this suggests that species richness in grassland swards is influenced by factors other than the level of Olsen P. Therefore, the ecological effects of P fertilizer need be considered additionally in relation to other components of soil fertility and nutrient availability.

Data from the Park Grass Experiment at Rothamsted provide convincing evidence for this view. This experiment, started in 1856 on a grass sward that had been established at least 200 years previously, can be described as the longest-running experiment in plant ecology anywhere in the world. Species richness on plots where Olsen P has been built up since the start of experiment can be compared directly with that on plots that have received no P. As well as P, other experimental inputs have included N and K as fertilizers and organic manures applied annually, with lime applied occasionally. These treatments have lead to an increase in plant biomass and where N was applied as ammonium sulfate to substantial decreases in soil pH. The soil is a silty clay loam and had a pH of about 5.5 in the 1850s. Each year, the sward is cut for hay in June/July and the aftermath cut for silage in September/October. It is never grazed. Thus, fertilizers have been the only nutrient additions other than those from the atmosphere. On some plots the fertilizer inputs have remained unchanged since 1856. Fertilizers are applied once each year, with 33 kg P ha^{-1} and 225 kg K ha^{-1} being applied in late winter. Where P and K are applied the plant available P and K in soil are well above the yield limiting value (~25 mg kg^{-1} Olsen P). Where no P or K is applied, plant available P and K in soil are very small (2–5 mg kg^{-1} Olsen P). Nitrogen is applied in early spring, and a comparison is made of 48 and 96 kg N ha^{-1}, applied as ammonium sulfate and sodium nitrate. The ammonium sulfate has greatly acidified the soil, the pH is now 3.7. A test of liming was first introduced in 1903, and a more comprehensive one in 1965. The latter aims to achieve and maintain pH values of 7, 6 and 5 on three subplots while that on a fourth subplot depends on the acidifying effect of the various inputs. Each year from 1991 to 2000 individual species of grasses, forbs and legumes were separated and their contribution to dry matter yield in swards from each experimental plot, was made just prior to them being cut for hay (Table 9.7; Crawley *et al.* 2005). Compared to with untreated plots, applying only P or PNaMg fertilizers decreased species richness only slightly, from 39 to 35 species

Table 9.7 Effects of soil phosphorus and acidity and applied nitrogen fertilizer on the total number of species identified in, and (in parentheses) the number of species contributing 10% or more to the total dry matter yield, of a permanent grass sward, Park Grass, Rothamsted. Average of ten years data, between 1991 and 2000. (Crawley *et al.* 2005.)

Treatment	Soil pH	Number of species	Soil pH	Number of species
None	7.2	39 (5)	5.2	36 (3)
P/PNaMg	7.0	35 (4)	5.2	30 (3)
N*1	7.1	32 (4)	5.8	34 (4)
N*1PK	6.7	25 (5)	5.4	27 (2)
N1	7.1	33 (4)	4.1	10 (2)
N2P	6.9	22 (3)	3.7	10 (2)
N2PK	6.9	22 (5)	3.7	3 (3)

N*1 = 48 kg N ha^{-1} as sodium nitrate
N1 = 48 kg N ha^{-1} as ammonium sulfate
N2 = 96 kg N ha^{-1} as ammonium sulfate
P = 33 kg P ha^{-1} each year since 1856
K = 225 kg K ha^{-1} each year since 1856

$plot^{-1}$ on soils with pH about 7 and from 36 to 30 species $plot^{-1}$ on soils with pH 5.2. In both cases the number of dominant species was small. Species richness was decreased from 39 to about 32 species $plot^{-1}$ where only N (48 kg ha^{-1}) was applied either as sodium nitrate or ammonium sulfate and soil pH maintained at about 7. But species richness declined to 10 species $plot^{-1}$ where the application of 48 kg N ha^{-1} as ammonium sulfate decreased soil pH to 4.1. Applying N at 96 kg ha^{-1} with P or PK fertilizer showed the greatest reduction of species richness, irrespective of whether the N was applied as sodium nitrate or as ammonium sulfate. Species richness declined to a little over 20 species $plot^{-1}$ where soil pH was maintained at 6.9, but declined to 10 species $plot^{-1}$ with P fertilizer and 3 species $plot^{-1}$ with PK fertilizer where soil pH was 3.7. The data from this experiment suggest that small amounts of applied N fertilizer and a high level of plant-available soil P (200 mg kg^{-1} Olsen P) cause only a small decline in species richness in a permanent grass sward that is only harvested for hay in June/July with any subsequent growth removed in late autumn. It is the combination of N and P or PK fertilizer, especially on acid soils, that markedly reduces species richness.

An interesting feature of this set of data from the Park Grass Experiment are the changes in the species that contribute 10% or more to the dry matter yield as a result of the differences in nutrient inputs on the most acid soils for each treatment. On the untreated soil with pH 5.2 there are 36 species present but only two contribute more than 10% to dry matter yield. These are *Agrostis capillaries* (45%) and *Festuca rubra* (30%). On the ammonium sulfate treatments where only N is applied and soil pH is 4.1 there are 10 species $plot^{-1}$ and two dominant species, *A. capillaries* which contributes 65% of the biomass and *Anthoxanthum odoratum* which contributes 30% of the biomass. Applying P in addition to N as ammonium sulfate and allowing soil pH to fall to 3.7 also results in 10 species $plot^{-1}$ but *A. capillaris* now contributes only 30% and *A. odoratum* 70% of the biomass. On the treatment where 144 kg N ha^{-1} is given with PK, soil pH is still 3.7 but the sward is 100% *Holcus lanatus*.

SUMMARY

The global demand for P fertilizers to produce crops for food for man and animals is currently about 36.8 Mt P_2O_5 and is expected to increase by 2.6% per annum over the next five years. About 85% of all PR mined is processed for use in agriculture. Estimates vary as to how long known PR reserves will last, but it is universally recognized that they are finite and that this precious resource must be used as efficiently as possible.

It is now envisaged that soil P is held by a continuum of bonding energies to the surfaces of and within the matrix of soil components. Associated with the varying strengths of bonding, soil P can be considered to exist in four P pools of contrasting availability to plants (Figure 9.2). Much evidence from field experiments shows that there is a reversible transfer of P between the first three pools. Phosphorus removed from the soil solution by uptake by plant roots is replenished by P from the readily available pool. The widely and long held view that a large proportion of applied fertilizer P becomes irreversibly fixed in the soil is no longer tenable. Furthermore, the efficiency with which fertilizer P is acquired by crops over the long term is much higher than is generally appreciated.

In very broad terms, soils used for agriculture may be divided into those adequately supplied with P and managed intensively, as in many developed countries, and those with too little plant-available P on which extensive agriculture is practiced, as in many developing countries. The concept of exchanging pools of soil P relates well to the known responses of crops to soil and fertilizer P. It also allows the concept of critical levels of plant available P related to soil type and farming system to be developed, which is particularly useful for the management of intensive agricultural systems.

The need for crops to utilize P more efficiently has led to a re-examination of the way in which plants adapt, and even thrive, on low-P soils. The morphological, physiological, biochemical and molecular responses of plants to environments with low P availability are being examined by scientists of various disciplines. These responses have been successfully exploited to improve the use of soil and fertilizer P in crop nutrition in soils of both adequate and low P availability.

Environmental issues relating to the use of P fertilizers concern the fate of radionuclide and Cd impurities and P losses from soils to watercourses. It is suggested that improved management practices, the development of P-efficient crops and the implementation of Codes of Good Practice for agriculture should assuage concerns that use of P fertilizers may affect the environment detrimentally.

REFERENCES

Ae N, Arihara J, Okada K, Yoshihara T, Johansen C (1990) Phosphorus uptake by pigeon pea and its role in cropping systems of the Indian subcontinent. Science 248: 477–480

Amtmann A, Hammond JP, Armengaud P, White PJ (2006) Nutrient sensing and signalling in plants: potassium and phosphorus. Adv Bot Res 43: 209–257

Barber SA (1979) Soil phosphorus after 25 years of cropping with five rates of phosphorus application. Commun Soil Sci Plant Anal 10: 1459–1468

Barber SA (1995) Soil Nutrient Bioavailability: A Mechanistic Approach. Wiley, New York

Barrow NJ (1980) Evaluation and utilisation of residual phosphorus in soils. In: Khasawneh FE, Sample EC, Kamprath EJ (eds), The Role of Phosphorus in Agriculture. ASA/CSSA/SSSA, Madison, WI, pp 333–359

Barrow NJ (1983a) A mechanistic model for describing the sorption and desorption of phosphate by soil. J Soil Sci 34: 733–750

Barrow NJ (1983b) On the reversibility of phosphate sorption by soils. J Soil Sci 34: 751–758

Bates TR, Lynch JP (1996) Stimulation of root hair elongation in *Arabidopsis thaliana* by low phosphorus availability. Plant Cell Environ 19: 529–538

Bates TR, Lynch JP (2001) Root hairs confer a competitive advantage under low phosphorus availability. Plant Soil 236: 243–250

Batjes NH (1997) A world data set of derived properties by FAO-UNESCO soil unit for global modelling. Soil Use Manag 13: 9–16

Beck MA, Sanchez PA (1996) Soil P budget and movement after 13 years of fertilized cultivation in the Amazon basin. Plant Soil 184: 23–31

Blake L, Johnston AE, Poulton PR, Goulding KWT (2003) Phosphorus fractions following positive and negative phosphorus balances for long periods. Plant Soil 254: 245–261

Bundy LG, Tunney H, Halvorson AD (2005) Agronomic aspects of phosphorus management. In: Sims JT, Sharpley AN (eds), Phosphorus: Agriculture and the Environment. Agronomy Monograph 46, ASA/CSSA/SSSA, Madison, WI, pp 685–728

Cassman KG, Peng S, Olk DC, Ladha JK, Reichardt W, Dobermann A, Singh U (1998) Opportunities for increasing nitrogen use efficiency from improved resource management in irrigated rice systems. Field Crop Res 56: 7–38

Chang SC, Jackson ML (1958) Soil phosphorus fractionation in some representative soils. J Soil Sci 9: 109–119

Chen Y, Barak P (1982) Iron nutrition of plants in calcareous soils. Adv Agron 35: 217–240

Claassen N, Syring KM, Jungk A (1986) Verification of a mathematical model by simulating potassium uptake from soil. Plant Soil 95: 209–220

Coleman R (1942) Utilization of adsorbed phosphate by cotton and oats. Soil Sci 54: 237–246

Crawley MJ, Johnston AE, Silvertown J, Dodd M, De Mazancourt C, Heard MS, Henman DF, Edwards GR (2005) Determinants of species richness in the Park Grass experiment. Am Nat 165: 179–192

Critchley CNR, Chambers BJ, Fowbert JA, Bhogal A, Rose SC, Sanderson RA (2002) Plant species richness, functional type and soil properties of grasslands and allied vegetation in English Environmentally Sensitive Areas. Grass Forage Sci 57: 82–92

Crowther EM, Warren RG, Nagelschmidt G, Cooke EH (1951) The production and agricultural value of silicophosphate fertilizers. Part V. Laboratory and pot culture experiments. Permanent records of research and development. Monograph 11, Ministry of Supply, London, pp 108–283

Curtin D, Syers JK (2001) Lime-induced changes in indices of soil phosphate availability. Soil Sci Soc Am J 65: 147–152

Dean LA (1938) An attempted fractionation of soil phosphorus. J Agric Sci 28: 234–246

Dechassa N, Schenk MK, Claassen N, Steingrobe B (2003) Phosphorus efficiency of cabbage (*Brassica oleraceae* L. var. *capitata*), carrot (*Daucus carota* L.), and potato (*Solanum tuberosum* L.). Plant Soil 250: 215–224

De Kok LJ, Schnug E (2007) (eds) Protecting Water Bodies from Negative Impact of Agricultural Loads and Fate of Fertilizer Derived Uranium. Backhuys, BV Leiden, The Netherlands

Delhaize E, Ryan PR, Randall PJ (1993) Aluminium tolerance in wheat (*Triticum aestivum* L.) II Aluminium-stimulated excretion of malic acid from root apices. Plant Physiol 103: 695–702

Dinkelaker B, Römheld V, Marschner H (1989) Citric acid excretion and precipitation of calcium citrate in the rhizosphere of white lupin (*Lupinus albus* L) Plant Cell Environ 12: 285–292

Drew MC (1975) Comparison of the effects of a localized supply of phosphate, nitrate, ammonium and potassium on the growth of the seminal root system, and the shoot, in barley. New Phytol 75: 479–490

Dyer B (1894) On the analytical determination of probably availably "mineral" plant food in soils. J Chem Soc Trans 65: 115–167
Dyer B (1902) Results of investigations on the Rothamsted soils. USDA Bull Off Exp Stns 106
Evans TD, Syers JK (1971) Application of autoradiography to study the fate of 33P labelled orthophosphate added to soil crumbs. Soil Sci Soc Am Proc 35: 906–909
Fairhurst TR, Lefroy R, Mutert E, Batjes N (1999) The importance, distribution and causes of phosphorus deficiency as a constraint in crop production in the tropics. Agrofor Forum 9: 2–8
Falk WE, Wymer D (2006) Uranium in phosphate fertilizer production In: Merkel BJ, Hasche-Berger A (eds), Uranium in the Environment. Springer, Berlin/Heidelberg, pp 857–866
Fixen PE (2007) Potential biofuels influence on nutrient use and removal in the U.S. Better Crops 91: 12–14
Föhse D, Jungk A (1983) Influence of phosphate and nitrate supply on root hair formation of rape, spinach and tomato plants. Plant Soil 74: 359–368
Gahoonia TS, Nielsen NE (1997) Variation in root hairs of barley cultivars doubled soil phosphorus uptake. Euphytica 98: 177–182
Gahoonia TS, Nielsen NE (2004) Barley genotypes with long root hairs sustain high grain yields in low-P field. Plant Soil 262: 55–62
Gahoonia TS, Nielsen NE, Lyshede OB (1999) Phosphorus acquisition in the field at three levels of P fertilization. Plant Soil 211: 269–281
Gardner WK, Parbery DG, Barber DA (1982) The acquisition of phosphorus by *Lupinus albus* L. II. The effect of varying phosphorus supply and soil type on some soil characteristics of the soil/root interface. Plant Soil 68:33–41
George TS, Richardson AE (2008) Potential and limitations to improving crops for enhanced phosphorus utilization. In: White PJ, Hammond JP (eds), The Ecophysiology of Plant-Phosphorus Interactions. Springer, Dordrecht, The Netherlands, pp 247–270
Gerke J, Biessner L, Römer W (2000) The quantitative effect of chemical phosphate mobilization by carboxylate anions on P uptake by a single root. 1. The basic concept and determination of soil parameters. J Plant Nutr Soil Sci 163: 207–212
Gillingham AG, Syers JK, Gregg PEH (1980) Phosphorus uptake and return in grazed, steep hill pastures. I. Above-ground components of the phosphorus cycle. New Zeal J Agric Res 23: 323–330
Greenwood DJ, Karpinets TV, Stone DA (2001) Dynamic model for the effects of soil P and fertilizer P on crop growth, P uptake and soil P in arable cropping: model description. Ann Bot 88: 279–291
Greenwood DJ, Stellacci AM, Meacham MC, Broadley MR, White PJ (2005) Phosphorus response components of different *Brassica oleracea* genotypes are reproducible in different environments. Crop Sci 45: 1728–1735
Greenwood DJ, Stellacci AM, Meacham MC, Mead A, Broadley MR, White PJ (2006a) Relative values of physiological parameters of P response of different genotypes can be measured in experiments with only two P treatments. Plant Soil 281: 159–172
Greenwood DJ, Stellacci AM, Meacham MC, Mead A, Broadley MR, White PJ (2006b) *Brassica* cultivars: P response and fertilizer efficient cropping. Acta Hort 700: 97–102
Halvorson AD, Black AL (1985) Long-term dryland crop responses to residual phosphate fertilizers. Soil Sci Soc Am J 49: 928–933
Haynes RJ (1982) Effects of lime on phosphate availability in acid soil: a critical review. Plant Soil 68: 289–308
Haynes RJ (1984) Effect of lime, silicate and phosphate applications on the concentrations of extractable aluminium and phosphate in a spodosol. Soil Sci 138: 8–14
Heckrath G, Brookes PC, Poulton PR, Goulding KWT (1995) Phosphorus leaching from soils containing different phosphorus concentrations in the Broadbalk experiment. J Environ Qual 24: 904–910
Heffer P, Prud'homme MPR, Muirhead B, Isherwood KF (2006) Phosphorus Fertilization: Issues and Outlook. Proceedings 586, International Fertiliser Society, York
Hemwall JB (1957) The fixation of phosphorus by soils. Adv Agron 9: 95–112

Hermans C, Hammond JP, White PJ, Verbruggen N (2006) How do plants respond to nutrient shortage by biomass allocation? Trends Plant Sci 11: 610–617
Hetrick BAD (1991) Mycorrhizas and root architecture. Experientia 47: 355–362
Hilton J (2006) Phosphogypsum Management and Opportunities for Use. Proceedings 587, International Fertiliser Society, York
Hinsinger P, Plassard C, Tang CX, Jaillard B (2003) Origins of root-mediated pH changes in the rhizosphere and their responses to environmental constraints: a review. Plant Soil 248: 43–59
Hoffland E (1992) Quantitative evaluation of the role of organic acid exudation in the mobilization of rock phosphate by rape. Plant Soil 140: 279–289
Holford ICR, Scheirtzer BE, Crocker GJ (1994) Long-term effects of lime on soil phosphorus solubility and sorption in eight soils. Aust J Soil Sci 32: 795–803
Horst WJ, Waschkies C (1987) Phosphatversorgung von Sommerweisen (*Triticum aestivum* L) in Mischkultur mit Weisser Lupine (*Lupinus albus* L). Z Pflanz Bodenkunde 150: 1–8
Horst WJ, Kamh M, Jibrin JM, Chude VO (2001) Agronomic measures for increasing P availability to crops. Plant Soil 237: 211–223
Howeler RH, Sieverding E, Saif S (1987) Practical aspects of mycorrhizal technology in some tropical crops and pastures. Plant Soil 100: 249–283
Huffman EO (1962) Reactions of Phosphate in Soils: Recent Research at TVA. Proceedings 71, International Fertiliser Society, York
Huffman EO (1968) Behaviour of fertilizer phosphorus. In: 9th International Congress of Soil Science Transactions II, pp 745–754
Jackson ML (1963) Aluminium bonding in soil: a unifying principle in soil science. Soil Sci Soc Am Proc 27: 1–10
Jarvie HP, Neal C, Withers PJA (2006) Sewage effluent phosphorus: a greater risk to river eutrophication than agricultural phosphorus? Sci Total Environ 360: 246–253
Johnston AE (1975) The Woburn Market Garden Experiment, 1942–1969. II. The effects of treatments on soil pH, soil carbon, nitrogen, phosphorus and potassium. In: Rothamsted Experimental Station Report for 1974, pp 102–132
Johnston AE (1994) The Rothamsted Classical Experiments. In: Leigh RA, Johnston AE (eds), Long-Term Experiments in Agricultural and Ecological Sciences. CAB International, Wallingford, pp 9–37
Johnston AE (2005) Phosphorus nutrition of arable crops. In: Sims JT, Sharpley AN (eds), Phosphorus: Agriculture and the Environment. Agronomy Monograph 46, American Society of Agronomy, Madison, WI, pp 495–520
Johnston AE, Dawson CJ (2005) Phosphorus in Agriculture and in Relation to Water Quality. Agricultural Industries Confederation, Peterborough
Johnston AE, Jones KC (1995) The Origin and Fate of Cadmium in Soil. Proceedings 366, International Fertiliser Society, York, pp 2–40
Johnston AE, Poulton PR (1977) Yields on the Exhaustion Land and changes in the N P K contents of the soil due to cropping and manuring, 1852–1975. In: Rothamsted Experimental Station Report for 1976, Part 2, pp 53–85
Johnston AE, Poulton PR (1992) The role of phosphorus in crop production and soil fertility: 150 years of field experiments at Rothamsted, United Kingdom. In: Schultz JJ (ed), Phosphate Fertilizers and the Environment. IFDC, Muscle Shoals, AL, pp 45–64
Johnston AE, Poulton PR (2005) Soil Organic Matter: Its Importance in Sustainable Agricultural Systems. Proceedings 565, International Fertiliser Society, York
Johnston AE, Syers JK (1998) (eds) Nutrient Management for Sustainable Crop Production in Asia. CAB International, Wallingford
Johnston AE, Syers JK (2006) Changes in Understanding the Behaviour of Soil and Fertiliser Phosphorus: Implications for Their Efficient Use in Agriculture. Proceedings 589, International Fertiliser Society, York
Johnston AE, Poulton PR, Syers JK (2001) Phosphorus, Potassium and Sulphur Cycles in Agricultural Soils. Proceedings 465, International Fertiliser Society, York

Jungk A (1994) Phosphorus supply of plants - how is it accomplished? Proc Natl Sci Council ROC B 18: 187–197
Jungk A (2001) Root hairs and the acquisition of plant nutrients from soil. J Plant Nutr Soil Sci 164: 121–129
Jungk A, Claassen N (1989) Availability in soil and acquisition by plants as the basis for phosphorus and potassium supply to plants. Z Pflanz Bodenkunde 152: 151–157
Kamprath EJ, Watson ME (1980) Conventional soil and tissue tests for assessing phosphorus status of soils. In: The Role of Phosphorus in Agriculture. ASA/CSSA/SSSA, Madison, WI, pp 433–469
Khasawneh FE, Sample EC, Kamprath EJ (1980) (eds) The Role of Phosphorus in Agriculture. American Society of Agronomy, Madison, WI
Kirkby EA (1981) Plant growth in relation to nitrogen supply. In: Clarke FE, Rosswall T (eds), Terrestrial Nitrogen Cycles, Processes, Ecosystem Strategies and Management Impacts. Ecology Bulletins Stockholm 33, pp 249–267
Kirkby EA, Römheld V (2006) Physiological Aspects of Plant Phosphorus in Relation to Its Acquisition from the Soil. Proceedings 588, International Fertiliser Society, York
Kratz S, Schnug E (2006) Rock phosphates and P fertilizers as source of U contamination in agricultural soils. In: Merkel BJ, Hasche-Berger A (eds), Uranium in the Environment. Springer, Berlin, pp 57–68
Kristoffersen AO, Greenwood DJ, Sogn TA, Riley H (2006) Assessment of the dynamic phosphate model PHOSMOD using data from field trials with starter fertilizer to cereals. Nutr Cycling Agroecosyst 74: 75–89
Kurtz LT (1953) Phosphorus in acid and neutral soils. In: Pierre WH, Norman AG (eds), Phosphorus in Crop Nutrition IV. American Society of Agronomy, Academic, New York, pp 59–88
Lambers H, Chapin FS III, Pons TL (1998) Plant Physiological Ecology. Springer, New York
Lambers H, Shane MW, Cramer MD, Pearse SJ, Veneklaas EJ (2006) Root structure and functioning for efficient acquisition of phosphorus: matching morphological and physiological traits. Ann Bot 98: 693–713
Lamont RE, Whitney DA (1991) Evaluation of starter fertilizer for grain sorghum production. J Fert Issues 8: 20–24
Larsen S (1967) Soil phosphorus. Adv Agron 19: 151–210
Leamer RW (1963) residual effects of phosphate fertilizer in an irrigated rotation in the south west. Soil Sci Soc Am Proc 27: 65–68
Le Bayon RC, Weisskopf L, Martinoia E, Jansa J, Frossard E, Keller F, Föllmi KB, Gobat J-M (2006) Soil phosphorus uptake by continuously cropped *Lupinus albus*: a new microcosm design. Plant Soil 283: 309–321
Leigh RA, Johnston AE (1986) An investigation of the usefulness of phosphorus concentrations in tissue water as indicators of the phosphorus status of field grown spring barley. J Agr Sci 107: 329–333
Leikam DF, Achorn FP (2005) Phosphate fertilizers: Production, characteristics, and technologies. In: Sims JT, Sharpley AN (eds), Phosphorus: Agriculture and the Environment. Agronomy Monograph 46, ASA, Madison, WI, pp 23–50
Lindsay WL (1979) Chemical Equilibrium in Soils. Wiley, New York
Lindsay WL, Moreno EC (1960) Phosphate phase equilibria in soils. Soil Sci Soc Am Proc 24: 177–182
Liu C, Muchhal US, Mukatira U, Kononowicz AK, Raghothama KG (1998) Tomato phosphate transporter genes are differentially regulated in plant tissues by phosphorus. Plant Physiol 116: 91–99
Lynch J (1995) Root architecture and plant productivity. Plant Physiol 109: 7–13
Lynch J (2007) Roots of the second green revolution. Aust J Bot 55: 493–512
Lynch JP, Brown KM (2008) Root strategies for phosphorus acquisition. In: White PJ, Hammond JP (eds), The Ecophysiology of Plant-Phosphorus Interactions. Springer, Dordrecht, The Netherlands, pp 83–116

Mackay AD, Barber SA (1985) Soil moisture effects on root growth and phosphorus uptake by corn. Agron J 77: 519–523
Mansell GP, Pringle RM, Edmeades DC, Shannon PW (1984) Effects of lime on pasture production on soils in the North Island of New Zealand. 3. Interaction of lime with phosphorus. New Zeal J Agric Res 27: 363–369
Marrs RH (1993) Soil fertility and nature conservation in Europe: theoretical considerations and practical management solutions. Adv Ecol Res 24: 241–300
Marschner H (1995) Mineral Nutrition of Higher Plants (2nd edition). Academic, London
Marschner P (2008) The effect of rhizosphere microorganisms on P uptake by plants. In: White PJ, Hammond JP (eds), The Ecophysiology of Plant-Phosphorus Interactions. Springer, Dordrecht, The Netherlands, pp 165–176
Mattingly GEG (1970) Residual value of basic slag, Gafsa rock phosphate and superphosphate in a sandy podzol. J Agric Sci 75: 413–418
McCollum RE (1991) Build-up and decline in soil phosphorus: 30-year trends on a Typic Umprabuult. Agron J 83: 77–85
McDowell R, Sharpley A, Brookes P, Poulton P (2001) Relation between soil test phosphorus and phosphorus release to solution. Soil Sci 166: 137–149
McGonigle TP, Miller MH (1999) Winter survival and extraradical hyphae and spores of arbuscular mycorrhizal fungi in the field. Appl Soil Ecol 12: 41–50
McKenzie RH, Roberts TL (1990) Soil and fertilizer phosphorus update. In: Proceedings of the Alberta Soil Science Workshop, pp 84–104
Mengel DB, Barber SA (1974) Rate of nutrient uptake per unit of corn root under field conditions. Agron J 66: 399–402
Mengel K, Kirkby EA (2001) Principles of Plant Nutrition (5th edition). Kluwer, Dordrecht, The Netherlands
Ministry of Agriculture, Fisheries and Food (MAFF) (2000) Fertiliser Recommendations for Agricultural and Horticultural Crops (RB209) (7th edition). HMSO, Norwich
Neeteson JJ, Schröder JJ, Smit AL, Bos JFFP, Verloop J (2006) Need and Opportunities to Reduce Phosphorus Inputs, Soil Supply and Loss from Agriculture in the Netherlands. Proceedings 595, International Fertiliser Society, York, pp 1–23
Neumann G, Römheld V (2006) The release of root exudates as affected by the plant physiological status. In: Pinto R, Varanini Z, Nannipieri P (eds), The Rhizosphere: Biochemistry and Organic Substances at the Soil Plant Interface (2nd edition). CRC Press, Boca Raton, FL, pp 41–94
Nicholson FA, Jones KC, Johnston AE (1994) Effects of phosphate fertilizer and atmospheric deposition on long-term changes in the cadmium content of soils and crops. Environ Sci Technol 28: 2170–2175
Pierre WH, Norman AG (1953) (eds) Soil and Fertilizer Phosphorus in Crop Nutrition. Academic, New York
Posner AM, Barrow NJ (1982) Simplification of a model for ion adsorption on oxide surfaces. J Soil Sci 33: 211–217
Rae AL, Jarmey JM, Mudge SR, Smith FW (2004) Over-expression of a high-affinity phosphate transporter in transgenic barley plants does not enhance phosphate uptake rates. Funct Plant Biol 31: 141–148
Raghothama KG, Karthikeyan AS (2005) Phosphate acquisition. Plant Soil 274: 37–49
Raven JA (2008) Phosphorus and the future. In: White PJ, Hammond JP (eds), The Ecophysiology of Plant-Phosphorus Interactions. Springer, Dordrecht, The Netherlands, pp 271–283
Raven JA, Smith FA (1976) Nitrogen assimilation and transport in vascular land plants in relation to intracellular pH regulation. New Phytol 76: 415–431
Reisenauer HM (1966) Mineral nutrients in soil solution. In: Altman PL, Ditmer DS (eds), Environmental Biology. Fed Am Soc Exp Biol, Bethesda, pp 507–508
Rengel Z, Marschner P (2005) Nutrient availability and management in the rhizosphere: exploiting genotypic differences. New Phytol 168: 305–312

Roelofs RFR, Rengel Z, Cawthray GR, Dixon KW, Lambers H (2001) Exudation of carboxylates in Australian Proteaceae: chemical composition. Plant Cell Environ 24: 891–904
Rothbaum HP, McGaveston DA, Wall T, Johnston AE, Mattingly GEG (1979) Uranium accumulation in soils from long-continued applications of superphosphate. J Soil Sci 30: 147–153
Runge-Metzger A (1995) Closing the cycle: obstacles to efficient P management for improved global security. In: Tiessen H (ed), Phosphorus in the Global Environment: Transfers, Cycles And Management. Wiley, Chichester, pp 27–42
Russell EJ (1912) Soil Conditions and Plant Growth. Longmans Green, London
Rutherford PJ, Dudas MJ, Samek RA (1994) Environmental impacts of phosphogypsum. Sci Total Environ 149: 1–38
Ryan PR, Delhaize E, Jones DL (2001) Function and mechanism of organic anion exudation from plant roots. Annu Rev Plant Physiol Plant Mol Biol 52: 527–560
Sample EC, Soper RJ, Racz GJ (1980) Reactions of phosphate fertilizers in soils. In: Khasawneh FE, Sample EC, Kamprath EJ (eds), The Role of Phosphorus in Agriculture. ASA/CSSA/SSSA, Madison, WI, pp 263–310
Sanchez CA (2007) Phosphorus. In: Barker AV, Pilbeam DJ (eds), Handbook of Plant Nutrition. CRC Press, Boca Raton, FL, pp 51–90
Sas L, Rengel Z, Tang C (2001) Excess cation uptake and extrusion of protons and organic anions by *Lupinus albus* under phosphorus deficiency. Plant Sci 160: 1191–1198
Schachtman DP, Reid RJ, Ayling SM (1998) Phosphorus uptake by plants: from soil to cell. Plant Physiol 116: 447–453
Schofield RK (1955) Can a precise meaning be given to "available" soil phosphorus? Soil Fert 28: 373–375
Scholten LC, Timmermans CWM (1996) Natural radioactivity in phosphate fertilizers. Fert Res 43: 103–107
Schweiger PF, Robson AD, Barrow NJ (1995) Root hair length determines beneficial effect of *Glomus* species on shoot growth. New Phytol 231: 247–254
Sewell PL, Ozanne PG (1970) The effect of modifying root profiles and fertilizer solubility on nutrient uptake. In: Miller TC (ed), Proceedings of the Australian Plant Nutrition Conference, Mt Gambier, CSIRO, Australia, pp 6–9
Shane MW, McCully ME, Lambers H (2004) Tissue and cellular phosphorus storage during development of phosphorus toxicity in *Hakea prostrata* (Proteaceae). J Exp Bot 55: 1033–1044
Silberbush M, Barber SA (1983) Sensitivity of simulated phosphorus uptake to parameters used by a mechanistic-mathematical model. Plant Soil 74: 93–100
Singh BR, Lal R (2005) Phosphorus management in low-input agricultural systems. In: Sims JT, Sharpley AN (eds), Phosphorus: Agriculture and the Environment, Agronomy Monograph 46, ASA/CSSA/SSSA, Madison, WI, pp 729–759
Singh DK, Sale PWG, Routley RR (2005) Increasing phosphorus supply in subsurface soil in northern Australia: rationale for deep placement and the effects with various crops. Plant Soil 269: 35–44
Smith SE, Read DJ (1997) Mycorrhizal Symbiosis. Academic, San Diego, CA
Steén IE (2006) Phosphorus for Livestock: Requirements: Efficient Use and Excretion. Proceedings 594, International Fertiliser Society, York
Stockdale EA, Watson CA, Edwards AC (2006) Phosphate Rock: Using Biological Processes to Increase Its Effectiveness as a Fertiliser. Proceedings 592, International Fertiliser Society York
Stone DA (1998) The effect of starter fertilizer injection on the growth and yield of drilled vegetable crops in relation to soil nutrient status. J Hort Sci Biotech 73: 441–445
Sumner ME, Farina MPW (1986) Phosphorus interactions with other nutrients and lime in field cropping systems. Adv Soil Sci 5: 201–236
Syers JK, Johnston AE, Curtis D (2007) Efficiency of Soil and Fertilizer Phosphorus Use: Reconciling Changing Concepts of Soil Phosphorus Behaviour with Agronomic Information. FAO, Vienna

Thomson CJ, Marschner H, Römheld V (1993) Effect of nitrogen fertilizer form on pH of the bulk soil and rhizosphere and on the growth phosphorus and micronutrient uptake of bean. J Plant Nutr 16: 493–506

Vance CP (2001) Symbiotic nitrogen fixation and phosphorus acquisition. Plant nutrition in a world of declining renewable resources. Plant Physiol 127: 390–397

Vance CP, Uhde-Stone C, Allan DL (2003) Phosphorus acquisition and use: critical adaptations by plants for securing a nonrenewable resource. New Phytol 157: 423–447

Van Kauwenbergh SJ (1997) Cadmium and Other Minor Elements in World Resources of Phosphate Rock. Proceedings 400, International Fertiliser Society, York

von Uexküll HR, Mutert E (1995) Global extent, development and economic impact of acid soils. Plant Soil 171: 1–15

Way JT (1850) On the power of soils to absorb manure. J Roy Agric Soc England 11: 313–379

Werner W, Scherer HW (1995) Quantity/intensity relation and phosphorus availability in south Brasilian latersols as affected by form and placement of phosphate and farmyard manure. In: Data *et al.* (eds), Plant Soil Interactions at Low pH. Kluwer, Dordrecht, The Netherlands, pp 129–133

White PJ, Hammond JP (2006) Updating the estimate of the sources of phosphorus in UK waters. Final Report on Defra project WT0701CSF.

White PJ, Hammond JP (2008) Phosphorus nutrition of terrestrial plants. In: White PJ, Hammond JP (eds), The Ecophysiology of Plant-Phosphorus Interactions. Springer, Dordrecht, The Netherlands, pp 51–81

White PJ, Broadley MR, Greenwood DJ, Hammond JP (2005) Genetic Modifications to Improve Phosphorus Acquisition by Roots. Proceedings 568, International Fertiliser Society, York

Wissuwa M (2003) How do plants achieve tolerance to phosphorus deficiency? Small causes with big effects. Plant Physiol 133: 1947–1958

Wissuwa M, Ae N (2001) Genotypic variation for tolerance to phosphorus deficiency in rice and the potential for its exploitation in rice improvement. Plant Breeding 120: 43–48

Withers PJA, Nash DM, Laboski CAM (2005) Environmental management of phosphorus fertilizers. In: Sims JT, Sharpley AN (eds), Phosphorus: Agriculture and the Environment. Agronomy Monograph 46, ASA/CSSA/SSSA, Madison, WI, pp 781–828

Wu QT, Morel JL, Guckert A (1989) Effect of nitrogen source on cadmium uptake by plants. Compt Rend Acad Sci 309: 215–220

Yan Y, Wu P, Ling H, Xu G, Xu F, Zhang Q (2006) Plant nutriomics in China: an overview. Ann Bot 98: 473–482

Zhang K, Greenwood DJ, White PJ, Burns IG (2007) A dynamic model for the combined effects of N, P and K fertilizers on yield and mineral composition; description and experimental test. Plant Soil 298: 81–98

Chapter 10
DIAGNOSING PHOSPHORUS DEFICIENCY IN CROP PLANTS

John P. Hammond and Philip J. White

DEFINING PLANT PHOSPHORUS AVAILABILITY

In addition to carbon, hydrogen and oxygen, plants require at least 11 mineral elements to complete their life cycle, with another four mineral elements being potentially beneficial (Marschner 1995). The supply of three of these elements, nitrogen (N), phosphorus (P), and potassium (K) is often such that it limits the growth of plants in agricultural systems. To avoid yield losses and/or poor crop quality, the supply of these elements is ensured by the use of fertilizers. The supply of these elements by fertilizers must be optimized for the crop and its growth conditions. This optimization is necessary to avoid supplying too much of an element, which could have a negative affect on the crop yield or the local environment, or supplying too little of an element, which could prevent the crop from reaching its maximum potential yield. Typically, increasing the supply of a limiting element will increase the maximum yield up to a point (Figure 10.1). After this point, the maximum yield will remain constant (or even decline again) with further increases in the supply of the element. Therefore, supplying elements in excess of the optimum becomes uneconomic, since no extra yield will be realized for any additional input. Understanding the nutritional status of the crop is therefore critical to optimizing the supply of these elements. This chapter will review the potential of current and future techniques for establishing the P requirements of crop plants.

Phosphorus in the soil-plant continuum

Not only is P an essential macronutrient required by all living organisms, but it is also one of the most unavailable and inaccessible nutrients present in the soil. Thus establishing a plant's requirement for P, and the ability of the soil to supply this requirement, is crucial in managing cropping systems (Holford 1997; Jain *et al.* 2007). Phosphorus is acquired by plants, in the form of inorganic phosphate (Pi), from the soil solution via their roots. The amount of Pi available to plants in the soil solution is determined by the interaction between the Pi in the soil solution, inorganic

P.J. White, J.P. Hammond (eds.) *The Ecophysiology of Plant-Phosphorus Interactions*,

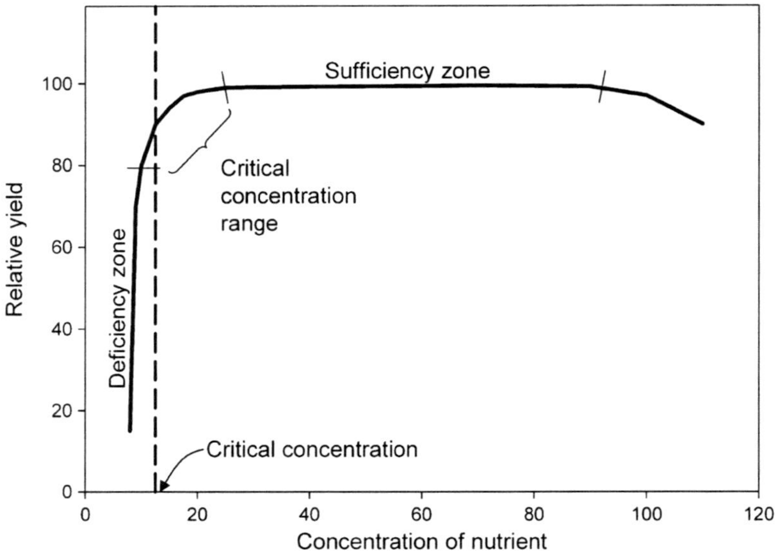

Fig. 10.1 Relationship between the relative crop yield and the concentration of the nutrient in the plant tissue or being supplied to the plant

and organic P compounds in the soil and by the forms of Pi taken up from the soil solution by plants (Raghothama 1999; Figure 10.2). The interactions between these P pools within the soil are governed by the dissolution of P from phosphate rocks, the solubilization and precipitation of P with Ca carbonates, Al and Fe hydrous oxides and Al silicates and sorption onto and desorption of P from the surfaces of soil and clay particles (Frossard *et al.* 1995; Richardson 2001; Kirkby and Johnston 2008; Figure 10.2).

Uptake of Pi by plants from the soil solution, primarily in the form of $H_2PO_4^-$ (Ullrich-Eberius *et al.* 1981), causes an imbalance in the equilibrium between the different pools of P within the soil (Figure 10.2). This imbalance quickly results in the depletion of Pi from the rhizosphere, leading to the formation of a concentration gradient of Pi from the bulk soil to the root. Desorption of Pi from soil particles and minerals acts to replenish soil solution Pi (Jungk *et al.* 1993). However, this occurs slowly and a depletion zone forms around the roots of plants. The low solubility of many P salts results in a poor supply of Pi to the plant roots, potentially leading to P deficiency in the plant.

In many plants, a prolonged reduction in the availability of Pi reduces their yield potential. However, plants have evolved to cope with low P availability, and the mechanisms they have developed act both to increase acquisition of P from the soil and to improve its use internally (George and Richardson 2008; Lynch and Brown 2008; Vance 2008; White and Hammond 2008). Visually, P deficiency results in a reduced stature, acute leaf angles, suppression of tillering, prolonged dormancy, early senescence and decreased size and number of flowers and buds (Bould *et al.*

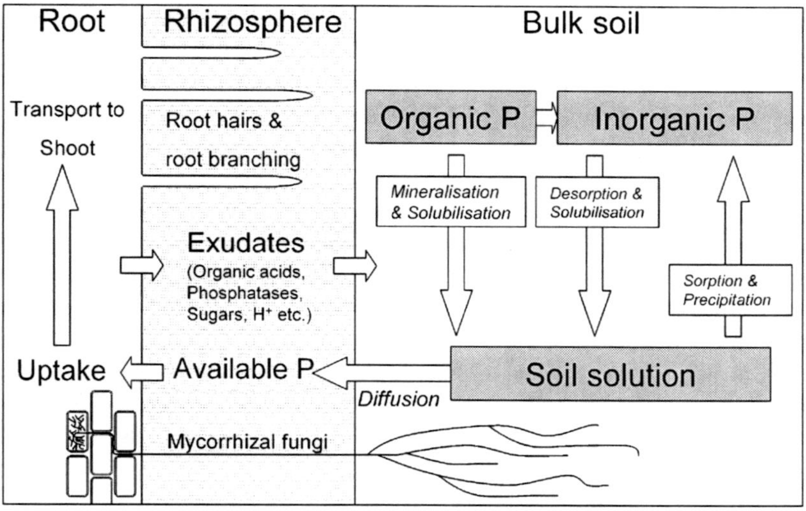

Fig. 10.2 Interactions between the organic, inorganic and soil solution pools of phosphorus (P) in the bulk soil and rhizosphere interactions with the plant root

1983; Bergmann 1992; Marschner 1995; Mengel and Kirkby 2001). Symptoms of P deficiency occur first in older leaves. The development of dark green or blue-green foliage is among the first symptoms of P deficiency. Red, purple or brown pigments develop in leaves, especially along veins. This is a consequence of anthocyanin production, which is induced by increased leaf sucrose concentrations (Müller *et al.* 2005; Teng *et al.* 2005; Amtmann *et al.* 2006; Solfanelli *et al.* 2006) and is thought to protect nucleic acids from UV damage and chloroplasts from photoinhibitory damage caused by P-limited photosynthesis (Hoch *et al.* 2001). The majority of these adaptations are not observable until P deficiency is advanced. Consequently, remedial action to correct advanced P deficiency, diagnosed by visual symptoms, is often too late to overcome the yield reduction. Thus, *a priori* knowledge of the soils ability to supply adequate amounts of P or the P status of the crop is required to manage crop P nutrition effectively through the application of inorganic Pi fertilizers. This maintains an adequate supply of Pi to the roots of crop plants, thereby ensuring crop yield and quality is maintained.

Worldwide, the use of P fertilizers in agriculture accounts for 80–90% of mineral P use (Bieleski and Ferguson 1983; Steen 1998). The non-renewable nature of P means that economically recoverable sources of phosphate rocks will be exhausted within the next 50 to 150 years (Runge-Metzger 1995; Denison and Kiers 2005; Cohen 2007). A compounding factor in the excessive use of P fertilizers is the nature of the P-soil interactions. It is estimated that up to 80% of applied P is bound rapidly to the soil (Holford 1997). To compensate for this, farmers frequently apply P fertilizers in excess of the recommended application to ensure crop production (Goldstein 1992). This can often lead to P in the soil exceeding the requirements of the crop. This situation is particularly common in the USA and Western Europe, where there is great concern that P from agriculture may reach watercourses and lead to the processes of eutrophication (Sharpley 1995; Raghothama 1999; Withers *et al.* 2001; White and Hammond 2006).

Consequently, to prevent damage to the environment, and to reduce the consumption of this non-renewable resource, improved management of P fertilizer use must be implemented. However, modifications in agricultural practices to minimize the impact on the environment must allow economically viable farming to continue (Higgs *et al.* 2000). The management of P fertilizer use at the farm-level seeks to reduce the use of P fertilizers, preventing excess P from entering the environment. This can be achieved by growing crops that require less P fertilization (i.e. crops that can acquire or use P more efficiently), and the use of cultural techniques to maintain P levels in the field, such as crop rotation and intercropping (Frossard *et al.* 2000; Vance 2001).

Modeling plant responses to phosphorus supply

A better understanding of how plants respond to P availability and the factors affecting the availability of soil P to plants will help prevent excess additions of P fertilizers, allowing a balance to be reached whereby the available P matches the crop's requirements (Frossard *et al.* 2000). Key components to these scientific endeavors are the mathematical and statistical approaches used to describe the responses of plants to increasing supplies of P. Typically, these describe mathematically how the yield of the crop changes with increasing availability of P (or any other element) and are commonly fitted to data using a least squares approach. Many equations have been proposed to describe crop yields as a function of the availability of mineral elements, but a full description of these is beyond the scope of this chapter; the reader is referred to Black (1993), which provides a comprehensive coverage of this topic. In the embryonic years of agricultural chemistry, the 'law of the minimum' emerged as the first theory describing plant development in relation to the availability of mineral elements (Sprengel 1828, 1837; Liebig 1855). The 'law of the minimum' states that plant growth is limited by a single factor at any one time. Thus, crop yield would increase as a straight line, from the control yield, with the addition of fertilizer, up to a point when another factor, for example another element or water availability, became limiting. Further additions of the fertilizer would not give rise to any further increase in yield, only an increase in the availability of the next limiting factor would cause an increase in yield. This constant-return relationship, in which increase in yield per unit of added fertilizer is constant, has been observed for the yield responses of some crops to the supply of certain mineral elements (Pinkerton 1991; Rubio *et al.* 2003), but more commonly a curved relationship is observed in which increases in yield decline as increasing units of fertilizer are added (Black 1993). These relationships have a precise upper limit. Mitscherlich (1909) was one of the first to consider these curved responses and applied an exponential growth equation to describe his data on oat yields in response to P fertilizer applications. Since then many mathematical equations have been proposed to describe how a crop responds to the addition of fertilizers. These include many modifications of Mitscherlich's exponential growth equation, polynomials (e.g. linear or quadratic), inverse polynomials, sigmoidal, hyperbola and linear-plateau or split-line equations (Greenwood *et al.* 1971, 1980, 2001; Mead and Pike 1975; Black 1993; Claassen and Steingrobe 1999; Rayment 2005). Mitscherlich's equation and the polynomial equations have been used widely due to the ease with which they can be fitted to data using common computer packages, but they can overestimate yields (Rayment 2005). In contrast the linear-plateau and split-line equations, which relate to the law of the minimum, require an iterative process to resolve and fit. These generally give lower estimates of optimal fertilizer rates when compared to polynomial equations, but require larger data sets for fitting, which has also limited their use (Rayment 2005).

These equations form the basis for many empirical (and semi-mechanistic) models used to predict the yield potential for given inputs of nutrients. These models can be used to determine the optimal fertilizer applications required to obtain the maximum yield, whilst minimizing the impact of excess fertilizer applications on the environment. An alternative to these empirical models are mechanistic models, which aim to represent the physical, chemical, and biological processes that P, in all its forms, undergoes both spatially and temporally in the soil-plant continuum. However, only a few of these mechanistic models have been developed sufficiently to account for most of these process and are capable of estimating crop performance relative to soil or plant test results (Bar-Yosef 2003; Karpinets *et al.* 2004).

These models are dependent on accurate information on the crop, weather, soil, and current availability of P to the plant. Chemical analyses of soil and plant material are most commonly used to assess the P fertilizer requirements of a field site, but new technologies might offer more precise alternatives in the future (Campbell 1994; van Raij 1998; Smethurst 2000; Hammond *et al.* 2003; Clark *et al.* 2005).

ANALYSIS OF SOIL SAMPLES TO PREDICT CROP P STATUS

The regression of soil nutrient content against crop yield, using the equations mentioned above, can be used empirically to determine the nutrient levels required to obtain maximum yields. The use of soil analysis in this way has been the basis of fertilizer recommendations for many decades (Smethurst 2000). In terms of P, the most important concept is the availability of P to plants. This concept was originally proposed by Dyer (1894) in his examination of soil from Hoos field, Rothamsted. Dyer ascertained that the P in a soil, as determined by extraction with strong acids, was not necessarily all available to the plant. This is because plants take up Pi from the soil solution, and when depleted, this is replenished by the exchangeable P held in the soil (Black 1993; Holford 1997). These two soil properties, the concentration of Pi in the soil solution and the exchangeable P in the soil, are related to each other by the buffering capacity of the soil, which defines the ability of the soil solution Pi concentration to increase in response to added fertilizer or replenishment from exchangeable P pool or to decrease following plant P uptake (Holford 1997). Thus, to best describe a soil's ability to supply P to plants, estimates of the soil solution Pi concentration, exchangeable P and buffering capacity of the soil is needed (Black 1993). Determining these parameters in routine soil testing are not practical (Sibbesen and Sharpley 1997), but a range of extractants and procedures have been developed to estimate these soil parameters and provide useful approximations of a soil's ability to supply P to a crop.

Soil test extractants for P

In his pioneering study, Dyer (1894) hypothesized that since 'plants help themselves to a part of their mineral food' an extraction solution with a similar pH to root sap, in this case a 1% solution of citric acid, would be a good indicator of the P available to plants (Dyer 1894). Since then, many soil analysis procedures have been developed to measure the amount of Pi in the soil that is available for uptake by plants using analogous approaches (Table 10.1). The majority of soil extractants for P contain either dilutions of strong acids, weak acids, complexing ions, or bicarbonate ions. Dilutions of strong acids such as, HCl, HNO_3 and H_2SO_4, principally act as solvents, providing sufficient H ions to dissolve primarily Ca phosphates, but also have the ability to solubilize some Al and Fe phosphates (Thomas and Peaslee 1973). Weak acids, including citrate, lactate and acetic acid are likely to act by anion replacement, replacing Pi absorbed to Ca, Al, or Fe. They also help reduce or prevent the re-absorption of Pi as do sulfate ions (Olsen *et al.* 1954; Thomas and Peaslee 1973; Kamprath and Watson 1980). Complexing ions, such as fluoride, are effective at complexing Al ions, thus releasing Pi from Al-Pi (Chang and Jackson 1957) and also precipitating calcium to release Pi from Ca-Pi (Thomas and Peaslee 1973). Finally, solutions containing bicarbonate ions act through the hydrolysis of Al-Pi and Fe-Pi with OH ions to release Pi, and also through the precipitation of Ca as $CaCO_3$, thus lowering the Ca activity of the soil and making the Ca-Pi more extractable (Thomas and Peaslee 1973). Since these extractants target different species of P in the soil, the use of some extractants is more appropriate on specific soil types. For example, the Olsen method was developed for use on calcareous soils, as the bicarbonate ion is effective at neutralizing Ca and, although it can be used on acidic soils, the results correlate less well with crop yield (Bray and Kurtz 1945; Kamprath and Watson 1980; Fixen and Grove 1990). Acid extractants containing NH_4F are less well suited to calcareous soils, as the soluble P may be precipitated by CaF_2 (Smillie and Syers 1972). Acid based extractants have also been shown to overestimate the P available in rock phosphate amended soils (Nuernberg *et al.* 1998), or underestimate the P available in soils with high clay or hydrous Al or Fe oxide contents (Chien 1978; van Raij *et al.* 1986; Kuo 1996; Bissani *et al.* 2002). To avoid some of these limitations and simplify the determination of soil P status,

Table 10.1 Common soil extraction methods used to determine the amount of soil P available to the crop

Soil test	Chemical composition[a]	Soil: extractant ratio[a]	Reference
AB-DTPA	1 M NH_4HCO_3, 0.005 M DTPA, pH 7.6	1:2	Soltanpour and Schwab 1977
Bray I	0.025 N HCl, 0.03 N NH_4F	1:10	Bray and Kurtz 1945
Bray II	0.1 N HCl, 0.03 N NH_4F	1:17	Bray and Kurtz 1945
Citric acid	1% Citric acid	1:10	Dyer 1894
Colwell	0.5 M $NaHCO_3$, pH 8.5	1:100	Colwell 1963
Egnér	0.02 N Ca lactate, 0.02 N HCl	1:20	Egnér *et al.* 1960
Fe strip	0.01 M $CaCl_2$	1:40	Menon *et al.* 1988
Mehlich-1	0.05 N HCl, 0.025 N H_2SO_4	1:5	Mehlich 1954
Mehlich-3	0.2 N CH_3COOH, 0.015 N NH_4F, 0.25 N NH_4NO_3, 0.013 N HNO_3, and 0.001 M EDTA	1:10	Mehlich 1984
Morgan	0.52 N CH_3COOH, 0.72 N NaOAc, pH 4.8	1:10	Morgan 1941
Modified Morgan	1.25 N CH_3COOH, 0.62 N NH_4OH, pH 4.8	1:5	McIntosh 1969
Olsen	0.5 M $NaHCO_3$, pH 8.5	1:20	Olsen *et al.* 1954
Pw	demineralized water	1:60	Sissingh 1971
Resin	Anion-exchange resin	1:1	Amer *et al.* 1955

[a]Values are those given in the original descriptions of the methods, more recent or modified versions may differ from these values

ion exchange techniques have been developed that act as sinks, absorbing ions from soil suspensions, thus mimicking the continuous dissolution of P from the exchangeable pool by plant roots (Amer *et al.* 1955; Yang *et al.* 1991; van Raij 1998). These methods include the use of anion exchange resin or membranes and iron-oxide impregnated filter paper (Amer *et al.* 1955; Menon *et al.* 1988, 1997). Since these methods rely on P sorption and desorption reactions with the soil instead of releasing P from the soil using chemical extractants, they are less susceptible to differences in soil properties (e.g. pH, clay content, soil mineralogy; van Raij 1998; Myers *et al.* 2005), but they are not without their problems, such as adhesion of soil particles to the iron-oxide strips and the lack of standardization for these procedures in commercial laboratories (Myers *et al.* 1995, 2005). Most recently the use of Near Infrared Reflectance (NIR) spectroscopy has been evaluated for estimating P availability in soils (van Vuuren *et al.* 2006). This technique has been used for measurements in the laboratory (Cohen *et al.* 2007; Morón and Cozzolino 2007) and continuous measurements in the field, enabling precision soil P maps to be developed that could potentially be used for precision application of P fertilizers (Maleki *et al.* 2007). This technique would allow higher throughput of samples, but more work is needed on calibrating and validating it for different systems and geo-climatic regions (van Vuuren *et al.* 2006).

Following the extraction process the amount of Pi present in the soil extract is determined either colormetrically (Murphy and Riley 1962) or using inductively coupled plasma (ICP) spectroscopic techniques (Pittman *et al.* 2005; Sikora *et al.* 2005). The result is commonly converted into an index suitable for making P fertilizer recommendations or determining the probability of getting a profitable response from the addition of P fertilizer. These indexes are based on the equations and models described in the preceding section, which relate soil test P values to plant P uptake or crop yield and are the result of extensive calibrations (Black 1993). The calibration of these indices is a major investment for establishing a new soil test, since the soil test must be calibrated against many crops, soil types and geo-climatic regions (Smethurst 2000). This approach takes time and significant resources to obtain good calibrations between P extracted from the soil sample and crop yield (Alva 1993). It is not feasible to account for all these potential variables experimentally, resulting in approximations being drawn from similar conditions (Alva 1993; Kara *et al.* 1997; van Raij 1998; Smethurst 2000).

Given the diversity of analytical procedures, soil types and cropping systems, the consistency of analytical procedures between soil testing laboratories must be closely monitored, especially with the increasing use of these procedures for environmental management of P (Grava 1975; Kleinman *et al.* 2001; Cantarella *et al.* 2006). Differences in field location (i.e. geo-climatic differences), soil sampling, storage, transportation, drying, soil test extractants and local variations in procedures contribute to variations observed between soil testing laboratories (Kleinman *et al.* 2001; Rayment 2005; Cantarella *et al.* 2006). A comparison of 16 soil test procedures for P on 135 European soils revealed significant differences in soil test P levels between extractants and laboratories (Neyroud and Lischer 2003). A comparison of 24 US soils using a variety of extraction procedures by nine laboratories showed that Olsen P had the greatest variability, but in general soil test data were highly correlated between laboratories (Kleinman *et al.* 2001).

There is still a continuing need for further development of soil testing procedures, in particular to meet the ever increasing regulations imposed on fertilizer use in agriculture to meet strict environmental regulations. However, the variability resulting from soil type, climate and croping systems is still a potential problem for interpreting results from soil tests. Measuring the P or Pi content of the crop however, might address these problems (Campbell 1994; Bollons and Barraclough 1997; van Raij 1998).

ANALYSIS OF PLANT TISSUES TO DETERMINE CROP P STATUS

Since plants are well evolved to acquire mineral elements across a range of external availabilities, measuring the concentrations of elements in their tissues provides a valuable insight into their nutritional status. The use of plant tissue analysis to determine crop P fertilizer requirements can thus provide an alternative to soil analysis. Analysis of plant tissue samples performed during the growing season can also complement soil analysis results, providing additional information for supplemental fertilizer applications.

Qualitatively, assessments can be made based on visual symptoms of the crop; however, visual symptoms may not become apparent in time for remedial action to be taken. For plant analysis to provide a quantitative measure of plant nutritional status a comparison must be made. This comparison may be semi-quantitative, for example, between plants growing poorly in a field and plants growing vigorously in the same field. More commonly, the comparison is between the level of the element in the crop, as determined by plant analysis, and pre-defined values (or ranges of values) that correspond to the amount of element in the plant required to provide an optimal yield or indicate sufficiency (Black 1993). These values are most commonly derived experimentally, with the crop grown at different fertilizer supply rates. The final absolute or relative yields are then plotted against the concentration of the element in the plant tissue using the equations described previously (Figure 10.1). These plots can be divided into three parts. The first phase, where the yield increases with increasing tissue element concentration is termed the deficiency zone or minimum concentration range. This zone ends with the critical concentration range. This has various definitions depending on how it is calculated, but usually represents a tissue element concentration that, under experimental conditions, gives rise to a percentage, commonly 90–95%, of the maximum yield (Macy 1936; Black 1993). Following this, the tissue element concentration continues to increase with increasing supply, but the yield remains constant. This is referred to as the zone of luxury consumption or sufficiency zone. Beyond this, the yield may decrease if the concentration of the element in plant tissues becomes toxic (Figure 10.1).

Typically, tissue P concentrations are determined on dried tissue samples, which are digested with either nitric or sulfuric acid. The filtered samples are then analyzed using atomic mass spectroscopy or ICP methods (Jones and Case 1990). The introduction of microwave digestion techniques has enabled more complete digestion of samples. Tissue Pi concentrations can be determined by digesting dried plant samples with 2% acetic acid. However, drying samples can result in the conversion of organic P to Pi, increasing the tissue Pi concentrations (Bollons and Barraclough 1997; Major and Barraclough 2003). Alternatively, tissue sap can be extracted from fresh (or freeze-thawed) samples and the Pi concentration measured using the molybdate-blue method (Murphy and Riley 1962; Burns and Hutsby 1984) or test strips and a hand help reflectometer (Major and Barraclough 2003)

As with soil analysis, regression of tissue P concentration against crop yield is required to inform P fertilizer application requirements. The data for these regressions are frequently obtained from experimental studies. For these data to be informative for P fertilizer recommendations the experiments should be field based under similar conditions to the test crop. However, there are large genotypic and environmental components to plant P status, so extrapolation beyond the test conditions could introduce errors into P fertilizer recommendations (Claassens 1990; Black 1993). For example, short periods of P deficiency can lead to an increase in uptake of Pi when P is re-supplied (Clarkson and Scattergood 1982; Claassens 1990). Changes in water potential arising from cold, drought, heavy rain or salinity may restrict or increase the supply of P to the root surface (White and Haydock 1970; Leigh and Johnston 1986; Claassens 1990; Holford 1997). Interactions of P with other minerals in the soil can also cause changes in tissue P status, for example N fertilization increased the shoot %P in wheat plants (Barraclough *et al.* 2000) but resulted in lower shoot %P in barley plants (Leigh and Johnston 1986). Recently, an integrated model for the combined effects of N, P, and K fertilizers on yield and mineral composition of crops has been developed (Zhang *et al.* 2007). This model suggests that %P and %N are strongly correlated and that fertilizer P depresses shoot %K.

To address the complexity of interactions between mineral elements affecting plant nutritional status it is possible to consider the ratios between elements. Much research has been focused on this approach. The element-ratios have been further developed into indices and form the backbone for the Diagnosis and Recommendation Integrated System (DRIS), which also incorporates environmental and soil data to make recommendations for crop fertilizer management (Beaufils 1971, 1973; Walworth and Sumner 1987; Hallmark and Beverly 1991). It has been successfully developed for many crops including corn (Elwali *et al.* 1985; Soltanpour *et al.* 1995), sugarcane (Sumner 1979), grassland (Bailey *et al.* 1997a,b, 2000), grapes (Robinson and McCarthy 1985) and citrus (Beverly *et al.* 1984). Element ratios are less susceptible to crop ontogeny, tissue sampling and seasonal changes, thus they can be more flexible. For example, the tissue used for analysis can affect the results; as P is recycled within the plant, new and old leaves may suffer transient P shortages which, if used for analysis could indicate the plant was P-deficient (Knowles *et al.* 1990; Campbell 1994). Critical concentrations are less flexible in the timing and identity of tissues since data is usually available only for specific tissues at a particular time in the season (Black 1993). However, element-ratios alone may not clearly indicate whether one element is limiting or whether there is an excess of another element, which limits their use practically, except as part of DRIS.

Previous studies have also demonstrated that storage P pools can be correlated to fertilizer application rates. The tissue Pi content (acetic acid extractable) in stem tissue from durum wheat has been shown to increase with the addition of P fertilizers (Knowles *et al.* 1990). Bollons and Barraclough (1997) demonstrated that it is possible to determine the P status of plants by analyzing their Pi concentration expressed on a tissue water basis (fresh weight minus dry weight). This was also shown to increase with P fertilizer applications.

Given the genotypic and environmental impacts on plant P status and the inability to account for all eventualities in experimental systems, diagnosing plant P status and making fertilizer P recommendation based on plant analysis still requires further research. However, analyzing the plant at a more molecular level may provide a better indication of plant P status.

Plants are likely to detect P deficiency and acclimatize physiologically to low P availability before visible signs of P deficiency develop. By monitoring the early changes in gene expression in response to P deficiency, it may be possible to identify and rectify any problem before it results in reduced yield, although in the case of P, this may be in the following season. The ability to monitor these changes at an early stage comes from recent advances in molecular biology, and in particular reporter gene technology. This has led to the concept of 'Smart Plants'.

SMART PLANTS

The natural abilities of plants to adapt and acclimate to growth under low P conditions provides opportunities for developing crops that can acquire or use P more efficiently. The use of modern molecular biological techniques to dissect the genetics underlying these adaptations and acclimatory mechanisms will prove invaluable in this process (Hammond *et al.* 2003, 2004a,b; Wu *et al.* 2003; Franco-Zorrilla *et al.* 2005; Misson *et al.* 2005; Amtmann *et al.* 2006; Hermans *et al.* 2006; Jain *et al.* 2007; Karthikeyan *et al.* 2007; Morcuende *et al.* 2007; Müller *et al.* 2007; Hammond and White 2008; White and Hammond 2008). The identification of genes influencing P acquisition and use efficiency can then be used to assist traditional breeding programs, or to modify the crops genetically. The development of crops efficient in the use and acquisition of P will not only serve to reduce P fertilizer inputs, but may also benefit farmers in developing countries who cannot afford expensive P fertilizers (Frossard *et al.* 2000; Vance 2001).

An alternative purpose for identifying genes that respond rapidly and specifically to P deficiency is to use them as indicators of P deficiency. The concept of Smart Plants has been developed to utilize these genes as markers for P deficiency. Smart Plants are plants that are able to inform us of their physiological status. Smart Plants are able to do this because they have been transformed with an appropriate promoter-marker gene construct. By combining the promoter elements from genes up-regulated during P deficiency with reporter genes such as *GFP* from the jellyfish *Aequorea victoria* (Chalfie *et al.* 1994; Haseloff *et al.* 1997) or the β-glucuronidase (*GUS*) gene (Jefferson *et al.* 1986), it is possible to monitor whether a plant is experiencing low P availability at a cellular level in real time. This enables the Smart Plant to inform us of its physiological P status.

To monitor a plant's response to P deficiency, promoters that function immediately upon P deficiency are required. Promoters that function later in the response may respond too late for any corrective measures to be taken. The promoter must also be specific in its response to P deficiency and not be activated by other stimuli. Ideally, the marker gene should have a short turnover time and be easily assayed *in situ* to make the diagnosis of plant P status as easy and as rapid as possible.

If Smart Plants are to be used to assay crop P status, it is important that the expression of the marker gene responds solely to P and increases as tissue P concentration declines, but before P deficiency affects plant growth. The Smart Plant concept has been shown to work in a proof-of-concept study using the model plant *Arabidopsis thaliana* (Hammond *et al.* 2003, 2004b). Shoot P concentration of *A. thaliana* plants decreased without an effect on growth between 24 and 72 h after P withdrawal. Promoters for genes whose expression is increased during this time period are suitable for Smart Plant technology.

The proof-of-concept for the Smart Plant was tested using the promoter from the gene, At*SQD1*, to control the expression of the *GUS* marker gene. The *SQD1* gene encodes an enzyme that is involved in the biosynthesis of plant sulfolipids, such as sulfoquinovosyl diacylglycerol (SQDG). This enzyme catalyzes the transfer of SO_3^-, from an unknown donor, to a molecule of UDP-glucose, giving rise to UDP-sulfoquinovose (Essigmann *et al.* 1998). The UDP-sulfoquinovose is thought to be the head group donor for SQDG biosynthesis (Benning 1998). SQDG is a component of the thylakoid membranes of chloroplasts. Under P limiting conditions, biosynthesis of phospholipids is restricted due to the lack of available P. In order to avoid damage to the important photosynthetic membranes during P deficiency, plants substitute sulfolipids for phospholipids. Consequently, increased expression of the *SQD1* gene is observed with increasing P deficiency (Essigmann *et al.* 1998).

It was observed that GUS activity in the leaves of transgenic *A. thaliana* plants bearing the *SQD1::GUS* construct increased following the withdrawal of P (Figure 10.3; Hammond *et al.* 2003, 2004b). A large increase in the activity of GUS was not observed until 220h after the removal of P. Since any period of P deficiency that reduces crop growth incurs a yield penalty (Broadley *et al.* 2002), it is important to detect P deficiencies in crops before they become acute and impact on growth. The shoot dry weight of *A. thaliana* plants grown hydroponically declined significantly following the withdrawal of P from the nutrient solution compared to *A. thaliana* plants grown on full nutrient solution, however, this significant difference was not evident until 220h after the withdrawal of P (Hammond *et al.* 2004b). This is the same time-point at which the large increase in GUS activity was observed in Smart Plants following the withdrawal of P (Figure 10.3). Thus, the withdrawal of P had a significant affect on growth before it was possible to detect GUS activity driven by the promoter from At*SQD1*. This implies that the promoter from At*SQD1* may become active relatively late in the response to P deficiency and is probably not suitable for monitoring plant P status. However, using the more sensitive technique of quantitative PCR, it was possible to detect increases in At*SQD1* transcript abundance over 100h before it could be detected using the MUG assay (Hammond *et al.* 2003).

In practice, a grower may also need to know whether P deficiency had been relieved following remedial P fertilization. Therefore, it would be necessary for the expression of any marker gene in a Smart Plant to return rapidly to basal levels following the application of P. To investigate the effects of re-supplying P on GUS activity, P was re-supplied to transgenic plants 316h after the withdrawal of P (Figure 10.3). At the time of P re-supply, GUS activity was between two- and three-fold higher than in P-replete plants. GUS activity remained at a similar level 24h after P was re-supplied and returned to basal levels 48h after P re-supply. This demonstrates the rapid response of the At*SQD1* promoter to remedial P supply and the fast turnover time of the marker gene product. This would allow the grower to determine quickly whether remedial P fertilization had been successful.

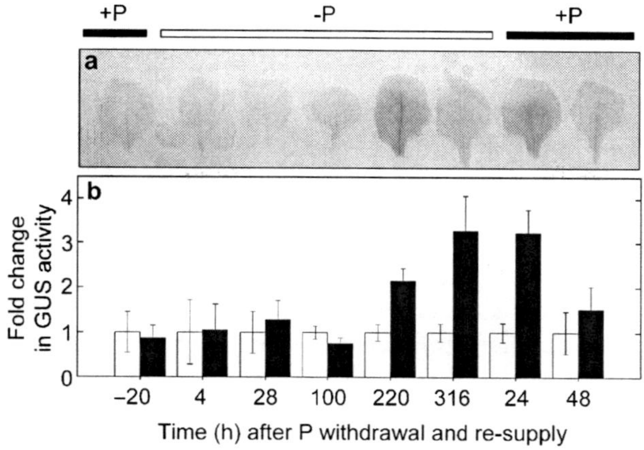

Fig. 10.3 Changes in the expression of the At*SQD1* gene following the withdrawal and re-supply of P. (a) Visualization of GUS activity (Stomp 1992) in the leaves of a transgenic *A. thaliana* line bearing a construct containing the *GUS* marker gene under the control of the promoter sequence for the P-sensitive gene, At*SQD1*. (b) GUS activity measured using the MUG assay (Martin *et al.* 1992) in the leaves of P-deficient (filled) and P-replete (open) *A. thaliana* plants containing the *GUS* marker gene under the control of the At*SQD1* promoter. GUS activity was expressed as nanomoles 4-MU min^{-1} mg^{-1} protein, and normalized to non-specific GUS activity in samples from P-replete plants (values from P-deficient plants divided by values from P-replete plants ± SEM). Plants were grown hydroponically and fully expanded leaves were harvested 20 h before, and 4, 28, 100, 220 and 316 h after the removal of P, and then 24 and 48 h after P was re-supplied

Since there is potential to develop Smart Plants to monitor many stresses, it is possible that many different Smart Plants would need to be grown simultaneously. However, this would be complex and consume valuable land. Alternatively, the promoters from genes responding specifically to different stresses could be stacked in a single construct in which each promoter controlled the expression of a different marker gene, for example other colored or fluorescent proteins, such as yellow fluorescent protein, cyan fluorescent protein and red fluorescent protein (Lansford *et al.* 2001; Visser *et al.* 2002) in a single Smart Plant. However, given the ethical and environmental issues associated with the use of genetically modified crops, non-GM alternatives may be required. One alternative approach would be to monitor the natural activities of enzymes involved in plant responses to P deficiency, such as acid phosphatases (Ascencio 1994). A further approach would be to identify native proteins whose expression increases in response to P withdrawal, to which antibodies can be developed. This would enable the use of antibody test sticks to monitor leaf or stem sap for the expression of these proteins, and thus determine the P status of the plant. The use of antibody test kits has already been developed for plant pathogen detection (Lyons 1997). The use of transcript profiling assays to monitor biotic and abiotic stresses is a another alternative to Smart Plants.

DIAGNOSTIC ARRAYS

In theory, the Smart Plant concept can be applied to any biotic or abiotic stress experienced by plants, provided that a suitable gene can be identified whose expression increases specifically and rapidly in response to the stress. This concept has recently been used to develop experimental plants that have the potential to monitor the bioavailability of Ni in contaminated soils (Krizek *et al.* 2003). An alternative and potentially more sensitive approach would be to use transcript profiling assays to monitor biotic and abiotic stresses (Hammond *et al.* 2004b; Boonham *et al.* 2007). For example, a microarray for the detection of viral pathogens in potatoes has been developed, offering the ability to detect and identify multiple viral pathogens with high sensitivity in a single assay (Boonham *et al.* 2003). Similarly, the sensitivity of quantitative PCR, with its ability to detect increases in At*SQD1* transcript abundance over 100h before it could be detected using the MUG assay, has already been shown (Hammond *et al.* 2003). Both quantitative PCR and microarrays offer the ability to monitor the expression of many genes simultaneously, and do not require the generation of transgenic plants. The reliance on the expression pattern of a single gene could result in false diagnosis of some stresses. Since P deficiency can be induced by a variety of factors, and many different signalling and metabolic pathways are implicated (Hammond *et al.* 2003, 2004a,b; Wu *et al.* 2003; Franco-Zorrilla *et al.* 2005; Misson *et al.* 2005; Amtmann *et al.* 2006; Hermans *et al.* 2006; Jain *et al.* 2007; Karthikeyan *et al.* 2007; Morcuende *et al.* 2007; Müller *et al.* 2007; Hammond and White 2008; White and Hammond 2008), it might be wise to identify a suite of genes responding differentially to P deficiency to allow plant P-stress to be monitored in all environments. The use of quantitative PCR or microarrays makes this approach possible. Furthermore, using whole genome microarrays might enable several stresses to be monitored simultaneously and the careful selection of probes could enable several crop species to be monitored using a 'universal' microarray.

Several issues need to be resolved before such techniques become widely available (Krizek *et al.* 2003). Firstly, the cost and practicalities of using quantitative PCR or microarrays to monitor plant stresses currently prevents such an approach being used for commercial agriculture, but these costs should diminish with time. Progress in this field is currently rapid, driven by the development of diagnostic technologies for human diseases, bio-security and microbial ecology (Cirino *et al.* 2004; Hahn *et al.* 2006; Petrik 2006; Sessitsch *et al.* 2006). Consequently, it is now possible to analyze multiple samples simultaneously (10–100) on one microarray, with each sample independently challenging hundreds to thousands of probes. Theoretically, it is therefore possible to assay samples for multiple abiotic and biotic stresses. Standardization of procedures would also need to be undertaken; especially if different technologies were being used to monitor changes in gene expression (Nuwaysir *et al.* 1999). Associated with this is the need for specialist laboratory equipment and specialist knowledge to undertake such diagnostic assays. Secondly, statistical techniques to normalize microarrays containing a few active genes (compared to many thousands of non-changing genes) and suitable techniques to distinguish between stressed and un-stressed plants based on the changes in expression of several genes would need to be developed. Again these are being developed for use in human diagnostics and should be applicable to plant based applications (Jaeger and Spang 2006; Chen 2007).

SUMMARY

To prevent damage to the environment, and to reduce the consumption of a non-renewable resource, improved management of P fertilizer use must be implemented. A key part of this process will be to understand the mechanisms that govern the availability of P to plants and the plant's responses to available P. Using this understanding, robust analytical techniques that have been calibrated and validated under field conditions can assist P fertilizer management strategies. New technologies are becoming available for use in precision agriculture, allowing detailed field maps of soil P to be made and used for precision fertilizer P applications. More detailed analyses of plant tissue samples are now also possible. In the future, the use of techniques to monitor plants at the molecular level will enable the plant to provide valuable insights into how they perceive their environment and indicate their requirements for fertilizer applications.

Acknowledgements Work of PJW and JPH on phosphorus nutrition is supported by grants HH3504SPO, HH3507SFV, HH3509SFV and WQ0119 from the UK Department for Environment, Food and Rural Affairs (Defra). Other work at SCRI is supported by the Scottish Executive Environment and Rural Affairs Department (SEERAD).

REFERENCES

Alva AK (1993) Comparison of Mehlich 3, Mehlich 1, ammonium bicarbonate-DTPA, 1.0M ammonium acetate and 0.2M ammonium chloride for extraction of calcium, magnesium, phosphorus, and potassium for a wide range of soils. Commun Soil Sci Plant Anal 24: 603–612

Amer F, Bouldin DR, Black CA, Duke FR (1955) Characterization of soil phosphorus by anion exchange resin adsorption and P^{32}-equilibration. Plant Soil 6: 391–408

Amtmann A, Hammond JP, Armengaud P, White PJ (2006) Nutrient sensing and signalling in plants: potassium and phosphorus. Adv Bot Res 43: 209–257

Ascencio J (1994) Acid phosphatase as a diagnostic tool. Commun Soil Sci Plant Anal 25: 1553–1564

Bailey JS, Beattie JAM, Kilpatrick DJ (1997a) The diagnosis and recommendation integrated system (DRIS) for diagnosing the nutrient status of grassland swards: I. Model establishment. Plant Soil 197: 127–135

Bailey JS, Cushnahan A, Beattie JAM (1997b) The diagnosis and recommendation integrated system (DRIS) for diagnosing the nutrient status of grassland swards: II. Model calibration and validation. Plant Soil 197: 137–147

Bailey JS, Dils RA, Foy RH, Patterson D (2000) The diagnosis and recommendation integrated system (DRIS) for diagnosing the nutrient status of grassland swards: III. Practical applications. Plant Soil 222: 255–262
Barraclough PB, Bollons HM, Chambers BJ, Bhogal A, Hatley D (2000) Development of on-farm plant tests for phosphate and potassium in wheat. Home Grown Cereals Authority Project Report No. 224
Bar-Yosef B (2003) Phosphorus dynamics. In: DK Benbi, R Nieder (eds) Handbook of Processes and Modeling in the Soil-Plant System. Haworth Reference Press, New York, pp 483–524
Beaufils ER (1971) Physiological diagnosis - a guide for improving maize production based on principles developed for rubber trees. Fert Soc S Afr J 1: 1–30
Beaufils ER (1973) Diagnosis and recommendation integrated system (DRIS): a general scheme for experimentation and calibration based on principles developed from research in plant nutrition. University of Natal, Soil Science Bulletin 1
Benning C (1998) Biosynthesis and function of the sulfolipid sulfoquinovosyl diacylglycerol Annu Rev Plant Phys 49: 53–75
Bergmann W (1992) Nutritional Disorders of Plants. Visual and Analytical Diagnosis. Gustav Fischer, Jena, Germany
Beverly RB, Stark JC, Ojala JC, Embleton TW (1984) Nutrient diagnosis of 'Valencia' oranges by DRIS. J Am Soc Hort Sci 109: 649–654
Bieleski RL, Ferguson IB (1983) Physiology and metabolism of phosphate and its compounds. In: A Läuchli, RL Bieleski (eds), Encyclopedia of Plant Physiology, New Series, Vol. 15A. Springer, Berlin, pp 422–449
Bissani CA, Tedesco MJ, Camargo FAD, Miola GL, Gianello C (2002) Anion-exchange resins and iron oxide-impregnated filter paper as plant available phosphorus indicators in soils. Commun Soil Sci Plant Anal 33: 1119–1129
Black CA (1993) Soil Fertility Evaluation and Control. Lewis, Boca Raton, FL
Bollons HM, Barraclough PB (1997) Inorganic orthophosphate for diagnosing the phosphorus status of wheat plants. J Plant Nutr 20: 641–655
Boonham N, Walsh K, Smith P, Madagan K, Graham I, Barker I (2003) Detection of potato viruses using microarray technology: towards a generic method for plant viral disease diagnosis. J Virol Methods 108:181–87
Boonham N, Tomlinson J, Mumford R (2007) Microarrays for rapid identification of plant viruses Annu Rev Phytopathol 45: 307–328
Bould C, Hewitt EJ, Needham P (1983) Diagnosis of Mineral Disorders in Plants. Volume 1: Principles. HMSO, London
Bray RH, Kurtz LT (1945) Determination of total, organic, and available forms of phosphorus in soils. Soil Sci 59: 39–45
Broadley MR, Burns A, Burns IG (2002) Relationships between phosphorus forms and plant growth. J Plant Nutr 25: 1075–1088
Burns IG, Hutsby W (1984) Development and evaluation of rapid tests for the estimation of phosphate and potassium in plant sap. Commun Soil Sci Plant Anal 15: 1463–1480
Campbell LC (1994) Plant and soil analysis: An Australian perspective. Commun Soil Sci Plant Anal 25: 767–780
Cantarella H, Quaggio JA, van Raij B, de Abreu MF (2006) Variability of soil analysis in commercial laboratories: implications for lime and fertilizer recommendations. Commun Soil Sci Plant Anal 37: 2213–2225
Chalfie M, Tu Y, Euskirchen G, Ward WW, Prasher DC (1994) Green fluorescent protein as a marker for gene expression. Science 263: 802–805
Chang SC, Jackson ML (1957) Fractionation of soil phosphorus. Soil Sci 84: 133–144
Chen JJ (2007) Key aspects of analyzing microarray gene-expression data. Pharmacogenomics 8: 473–482
Chien SH (1978) Reactions of phosphate rocks, rhenania phosphate, and super-phosphate with an acid soil. Soil Sci Soc Am J 42: 705–708

Cirino NM, Musser KA, Egan C (2004) Multiplex diagnostic platforms for detection of biothreat agents. Expert Rev Mol Diagn 4: 841–857

Clarkson DT, Scattergood CB (1982) Growth and phosphate transport in barley and tomato plants during the development of, and recovery from, phosphate-stress. J Exp Bot 33: 865–875

Claassens AS (1990) Factors influencing phosphorus uptake by plants: The dangers of misinterpretation of foliar phosphorus analysis. Commun Soil Sci Plant Anal 21: 1301–1312

Claassen N, Steingrobe B (1999) Mechanistic simulation models for a better understanding of nutrient uptake from soil. In: Z Rengel (ed), Mineral Nutrition of Crops. Fundamental Mechanisms and Implications. Haworth Press, New York, pp 327–367

Clark LJ, Gowing DJG, Lark RM, Leeds-Harrison PB, Miller AJ, Wells DM, Whalley WR, Whitmore AP (2005) Sensing the physical and nutritional status of the root environment in the field: a review of progress and opportunities. J Agr Sci 143: 347–358

Cohen D (2007) Earth's natural wealth: an audit. New Scientist 2605: 35–41

Cohen M, Mylavarapu RS, Bogrekci I, Lee WS, Clark MW (2007) Reflectance spectroscopy for routine agronomic soil analyses. Soil Sci 172: 469–485

Colwell J (1963) The estimation of the phosphorus fertiliser requirements of wheat in southern New South Wales by soil analysis. Aust J Exp Agr Anim Husb 3: 190

Denison RF, Kiers ET (2005) Sustainable crop nutrition: constraints and opportunities. In: MR Broadley, PJ White (eds), Plant Nutritional Genomics. Blackwell, Oxford, pp 242–286

Dyer B (1894) On the analytical determination of probable available "mineral" plant food in soils. J Chem Soc Trans 65: 115–167

Egnér H, Riehm H, Domingo WR (1960) Untersuchungen über die chemische Bodenanalyse als Grundlage für die Beurteilung des Nährstoffzustandes der Böden. II. Chemische Extraktionsmethoden zur Phosphor-und Kaliumbestimmung. Kungl Lantbrukshögsk Ann 26: 199–215

Elwali AMO, Gascho GJ, Sumner ME (1985) DRIS norms for 11 nutrients in corn leaves. Agron J 77: 506–508

Essigmann B, Güler S, Narang RA, Linke D, Benning C (1998) Phosphate availability affects the thylakoid lipid composition and the expression of *SQD1*, a gene required for sulfolipid biosynthesis in *Arabidopsis thaliana*. Proc Natl Acad Sci USA 95: 1950–1955

Fixen PE, Grove JH (1990) Testing soils for phosphorus. In: RL Westerman (ed) Soil Testing and Plant Analysis (3rd edition). Soil Science Society of America, Madison, WI, pp 141–180

Franco-Zorrilla JM, Martín AC, Leyva A, Paz-Ares J (2005) Interaction between phosphate-starvation, sugar, and cytokinin signaling in Arabidopsis and the roles of cytokinin receptors CRE1/AHK4 and AHK3. Plant Physiol 138: 847–857

Frossard E, Brossard M, Hedley MJ, Metherell A (1995) Reactions controlling the cycling of P in soils. In: H Tiessen (ed), Phosphorus in the Global Environment: Transfers, Cycles and Management, Wiley, Chichester, pp 107–137

Frossard E, Condron LM, Oberson A, Sinaj S, Fardeau JC (2000) Processes governing phosphorus availability in temperate soils. J Environ Qual 29: 15–23

George TS, Richardson AE (2008) Potential and limitations to improving crops for enhanced phosphorus utilization. In: White PJ, Hammond JP (eds), The Ecophysiology of Plant-Phosphorus Interactions. Springer, Dordrecht, The Netherlands, pp 247–270

Grava J (1975) Causes for variation in phosphorus soil tests. Commun Soil Sci Plant Anal 6: 129–138

Greenwood DJ, Wood JT, Cleaver TJ, Hunt J (1971) A theory of fertiliser response. J Agr Sci 77: 511–523

Greenwood DJ, Barnes A, Liu K, Hunt J, Cleaver TJ, Loquens SMH (1980) Relationships between the critical concentrations of nitrogen, phosphorus and potassium in 17 different vegetable crops and duration of growth. J Sci Food Agr 31: 1343–1353

Greenwood DJ, Karpinets TV, Stone DA (2001) Dynamic model for the effects of soil P and fertilizer P on crop growth, P uptake and soil P in arable cropping: model description. Ann Bot 88: 279–291

Goldstein AH (1992) Phosphate starvation inducible enzymes and proteins in higher plants. In: JL Wray (ed), Society for Experimental Biology Seminar Series 49: Inducible Plant Proteins. Cambridge University Press, Cambridge, pp 25–44

Hahn S, Mergenthaler S, Zimmermann B, Holzgreve W (2005) Nucleic acid based biosensors: the desires of the user. Bioelectrochemistry 67: 151–154

Hallmark WB, Beverly RB (1991) Review: an update in the use of the diagnosis and recommendation integrated system. J Fert Issues 8: 74–88

Hammond JP, White PJ (2008) Sucrose transport in the phloem: integrating root responses to phosphorus starvation. J Exp Bot 59: 93–109

Hammond JP, Bennett MJ, Bowen HC, Broadley MR, Eastwood DC, May ST, Rahn C, Swarup R, Woolaway KE, White PJ (2003) Changes in gene expression in Arabidopsis shoots during phosphate starvation and the potential for developing smart plants. Plant Physiol 132: 578–596

Hammond JP, Broadley MR, White PJ (2004a) Genetic responses to phosphorus deficiency. Ann Bot 94: 323–332

Hammond JP, White PJ, Broadley MR (2004b) Diagnosing phosphorus deficiency in plants. Asp Appl Biol 72: 89–98

Haseloff J, Siemering KR, Prasher DC, Hodge S (1997) Removal of a cryptic intron and subcellular localization of green fluorescent protein are required to mark transgenic *Arabidopsis* plants brightly. Proc Nat Acad Sci USA 94: 2122–2127

Hermans C, Hammond JP, White PJ, Verbruggen N (2006) How do plants respond to nutrient shortage by biomass allocation? Trends Plant Sci 11: 610–617

Higgs B, Johnston AE, Salter JL, Dawson CJ (2000) Some aspects of achieving sustainable phosphorus use in agriculture. J Environ Qual 29: 80–87

Holford ICR (1997) Soil phosphorus; its measurement, and its uptake by plants. Aust J Soil Res 35: 227–239

Hoch WA, Zeldin EL, McCown BH (2001) Physiological significance of anthocyanins during autumnal leaf senescence. Tree Physiol 21: 1–8

Jaeger J, Spang R (2006) Selecting normalization genes for small diagnostic microarrays. BMC Bioinformatics 7: 388

Jain A, Vasconcelos MJ, Raghothama KG, Sahi SV (2007) Molecular mechanisms of plant adaptation to phosphate deficiency. Plant Breeding Rev 29: 359–419

Jefferson RA, Burgess SM, Hirsh D (1986) Beta-glucuronidase from *Escherichia coli* as a gene-fusion marker. Proc Nat Acad Sci USA 83: 8447–8451

Jones JB, Case VW (1990) Sampling, handling, and analyzing plant tissue samples. In: Westerman RL (ed), Soil Testing and Plant Analysis (3rd edition). Soil Science Society of America, Madison, WI, pp 389–427

Jungk A, Seeling B, Gerke J (1993) Mobilization of different phosphate fractions in the rhizosphere. Plant Soil 156: 91–94

Kamprath EJ, Watson ME (1980) Conventional soil and tissue tests for assessing the phosphorus status of soils. In: FH Khasawneh, EC Sample, EJ Kamprath (eds), The Role of Phosphorus in Agriculture. ASA/CSSA/SSSA, Madison, WI, pp 433–469

Kara D, Özsavaşçi C, Alkan M (1997) Investigation of suitable digestion methods for the determination of total phosphorus in soils. Talanta 44: 2027–2032

Karpinets TV, Greenwood DJ, Ammons JT (2004) Predictive mechanistic model of soil phosphorus dynamics with readily available inputs. Soil Sci Soc Am J 68: 644–653

Karthikeyan AS, Varadarajan DK, Jain A, Held MA, Carpita NC, Raghothama KG (2007) Phosphate starvation responses are mediated by sugar signaling in *Arabidopsis*. Planta 225: 907–918

Kirkby EA, Johnston AE (2008) Soil and fertilizer phosphorus in relation to crop nutrition. In: White PJ, Hammond JP (eds), The Ecophysiology of Plant-Phosphorus Interactions. Springer, Dordrecht, The Netherlands, pp 177–223

Kleinman PJA, Sharpley AN, Gartley K, Jarrell WM, Kuo S, Menon RG, Myers R, Reddy KR, Skogley EO (2001) Interlaboratory comparison of soil phosphorus extracted by various soil test methods. Commun Soil Sci Plant Anal 32: 2325–2345

Knowles TC, Doerge TA, Clark LJ (1990) Diagnosing phosphorus deficiency in irrigated durum wheat using basal stem phosphate tissue analysis. Commun Soil Sci Plant Anal 21: 2053–2065

Krizek BA, Prost V, Joshi RM, Stoming T, Glenn TC (2003) Developing transgenic Arabidopsis plants to be metal-specific bioindicators. Environ Toxicol Chem 22: 175–181

Kuo S (1996) Phosphorus. In: DL Sparks (ed), Methods of Soil Analysis, Part 3, Chemical Methods. SSSA/ASA, Madison, WI, pp 869–919

Lansford R, Bearman G, Fraser SE (2001) Resolution of multiple green fluorescent protein color variants and dyes using two-photon microscopy and imaging spectroscopy. J Biomed Opt 6: 311–318

Leigh RA, Johnston AE (1986) An investigation of the usefulness of phosphorus concentrations in tissue water as indicators of the phosphorus status of field-grown spring barley. J Agr Sci 107: 329–333

Liebig J von (1855) Principles of Agricultural Chemistry with Special Reference to the Late Researches Made in England. Walton & Maberly, London

Lynch JP, Brown KM (2008) Root strategies for phosphorus acquisition. In: White PJ, Hammond JP (eds), The Ecophysiology of Plant-Phosphorus Interactions. Springer, Dordrecht, The Netherlands, pp 83–116

Lyons NF (1997) Use of antibody coated mobile particles for the detection of plant pathogens. Final Report for MAFF Project HH1738SX

Macy P (1936) The quantitative mineral nutrient requirements of plants. Plant Physiol 11: 749–764

Major BJ, Barraclough PB (2003) Measuring inorganic orthophosphate in oilseed rape using reflectoquant technology. Commun Soil Sci Plant Anal 34: 2839–2851

Maleki MR, Mouazen AM, Ramon H, De Baerdemaeker J (2007) Optimization of soil VIS-NIR sensor-based variable rate application system of soil phosphorus. Soil Till Res 94: 239–250

Marschner H (1995) Mineral Nutrition of Higher Plants (2nd edition). Academic, London

Martin T, Wöhner R-V, Hummel S, Willmitzer L, Frommer WB (1992) The GUS reporter system as a tool to study plant gene expression. In: SR Gallagher (ed), GUS Protocols: Using the GUS Gene as a Reporter of Gene Expression. Academic, San Diego, CA, pp 23–43

McIntosh JL (1969) Bray and Morgan soil extractants modified for testing acid soils from different parent materials. Agron J 61: 259–265

Mead R, Pike DJ (1975) Review of response surface methodology from a biometric viewpoint. Biometrics 31: 803–851

Mehlich A (1954) Determination of P, K, Na, Ca, Mg, and NH_4. Soil Test Division, Mimeo. North Carolina Department of Agriculture, Raleigh, NC

Mehlich A (1984) Mehlich 3 soil test extractant: A modification of Mehlich 2 extractant. Commun Soil Sci Plant Anal 15: 1409–1416

Mengel K, Kirkby EA (2001) Principles of Plant Nutrition (5th edition). Kluwer, Dordrecht, The Netherlands

Menon RG, Hammond LL, Sissingh HA (1988) Determination of plant-available phosphorus by the iron hydroxide impregnated filter-paper (P_i) soil test. Soil Sci Soc Am J 52: 110–115

Menon RG, Chien SH, Chardon WJ (1997) Iron oxide-impregnated filter paper (P_i test). II. A review of its application. Nutr Cycl Agroecosys 47: 7–18

Misson J, Raghothama KG, Jain A, Jouhet J, Block MA, Bligny R, Ortet P, Creff A, Somerville S, Rolland N, Doumas P, Nacry P, Herrerra-Estrella L, Nussaume L, Thibaud M-C (2005) A genome-wide transcriptional analysis using *Arabidopsis thaliana* Affymetrix gene chips determined plant responses to phosphate deprivation. Proc Natl Acad Sci USA 102: 11934–11939

Mitscherlich EA (1909) Das Gesetz des Minimums und das Gesetz des abnehmenden Bodenertrages. Landwirt Jahrb Schweiz 38: 537–552

Morcuende R, Bari R, Gibon Y, Zheng WM, Pant BD, Bläsing O, Usadel B, Czechowski T, Udvardi MK, Stitt M, Scheible WR (2007) Genome-wide reprogramming of metabolism and regulatory networks of *Arabidopsis* in response to phosphorus. Plant Cell Environ 30: 85–112

Morgan MF (1941) Chemical soil diagnosis by the Universal Soil Testing system. Connecticut Agric. Exp. Stn Bull 525: 22

Morón A, Cozzolino D (2007) Measurement of phosphorus in soils by near infrared reflectance spectroscopy: effect of reference method on calibration. Commun Soil Sci Plant Anal 38: 1965–1974

Müller R, Morant M, Jarmer H, Nilsson L, Nielsen TH (2007) Genome-wide analysis of the Arabidopsis leaf transcriptome reveals interaction of phosphate and sugar metabolism. Plant Physiol 143: 156–171

Müller R, Nilsson L, Nielsen LK, Nielsen TH (2005) Interaction between phosphate starvation signalling and hexokinase-independent sugar sensing in *Arabidopsis* leaves. Physiol Plant 124: 81–90

Murphy J, Riley J (1962) A modified single solution for the determination of phosphate in natural waters. Anal Chim Acta 27: 31–35

Myers RG, Pierzynski GM, Thien SJ (1995) Improving the iron oxide sink method for extracting soil phosphorus. Soil Sci Soc Am J 59: 853–857

Myers RG, Sharpley AN, Thien SJ, Pierzynski GM (2005) Ion-sink phosphorus extraction methods applied on 24 soils from the continental USA. Soil Sci Soc Am J 69: 511–521

Neyroud JA, Lischer P (2003) Do different methods used to estimate soil phosphorus availability across Europe give comparable results? J Plant Nutr Soil Sci 166: 422–431

Nuernberg NJ, Leal JE, Sumner ME (1998) Evaluation of an anion-exchange membrane for extracting plant available phosphorus in soils. Commun Soil Sci Plant Anal 29: 467–479

Nuwaysir EF, Bittner M, Trent J, Barrett JC, Afshari CA (1999) Microarrays and toxicology: The advent of toxicogenomics. Mol Carcinogen 24: 153–159

Olsen SR, Cole CV, Watanabe FS, Dean LA (1954) Estimation of Available Phosphorus by Extraction with Sodium Bicarbonate (Circular 39). USDA, Washington DC

Petrik J (2006) Diagnostic applications of microarrays. Transfusion Med 16: 233–247

Pinkerton A (1991) Critical phosphorus concentrations in oilseed rape (*Brassica napus*) and indian mustard (*Brassica juncea*) as affected by nitrogen and plant-age. Aust J Exp Agr 31: 107–115

Pittman JJ, Zhang H, Schroder JL, Payton ME (2005) Differences of phosphorus in Mehlich 3 extracts determined by colorimetric and spectroscopic methods. Commun Soil Sci Plant Anal 36: 1641–1659

Raghothama KG (1999) Phosphate acquisition. Annu Rev Plant Phys 50: 665–693

Rayment GE (2005) Statistical aspects of soil and plant test measurement and calibration in Australasia. Commun Soil Sci Plant Anal 36: 107–120

Richardson AE (2001) Prospects for using soil microorganisms to improve the acquisition of phosphorus by plants. Aust J Plant Physiol 28: 897–906

Robinson JB, McCarthy MG (1985) Use of petiole analysis for assessment of vineyard nutrient status in the Barossa district of South Australia. Aust J Exp Agr 25: 231–240

Rubio G, Liao H, Yan XL, Lynch JP (2003) Topsoil foraging and its role in plant competitiveness for phosphorus in common bean. Crop Sci 43: 598–607

Runge-Metzger A (1995) Closing the cycle: obstacles to efficient P management for improved global security. In: Tiessen H (ed), Phosphorus in the Global Environment: Transfers, Cycles and Management. Wiley, Chichester, pp 27–42

Sessitsch A, Hackl E, Wenzl P, Kilian A, Kostic T, Stralis-Pavese N, Sandjong BT, Bodrossy L (2006) Diagnostic microbial microarrays in soil ecology. New Phytol 171: 719–736

Sharpley A (1995) Identifying sites vulnerable to phosphorus loss in agricultural runoff. J Environ Qual 24: 947–951

Sibbesen E, Sharpley AN (1997) Setting and justifying upper critical limits for phosphorus in soils. In: H Tunney, OT Carton, PC Brookes, AE Johnston (eds), Phosphorus Loss from Soil to Water. CAB International, Cambridge, pp 151–176

Sikora FJ, Howe PS, Hill LE, Reid DC, Harover DE (2005) Comparison of colorimetric and ICP determination of phosphorus in Mehlich 3 soil extracts. Commun Soil Sci Plant Anal 36: 875–887

Sissingh HA (1971) Analytical technique of the Pw method, used for assessment of the phosphate status of arable soils in the Netherlands. Plant Soil 34: 483–486

Smethurst PJ (2000) Soil solution and other soil analyses as indicators of nutrient supply: a review. Forest Ecol Manag 138: 397–411

Smillie GW, Syers JK (1972) Calcium chloride formation during extraction of calcareous soil with fluoride: II. Implications to the Bray-1 test. Soil Sci Soc Am Proc 36: 25–30

Solfanelli C, Poggi A, Loreti E, Alpi A, Perata P (2006) Sucrose-specific induction of the anthocyanin biosynthetic pathway in Arabidopsis. Plant Physiol 140: 637–646

Soltanpour PN, Schwab AP (1977) New soil test for simultaneous extraction of macro-nutrients and micro-nutrients in alkaline soils. Commun Soil Sci Plant Anal 8: 195–207

Soltanpour PN, Malakouti MJ, Ronaghi A (1995) Comparison of diagnosis and recommendation integrated system and nutrient sufficiency range for corn. Soil Sci Soc Am J 59: 133–139

Sprengel C (1828) Von den Substanzen der Ackerkrume und des Untergrundes. J Techn Ökonom 2: 423–474; 3: 42–99, 313–352, and 397–421

Sprengel C (1837) Die Bodenkunde oder die Lehre vom Boden. Immanuel Müller, Leipzig

Steen I (1998) Phosphorus availability in the 21st century: Management of a non-renewable resource. Phosphorus Potassium 217: 25–31

Stomp AM (1992) Histochemical localization of β-glucuronidase. In: SR Gallagher (ed), GUS Protocols: Using the GUS Gene as a Reporter of Gene Expression. Academic, San Diego, CA, pp 103–113

Sumner ME (1979) Interpretation of foliar analyses for diagnostic purposes. Agron J 71: 343–348

Teng S, Keurentjes J, Bentsink L, Koornneef M, Smeekens S (2005) Sucrose-specific induction of anthocyanin biosynthesis in Arabidopsis requires the MYB75/PAP1 gene. Plant Physiol 139: 1840–1852

Thomas GW, Peaslee DE (1973) Testing soils for phosphorus. In: LM Walsh, Beaton JD (eds), Soil Testing and Plant Analysis (Revised edition). Soil Science Society of America, Madison, WI, pp 115–132

Ullrich-Eberius CI, Novacky A, Fischer E, Lüttge U (1981) Relationship between energy-dependent phosphate uptake and the electrical membrane potential in *Lemna gibba* G1. Plant Physiol 67: 797–801

Vance CP (2001) Symbiotic nitrogen fixation and phosphorus acquisition. Plant nutrition in a world of declining renewable resources. Plant Physiol 127: 390–397

Vance CP (2008) Plants without arbuscular mycorrhizae. In: White PJ, Hammond JP (eds), The Ecophysiology of Plant-Phosphorus Interactions. Springer, Dordrecht, The Netherlands, pp 117–142

van Raij B (1998) Bioavailable tests: alternatives to standard soil extractions. Commun Soil Sci Plant Anal 29: 1553–1570

van Raij B, Quaggio JA, da Silva NM (1986) Extraction of phosphorus, potassium, calcium, and magnesium from soils by an ion-exchange resin procedure. Commun Soil Sci Plant Anal 17: 547–566

van Vuuren JAJ, Meyer JH, Claassens AS (2006) Potential use of near infrared reflectance monitoring in precision agriculture. Commun Soil Sci Plant Anal 37: 2171–2184

Visser NV, Hink MA, Borst JW, van der Krogt GNM, Visser AJWG (2002) Circular dichroism spectroscopy of fluorescent proteins. FEBS Lett 521: 31–35

Walworth JL, Sumner ME (1987) The diagnosis and recommendation integrated system (DRIS), Adv Soil Sci 6: 149–188

White PJ, Hammond JP (2006) Updating the estimate of the sources of phosphorus in UK waters. Final Report on Defra project WT0701CSF.

White PJ, Hammond JP (2008) Phosphorus nutrition of terrestrial plants. In: White PJ, Hammond JP (eds), The Ecophysiology of Plant-Phosphorus Interactions. Springer, Dordrecht, The Netherlands, pp 51–81

White RE, Haydock KP (1970) Phosphate concentration in siratro as a guide to its phosphate status in the field. Aust J Exp Agr Anim Husb 10: 426–430

Withers PJA, Edwards AC, Foy RH (2001) Phosphorus cycling in UK agriculture and implications for phosphorus loss from soil. Soil Use Manage 17: 139–149

Wu P, Ma L, Hou X, Wang M, Wu Y, Liu F, Deng XW (2003) Phosphate starvation triggers distinct alterations of genome expression in Arabidopsis roots and leaves. Plant Physiol 132: 1260–1271

Yang JE, Skogley EO, Georgitis SJ, Schaff BE, Ferguson AH (1991) Phytoavailability soil test - development and verification of theory. Soil Sci Soc Am J 55: 1358–1365

Zhang K, Greenwood DJ, White PJ, Burns IG (2007) A dynamic model for the combined effects of N, P and K fertilizers on yield and mineral composition; description and experimental test. Plant Soil 298: 81–98

Chapter 11
POTENTIAL AND LIMITATIONS TO IMPROVING CROPS FOR ENHANCED PHOSPHORUS UTILIZATION

Timothy S. George and Alan E. Richardson

INTRODUCTION

Phosphorus (P) is an essential element required for cellular function and when deficient has a significant impact on plant growth and fecundity. Poor availability of P in soil and consequent P-deficiency represents a major constraint to crop production globally (Runge-Metzger 1995). Soil P status is also a key factor that controls the competitive dynamics and species composition in different natural ecosystems (McGill and Cole 1981; Attiwill and Adams 1993), and thus may have significant impact on biodiversity (Wassen *et al.* 2005). Many plant species have evolved in P-limited environments and, as a consequence, are known to possess a number of adaptive features that can enhance the acquisition of P from soil (Raghothama 1999; Vance *et al.* 2003; Richardson *et al.* 2005). However, ongoing selection of crop cultivars, in nutrient replete environments, for traits such as yield and vigor (and thus an adaptation to optimal production systems), may have resulted in cultivars that have 'lost' adaptive traits that are required to cope with P-deficiency (Manske *et al.* 2000; Buso and Bliss 1988). Identification of such traits and their introduction into elite material from traditional cultivars, wild relatives and other species through modern approaches in breeding (e.g. marker-assisted selection and/or genetic manipulation) may provide new opportunities to improve the efficiency of P-uptake by crop plants.

Phosphorus is taken up by plants as orthophosphate (Pi). In many soils the acquisition of P by plants is limited by low concentrations of Pi in soil solution, slow diffusion of Pi in soil, and the limited capacity for Pi to be replenished at the soil-root interface (Bieleski 1973). The paradox of this is that most soils contain large amounts of P that, if made more available, could sustain crop production for long periods with less dependence on P inputs (Harrison 1987). Phosphorus in soil is comprised of organic and inorganic P forms (Sanyal and De Datta 1991). The majority of the inorganic P is either adsorbed to soil constituents such as clays, sesquioxides and organic matter, or occurs in a range of precipitated mineral forms. Organic P generally accounts for around 50% of soil P, and is largely comprised of monoesters with lesser amounts of diesters and phosphonates (Newman and Tate 1980; Hawkes *et al.* 1984; Condron *et al.* 1990). Monoester P occurs predominantly

P.J. White, J.P. Hammond (eds.) *The Ecophysiology of Plant-Phosphorus Interactions*,

as derivatives of inositol hexakisphosphates, whereas sugar phosphates and diester P, such as nucleic acids and phospholipids, constitute only a small fraction (Dalal 1977; Anderson 1980). In order to be available to plants, inorganic P must be either desorbed or solubilized, and organic P must be mineralized to release Pi. Once in the soil solution, Pi is acquired rapidly by plant roots such that its concentration in close proximity to the root surface is estimated to be of the order ~0.05–0.2 μM (Barber 1984), which is significantly less than elsewhere in the soil environment, where soil solution concentrations are typically in the range of 1–5 μM (Bieleski 1973). However, slow Pi diffusion through soil to roots limits P-supply to the root surface and can thus restrict P-acquisition. In order to overcome this, plants have become effective in accessing P from spatially unexploited reserves through either root growth or altering the physicochemical characteristics of the rhizosphere. The range of plant traits that are associated with enhanced P acquisition in soil are highlighted in this chapter, and the potential and limitations for improving the P-use efficiency of plants by deploying such traits into crop germplasm are discussed.

MECHANISMS ENHANCING P ACQUISITION BY PLANTS

Plant responses to P-deficiency are characterized by a change in the expression of a large number of genes (Hammond *et al.* 2003; Wasaki *et al.* 2003; Wu *et al.* 2003). Genes with rapid response (less than one day) to P deficiency ('early' genes) are generally non-specific and related to general stress responses. The more significant changes in gene expression in response to P deprivation occur over a period of several days ('late' genes), and involve genes that are more commonly associated with traits that directly affect P nutrition (Hammond *et al.* 2003). Of particular significance is the up-regulation of genes that encode for high-affinity Pi transporters, by which roots take up P from soil solution. However, once 'easily' available Pi is depleted from the soil solution, it is the rate of Pi replenishment from the soil that is generally considered to be the main constraint to crop production. Consequently, plants possess other physiological and morphological adaptations to cope with P-limitation, the genes for many of which are also up-regulated under conditions of P-deficiency. These adaptations include both strategies that conserve P and strategies that increase access to soil P reserves through modified root structure, function and/or association with microorganisms (Raghothama 1999; Vance *et al.* 2003; Jakobsen *et al.* 2005; Amtmann *et al.* 2006).

Physiological adjustment to P-deficiency

Internal physiological adjustments that allow plants to conserve P have been reviewed previously (Bieleski 1973; Schachtman *et al.* 1998; Raghothama 1999; Vance *et al.* 2003; Hammond *et al.* 2004; Ticconi and Abel 2004) and are discussed elsewhere in this volume (White and Hammond 2008). In general, these changes include a reduced rate of growth, remobilization of internal P and increased turnover of nucleic acids, the use of alternative metabolic pathways and induction of secondary metabolites (e.g. anthocyanins to ameliorate photo-inhibitory damage), and modification to phospholipid membranes (Vance *et al.* 2003). Some key regulatory genes associated with these pathways have been identified (Hamburger *et al.* 2002; Aung *et al.* 2006; Bari *et al.* 2006) and these and other genes (as they are identified) may provide new opportunities for developing plants that are more efficient for internal P utilization, thus increasing the harvestable product per unit of P taken up by crops.

Increasing the absolute amount of P taken up by plants, however, is likely to have greater potential for improving plant tolerance to soils of low P availability and provide more appropriate traits for inclusion in agricultural germplasm. Orthophosphate is taken up by roots and transported within plants via specific Pi-transporters. A number of genes encoding Pi-transporters have been cloned (Rausch and Bucher 2002) and members of the *Pht1* family are particularly important for Pi uptake (Mitsukawa *et al.* 1997; Mudge *et al.* 2002; Schünmann *et al.* 2004). Expression of specific *Pht1* genes is localized to root epidermal cells, primarily in root hair cells, and these PHT1 proteins show high affinity Pi transport. These P transporters are induced by P-deficiency and are energized by ATPase-mediated efflux of protons, which then drive Pi transport across the plasma-membrane against the steep electrochemical gradient that occurs for Pi between plant cells and the soil solution (Bieleski 1973; Schachtman *et al.* 1998). The effectiveness of high-affinity Pi transporters at the root surface is considered unlikely to pose any major limitation to P acquisition by plants (Bieleski 1973; Schachtman *et al.* 1998), although this relies on Pi being available for uptake in the immediate vicinity of the transporter. Whilst over-expression of high-affinity Pi transporters in cell suspensions has been shown to increase the rate of Pi uptake (Mitsukawa *et al.* 1997), there appears to be no improvement in the P nutrition of whole plants when grown in either solution culture at low external P concentrations or in soil (Rae *et al.* 2004). This demonstrates that plants are well adapted for obtaining Pi at the low Pi concentrations commonly found in soil solution and, along with other evidence, suggests that morphological and functional characteristics of roots, and their interaction with microorganisms, may be of greater importance for improving the acquisition of P by crop plants.

Associations with soil microorganisms

Microorganisms affect P availability to plants through a number of mechanisms, including direct enhancement of P supply, by extending the volume of soil accessible to the plant (e.g. by mycorrhizal associations), increasing mobilization of Pi from soil inorganic or organic P fractions and stimulation of root growth (Jakobsen *et al.* 2005). The microbial biomass also contains a substantial pool of immobilized P (~5% of soil P; Richardson 1994; Oberson and Joner 2005) which, through turnover, can contribute to plant P nutrition. Microbially-mediated turnover of P is likely to be of particular significance within the rhizosphere (Jakobsen *et al.* 2005) and numerous studies have shown the presence of a large range of bacterial and fungal species that have ability to solubilize various forms of mineral P. In some cases, specific phosphorus solubilizing bacteria and fungi have been developed for use as commercial inoculants (Leggett *et al.* 2001). However, their widespread use has been limited by inconsistent performance across different agro-ecological environments (Richardson 2001; Wakelin *et al.* 2004). Another approach has been to clone genes from soil microorganisms that improve their P acquisition and express these directly in plants (López-Bucio *et al.* 2000; Delhaize *et al.* 2001; Richardson *et al.* 2001a; Yip *et al.* 2003; George *et al.* 2004), which, as discussed below, has also met with inconsistent responses.

The most widely studied association between plants and microorganisms that affects plant P nutrition is that with mycorrhizal fungi, with arbuscular mycorrhizae (AM) fungi being particularly important for many agricultural plants (Smith and Read 1997). The primary benefit to the plant of mycorrhizal associations comes from the ability of the fungus to increase the effective exploitation of a larger soil volume. In some cases, mycorrhizae have also been reported to increase the utilization of soil organic P and to enhance the exploitation of nutrient-rich patches in soil (Tarafdar and Marschner 1994; Feng *et al.* 2003; Gavito and Olsson 2003). However, it is generally thought that mycorrhizal fungi only provide access to pools of soil P that are already available to plants (Bolan 1991; Joner *et al.* 2000; Jakobsen *et al.* 2005). Recent evidence has shown that expression of plant Pi transporters at the root epidermis are down-regulated following the colonization of roots by mycorrhizal fungi, and that there is up-regulation of specific Pi-transporters localized at the membrane where the plant and fungus interface (Liu *et al.* 1998; Karandashov *et al.* 2004). Such interactions demonstrate the complexity of the Pi-uptake efficiency phenotype in plants and that functional redundancy exists between the mechanisms that enable plants to acquire P.

Changes in root morphology

When plants are P deficient, they allocate relatively more photosynthate to root production which allows the root system to explore greater volumes of soil (Bradshaw *et al.* 1960; Hermans *et al.* 2006; Hill *et al.* 2006). A commonly observed phenotype of P-deficient plants is an increase in specific root length to achieve longer or more branched roots per unit of root dry matter (Christie 1975; Fitter 1985; Hill *et al.* 2006). This is often concurrent with an increase in root hair length and density (Itoh and Barber 1983; Föhse *et al.* 1991). Plant-available P is generally concentrated in the top few centimeters of soil profiles or in nutrient-rich patches around fertilizer granules or decomposing organic matter. Consequently, root systems which have a large proportion of roots in the surface layers of soil (Lynch 1995; Lynch and Brown 2001; Manske *et al.* 2000) and/or possess the morphological plasticity to proliferate roots within nutrient-rich patches (Hodge

2004), are more effective in P acquisition. Such traits of root foraging have been demonstrated to be useful as predictors of the P-fertilizer requirements of plants, for example, across a range of Australian pasture species (Hill *et al.* 2006).

Root hair density and length increase when plants are P-deficient (Bates and Lynch 1996; Gahoonia and Nielsen 1997). Despite root hairs having a smaller uptake capacity per unit length than larger diameter roots, due to their smaller surface area per unit area of root surface (Barber 1984), they are extremely effective in P-acquisition. Indeed a number of studies have specifically highlighted the importance of root hairs for P acquisition (e.g. Gahoonia and Nielsen 1997). Barley and Rovira (1970) demonstrated that root systems with root hairs absorbed 78% more P than those without. However, the benefit of increasing root hair density is limited when the P-depletion zones around each root hair begin to overlap, such that there is an effective optimum of root-hair density (Ma *et al.* 2001). Under field conditions, Gahoonia *et al.* (1999) have shown that wheat and barley lines with longer root hairs and greater root hair density have greater P uptake over a range of levels of P fertilization, but larger differences between long-haired and short-haired varieties occurred in more P-deficient conditions.

In addition to root hairs, the formation of dense root systems is an effective means by which plants can increase P acquisition and adjust to low soil P levels. The proteoid roots formed by white lupin (*Lupinus albus*), which are dense clusters of 'bottle-brush'-like rootlets (Gardner *et al.* 1981; Dinkelaker *et al.* 1995; Keerthisinghe *et al.* 1998; Neumann *et al.* 1999), are a specific example. Cluster root formation is associated with the locally induced release of organic anions and phosphatase enzymes (Neumann *et al.* 1999; Watt and Evans 1999), and is therefore effective in both the mobilization and subsequent capture of P in the soil. The effectiveness of cluster roots in P-acquisition is a good example of the functional compatibility of different mechanisms effecting P acquisition.

Modification of rhizosphere biochemistry

Acidification of the rhizosphere through efflux of protons (H^+), which is commonly associated with the exudation of organic anions, can alter both the solubility of sparingly-soluble inorganic P compounds and affect the kinetics of Pi adsorption/desorption reactions in the soil and subsequent availability of P to plants (Gahoonia and Nielsen 1992; Hinsinger 2001). However, of potentially greater significance to plant P nutrition is the exudation of compounds that directly influence P availability such as, low-molecular-weight organic anions and phosphatase enzymes (Marschner *et al.* 1986; Hocking 2001; Ryan *et al.* 2001; Richardson *et al.* 2005).

Many studies have shown that organic anions modify the chemistry of the rhizosphere and mobilize various forms of inorganic and organic P. This is achieved by an increase in the dissolution of sparingly-soluble P minerals; reduced sorption of P by alteration of the surface characteristics of soil particles; desorption of Pi from sorption sites (ligand exchange and ligand dissolution); and through the chelation of cations (e.g. Al and Fe in acidic soils or Ca in alkaline soils) that are commonly associated or complexed with Pi in soil (Bar-Yosef 1991; Jones and Darrah 1994; Lan *et al.* 1995; Jones 1998). Organic anions may also promote the growth of rhizosphere microorganisms that improve plant P acquisition. The importance of organic anions in increasing the availability of organic P, and its subsequent mineralization by phosphatases, has also been identified recently (Jones 1998; Otani and Ae 1999; Hayes *et al.* 2000a; Hens *et al.* 2003; Li *et al.* 2003).

It is evident that exudation of organic anions from plant roots is facilitated by transport proteins (Neumann *et al.* 1999; Ryan *et al.* 2001) and is not just a passive release. At concentrations commonly found in the rhizosphere (10–100 μM; Jones 1998) citrate has a greater potential for P mobilization than other organic anions. High rates of citrate exudation from cluster roots on white lupins are associated with a large capacity for P-mobilization in soil by this species (Hocking *et al.* 1997; Vance *et al.* 2003; Richardson *et al.* 2007). Crop plants vary in the nature and amounts of organic anions they exude from roots. For example, roots of pigeon pea (*Cajanus cajan*) exude malonate and piscidic acid which, although able to mobilize P bound to Fe and Al (Ae *et al.* 1991; Otani *et al.* 1996), is not as effective as citrate. Roots of chickpea (*Cicer arietinum*) exude malonate and citrate, although there is some uncertainty about the ability of this species to access P bound to Fe or Al (Wouterlood *et al.* 2004; Veneklaas *et al.* 2003; Pearse *et al.* 2007). Increased organic anion efflux from roots also occurs in other species grown under P-deficient conditions (Lipton *et al.* 1987; Hedley *et al.* 1982; Hoffland *et al.* 1989; Kirk *et al.* 1999), including lucerne (*Medicago sativa*), oil seed rape (*Brassica napus*) and rice (*Oryza sativa*). Roots of bread wheat (*Triticum aestivum*) exude malate, but this occurs as a response to Al toxicity, rather than P-deficiency, and is specifically activated by the presence of monomeric Al (Ryan *et al.* 2001).

A number of studies have demonstrated significant rates of organic P mineralization in proportion to soil phosphatase activity (Trasar-Cepeda and Carballas 1991; George *et al.* 2002; Lopez-Hernandez *et al.* 1998; Oehl *et al.* 2001). In natural ecosystems, mineralization of soil organic P is thought to provide the major proportion of P to plants (Fox and Comerford 1992; Polglase *et al.* 1992). Similarly, in organic-based farming systems, and where green-manure crops are used for fertilization, high rates of organic P cycling have been observed (Oberson *et al.* 1996, 2001; Nziguheba *et al.* 1998; Maroko *et al.* 1999; Oehl *et al.* 2004).

The hydrolysis of organic P is mediated by the action of phosphatase enzymes in the extracellular environment, a process which is necessary for the subsequent uptake of Pi by plant roots. At present there is no evidence for direct uptake of dissolved organic P compounds by plants, although organic P substrates may be hydrolyzed within the root apoplast (Duff *et al.* 1994; George *et al.* 2008). Extracellular phosphatase activity of plant roots is induced under conditions of P-deficiency and is associated with either root cell walls (McLachlan 1980; Dracup *et al.* 1984; Barrett-Lennard *et al.* 1993; Hayes *et al.* 1999; Hunter and McManus 1999) or is released directly into the rhizosphere (Tarafdar and Claassen 1988; Tadano *et al.* 1993; Li *et al.* 1997; Gaume *et al.* 2001). The cloning of genes encoding extracellular phosphatases from *A. thaliana* (Haran *et al.* 2000) and *L. albus* has provided direct evidence for extracellular secretion and regulation of phosphatase expression in response to P-deficiency (Wasaki *et al.* 2000; Miller *et al.* 2001).

Extracellular secretion of phosphatases from roots is correlated with the ability of plants to obtain P from organic P sources when grown under sterile conditions (Tarafdar and Claassen 1988; Hayes *et al.* 2000b; Richardson *et al.* 2000; George *et al.* 2008). For example, wheat and a range pasture species are able to utilize P from various monoester (e.g. glucose-6-phosphate) and diester (e.g. ribonucleic acid) forms, but show limited capacity to acquire P directly from *myo*-inositol hexakisphosphate (Richardson *et al.* 2000; George *et al.* 2008), despite inositol phosphates being an abundant form of organic P in many soils.

Direct hydrolysis of organic P and subsequent utilization of the mineralized Pi by roots has also been demonstrated in soil-grown plants. Depletion of various pools of extractable organic P from the rhizosphere has been linked with greater phosphatase activities around plant roots (Chen *et al.* 2002; George *et al.* 2002, 2006). However, the relative contribution of extracellular phosphatases derived from roots and from microorganisms in the utilization of soil organic P is unclear, as the numbers and activity of bacteria and fungi are greater within the rhizosphere than in the bulk soil (Chen *et al.* 2002; Richardson *et al.* 2005). There is some evidence that phosphatases derived from soil fungi have a greater affinity for organic P compounds compared to phosphatases derived from plant roots (Tarafdar *et al.* 2001). Either way, it is evident that mineralization of organic P occurs in the rhizosphere and could make an important contribution to the orthophosphate requirement of plants for growth.

STRATEGIES TO IMPROVE P-UPTAKE EFFICIENCY IN AGRICULTURAL CROPS

It is evident that plants have a wide range of mechanisms and physiological traits that facilitate increased availability and acquisition of P from soil. Despite this, there have been few successful attempts to increase the efficiency of P utilization by crop plants directly. This is largely due to the complex interactions between the range of P-acquisition mechanisms and their efficacy in different agro-ecological environments (Caradus 1995; Wissuwa 2003; George *et al.* 2005c). Irrespective of this, a number of promising approaches for directly improving the P acquisition of plants have been identified, including the identification of key morphological and physiological traits associated with P-uptake efficiency and the possible enhancement of specific P-acquisition processes through either targeted breeding or by gene technology.

Genotypic variation in the efficiency of P-use

Genotypic variation in the efficiency of P use has been identified in a number of cereal crops including maize, wheat, barley and rice (Fageria *et al.* 1988; Fageria and Baligar 1997a,b; Ciarelli *et al.* 1998; Manske *et al.* 2000; Wissuwa and Ae

2001a,b; Osborne and Rengel 2002; Wang *et al.* 2005; Zhu *et al.* 2005; Liao *et al.* 2008) and a range of other crop species such as soybean, brassica, cowpea and pigeon pea (Furlani *et al.* 2002; Greenwood *et al.* 2006; Vesterager *et al.* 2006; Marschner *et al.* 2007). Such studies have routinely been targeted to either some measure of P-use efficiency (see Gregory and George 2005; White *et al.* 2005, for definitions of various use-efficiencies) or specific traits such as root architecture (Lynch and Brown 2001), root hair formation (Gahoonia and Nielsen 1996; Gahoonia *et al.* 1999) or phosphatase activity (George *et al.* 2008). In almost all cases, and for different traits, significant genotypic variation has been found between varieties of the crop species tested. This is exemplified in Figure 11.1,

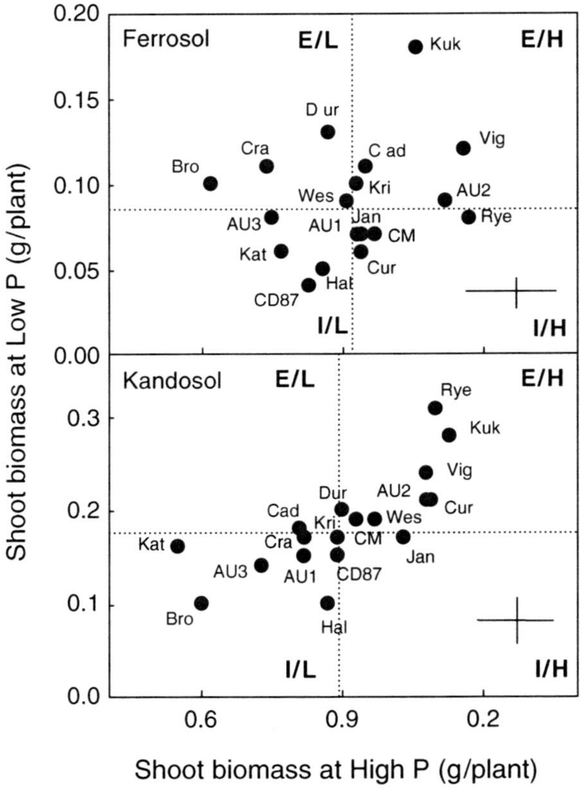

Fig. 11.1 Relationship between shoot biomass of plants grown in low-P and high-P treatments in two Australian soils (Ferrosol and Kandosol). Genotypes were classified as being P-efficient with high biomass potential (E/H), P-efficient with low biomass potential (E/L), P-inefficient with high biomass potential (I/H) or P-inefficient with low biomass potential (I/L). The dotted lines represent the median point of the data on both axes. For each panel, the vertical and horizontal bars show LSD ($P = 0.05$). The wheat genotypes are as follows: Brookton (Bro), Cadoux (Cad), CD87 (CD), Chuan-Mai 18 (CM), Cranbrook (Cra), Halberd (Hal), Janz (Jan), Katepwa (Kat), Krichauff (Kri), Kukri (Kuk), Vigour 18 (Vig), Westonia (Wes), three synthetic hexaploid wheats AUS29542 (AU1), AUS29563 (AU2), AUS29630 (AU3) and Rye cv.Ryesun (Rye), triticale cv. Currency (Cur) and durum wheat line 41 (Dur). (Taken from Liao *et al.* 2008.)

which shows significant differences in shoot biomass in a collection of 18 cereals (predominantly wheat genotypes) when grown in two contrasting Australian soils, either at low soil P (deficient) or with applied P. This study highlights the importance of identifying germplasm that is not only efficient under low P conditions (E; Figure 11.1) but also retains high biomass (yield) potential with applied P (E/H; Figure 11.1). It is also important that screens be conducted in soil rather than in artificial media or solution culture, where there may be a discrepancy in the relative ranking of different lines (Liao *et al.* 2008; Hayes *et al.* 2004). It is of further significance to consider that, for a multi-mechanistic trait such as response to P-deficiency, mechanisms (or markers and/or genes) identified for the trait may only be specific for the conditions used in the screen and not widely applicable to other environments. For example, variation in the activity of root-associated phosphatases, whilst effective in promoting growth of wheat plants *in-vitro*, appeared to show little relationship for predicting the growth response of wheat lines in a range of P-deficient soils (George *et al.* 2008). This complex genotype by environmental interaction is highlighted when attempts are made to assign P-efficiency traits to a single QTL (quantitative trait loci), as complex traits may not be attributable to a single mechanism (or QTL) in isolation (Wissuwa and Ae 2001b; Liao *et al.* 2008).

A common finding in many screens for P-efficiency is that traditional varieties or landraces have greater P-use efficiency than modern cultivars (Manske *et al.* 2000; Wissuwa and Ae 2001a). This is often attributed to the impact of breeding programs being performed under nutrient replete conditions (Buso and Bliss 1988). Moreover, it is considered that whilst such programs have produced cultivars with large harvest indexes, which are also relatively internally P-efficient, further gains in P limited yield are most likely to be achieved by improving the external P efficiency (Wissuwa and Ae 2001a). This has been done successfully for upland rice on volcanic soils in the Philippines (Figure 11.2), where near isogenic lines (NILs) generated from a backcross of a P-efficient landrace (Kasalath) to a P-inefficient variety (Nipponbare) resulted in lines with greater biomass and P uptake than the Nipponbare parental line under P-deficient conditions. Thus showing that there is potential for the introgression of traits for improved P-efficiency into modern high yielding varieties.

Root traits

Root growth, branching and root hair morphology are clearly important to the efficient acquisition of P from soil, and considerable genetic variation exists both across and within different plant species. For example, Lynch and van Beem (1993) showed that root architecture in bean (*Phaseolus vulgaris*) was responsive to low P availability and that variation in this trait between genotypes contributed to differing capacities for P uptake. Genotypes adapted to P-deficiency produced more adventitious roots in response to P stress and had greater capacity for root foraging in nutrient-rich surface layers of soil (Lynch and Brown 2001; Lynch 2005). Similar responses to root growth and development of lateral roots, and their contribution to

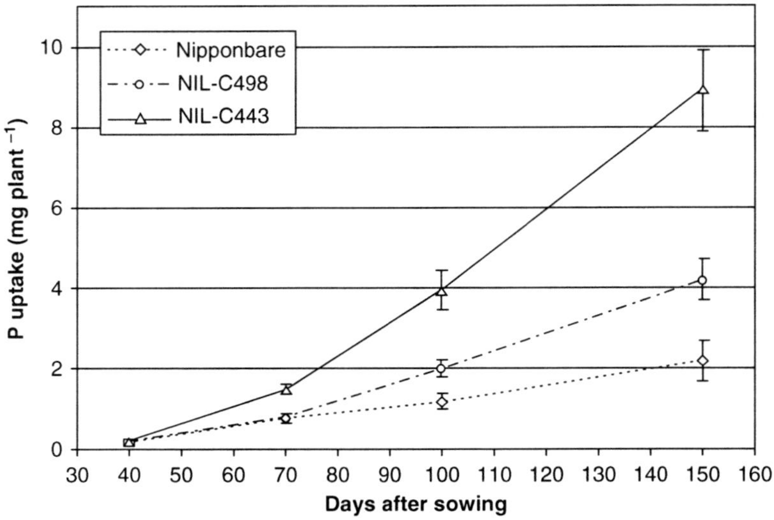

Fig. 11.2 Development of P uptake by a range of rice (*Oryza sativa*) lines over time when grown in a P-deficient Andisol from the Philippines. Included are near isogenic lines (NIL-C498 and NIL-C443) generated from a backcross of a P-efficient landrace (Kasalath) to a P-inefficient variety (Nipponbare). (Taken from Wissuwa and Ae 2001b.)

P-uptake have been reported in maize (*Zea mays*; Zhu and Lynch 2004) and wheat (Manske *et al.* 2000; Liao *et al.* 2008).

Variation in root hair length and morphology has likewise been reported for a number of crop species including white clover (*Trifolium repens*), barley (*Hordeum vulgare*), wheat, bean and soybean (Gahoonia and Nielsen 1997; Caradus 1979; Yan *et al.* 1995; Wang *et al.* 2004). In the case of white clover, high heritability of root hair length enabled the selection of this trait within a breeding program (Caradus 1995). Genotypic variation in root hair length affects the size of P-depletion zones around barley roots when grown in soil (Figure 11.3; Gahoonia and Nielsen 1997), and has been shown to contribute to improved P uptake under field conditions (Gahoonia *et al.* 1999).

Collectively, these observations suggest that selection for more P-efficient cultivars can be achieved through traditional breeding. However, efforts to develop successful breeding programs have been hampered by difficulties in developing robust and rapid screening methods for root traits under field conditions. The development of molecular markers that are appropriately linked to desired traits may offer new opportunities for success. Indeed, major QTLs for root hair length and P-uptake capacity have recently been identified in bean and maize (Yan *et al.* 2004; Zhu *et al.* 2005) and major QTLs associated with P-efficiency in rice have also been identified (Wissuwa and Ae 2001a; Wissuwa 2003, 2005). Alternatively, identification of genes that control Pi-uptake efficiency, or are associated with root development, may have direct application for plant improvement. For example, key regulatory

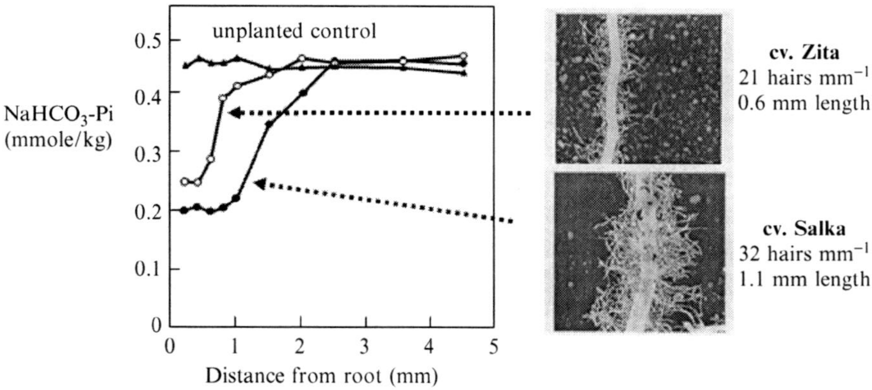

Fig. 11.3 Depletion of $NaHCO_3$-extractable inorganic P (mmole P/kg soil) from the rhizosphere of two cultivars of barley (*Hordeum vulgare*) with different root hair morphologies (cultivars Zita and Salka) as compared to an unplanted control soil. (Taken from Gahoonia and Nielsen 1997.)

genes that are involved in root proliferation, root branching and root hair development have been identified (e.g. Zhang and Forde 1998; Williamson *et al.* 2001) and analysis of whole plant genomes is providing new insights into pathways of genes that are associated with plant responses to P-deficiency (Hammond *et al.* 2003; Wasaki *et al.* 2003; Wu *et al.* 2003).

Extracellular organic anions

Given the importance of organic anions in P mobilization, it is possible that genetic variation in the capacity to exude organic anions may be exploited. Whilst information on intraspecific variation in organic-anion exudation from plants is relatively scarce, there is evidence that landraces of white lupin from acidic and alkaline soils show differences in organic anion exudation and relative capacity to access different forms of mineral P (Pearse *et al.* 2008). Differences in the ability of cultivars of pigeon pea to exude organic acids have similarly been identified (Subbarao *et al.* 1997; Ishikawa *et al.* 2002). Although it is difficult to know whether such differences can be exploited commercially to improve P uptake by plants, wide-based screening programs and selection of genotypes for increased exudation of organic-anions is worthy of pursuit.

Alternatively, the heterologous expression of genes for enzymes involved in organic anion synthesis in roots has been investigated as a means to increasing exudation of organic anions from roots. Over-expression of a bacterial citrate synthase (CS) gene in tobacco (*Nicotiana tabacum*) has been reported to increase citrate efflux from roots of transgenic lines compared to control plants (de la Fuente-Martínez *et al.* 1997). Enhanced efflux of citrate enabled the transgenic lines to access P from Ca-P that was otherwise unavailable to control plants (López-Bucio *et al.* 2000). However, using similar gene constructs and in some cases the same transgenic lines, Delhaize *et al.* (2001) could not confirm these results. Moreover, tobacco plants that over-expressed a tobacco CS, or were down-regulated for isocitrate dehydrogenase expression, showed no significant increase in citrate efflux even though in some cases the plants had greater internal citrate concentration (Delhaize *et al.* 2003). Notwithstanding this, it is apparent that there is potential to enhance organic acid exudation by targeting the citrate synthesis biosynthetic pathway. Over-expression of a plant gene for mitochondrial CS in *Arabidopsis thaliana* enhanced citrate efflux, with an associated small improvement in P acquisition (Koyama *et al.* 2000). Other approaches have attempted to increase organic-anion exudation by over-expressing other genes of the tricarboxylic acid (TCA) cycle such as, phosphoenol pyruvate carboxylase (PEPC) and malate dehydrogenase (MDH; Raghothama 1999). Over-expression of MDH in lucerne increased the efflux of organic anions from roots, but over-expression of PEPC did not (Tesfaye *et al.* 2001). Irrespective of the approach, it is worth noting that the increase in organic-anion efflux from transgenic plants needs to be of greater magnitude than that achieved thus far for *Arabidopsis* or lucerne to be of agronomic significance.

Genes that encode channels involved in the transport of organic anions from roots to the rhizosphere are a further option for a gene technology approach. Citrate-permeable channels in the plasma membrane of cluster roots of white lupin have been identified (Zhang *et al.* 2004), and a gene encoding a malate channel has been cloned from wheat (Sasaki *et al.* 2004). When expressed in transgenic barley this gene (*Almt1*) resulted in increased exudation of malate, albeit in an Al-activated manner (Delhaize *et al.* 2004). To be useful for P nutrition it is important that transporters be identified which are functional under conditions of P-deficiency, and that the expression of which is regulated appropriately in root tissues.

Release of extracellular phosphatase

A number of studies have investigated extracellular phosphatase activities of plant roots and significant genotypic variation across and within different species has been identified (Tadano *et al.* 1993; Li *et al.* 1997; Asmar 1997; Gaume *et al.* 2001; George *et al.* 2008). On this basis it has been proposed that variation in root phosphatase activities may be useful in the selection of genotypes for greater utilization of soil organic P (Asmar *et al.* 1995). George *et al.* (2008) screened a range of wheat lines for variation in the activity of root-associated phosphatases towards different organic P substrates (Figure 11.4). While some relationships were identified between the different activities and the ability of the plants to utilize specific organic P substrates *in vitro*, no clear relationships were found for the growth and P nutrition of the plants when grown in a range of soils (George *et al.* 2008). This suggests that variability in

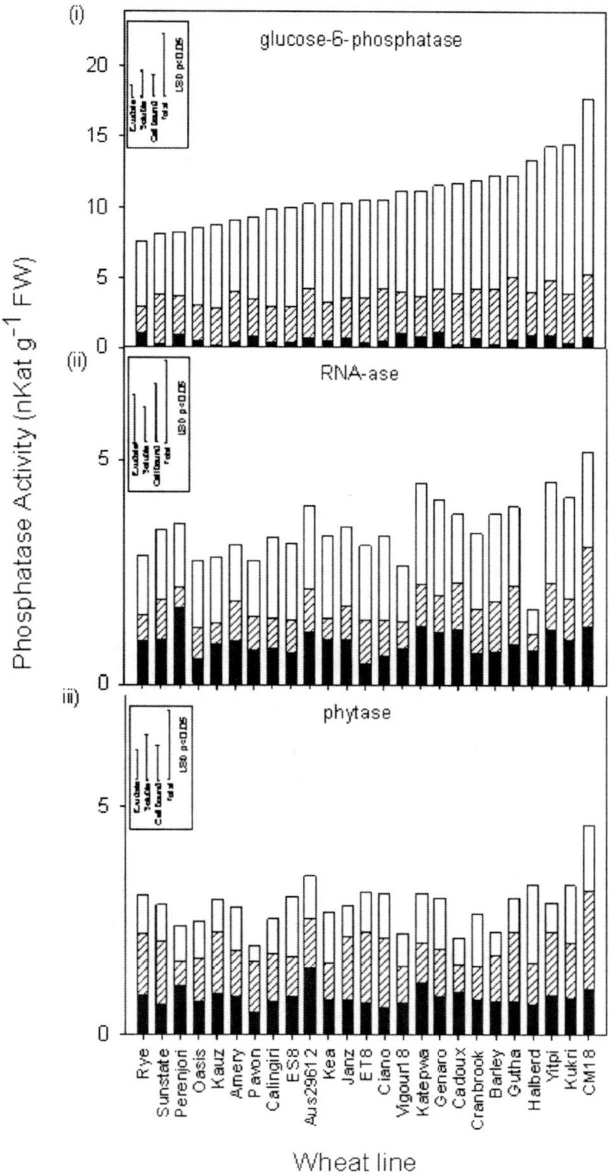

Fig. 11.4 Activities (nKat g^{-1} root FW) of phosphatases associated with roots of 23 cultivars of wheat, one of barley and one of rye. Phosphatase activities, assigned as exuded, soluble and cell-bound fractions (solid, hatched and open bars, respectively) were measured against (i) glucose 6-phosphate, (ii) ribonucleic acid and (iii) *myo*-inositol hexakisphosphate. Data are the mean of five replicates and are presented in ascending order with respect to total activity against glucose 6-phosphate. Differences between cultivars for each root phosphatase activity fraction, and the total activity, were established using ANOVA and for each panel are shown by LSD ($p < 0.05$) bars. (Adapted from George *et al.* 2008.)

phosphatase activities either has little significance in the P nutrition of soil-grown plants or, more likely, that any benefit from hydrolysis of organic P by either plant or microbial-derived phosphatases was common to all genotypes.

Research effort has also focused on improving the ability of plants to acquire P directly from common forms of soil organic P, such as inositol phosphates. A number of studies have developed transgenic plants with heterologous expression of microbial phytases (Richardson *et al.* 2001a; Zimmermann *et al.* 2003; Lung *et al.* 2005; Xiao *et al.* 2005). Previous studies demonstrated that many plants (including monocots and dicots) have limited capacity to utilize P from *myo*-inositol hexakisphosphate, and that this was associated, at least in part, with low levels of extracellular phytase (Findenegg and Nelemans 1993; Hayes *et al.* 2000b; Richardson *et al.* 2000, 2001b). Transgenic plants that over-express microbial phytase and release it from their roots have novel ability to hydrolyze P from *myo*-inositol hexakisphosphate and, when grown under controlled conditions, showed enhanced growth and P nutrition (Figure 11.5; George *et al.* 2004; Richardson *et al.* 2001a; Mudge *et al.* 2003; Zimmermann *et al.* 2003). However, when grown

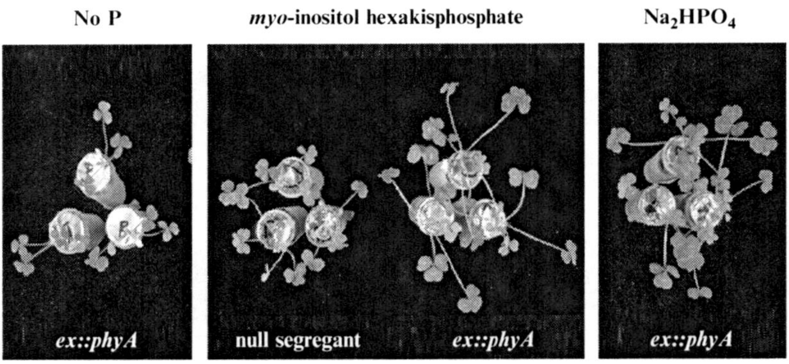

shoot dry wt (mg $plant^{-1}$)	28.1	40.7	51.8	47.9
shoot phosphorus (μg P $shoot^{-1}$)	48.3	103.5	299.0	305.3
exuded root phytase activity* (nKat g^{-1} root dry wt)	-	1.3	107.9	-

* activity for wild-type plants was 0.6 nKat g^{-1} root dry wt

Fig. 11.5 Growth, phosphorus nutrition and activity of phytase exuded from the roots of transgenic *Trifolium subterraneum*. Shown are plants that release the *Aspergillus niger* phytase (*ex::phyA*) as an extracellular enzyme and the corresponding null segregant transgenic control line. Plants were grown for 28 days in sterile agar either without added phosphorus (no P) or with phosphorus supplied as sodium phytate (*myo*-inositol hexakisphosphate) or disodium phosphate at 0.9 mM (with respect to phosphate). (Taken from George *et al.* 2004.)

in a range of soils these plants have generally shown limited capacity to access additional P over that of control plants (George *et al.* 2004, 2005b). Factors that may limit the capacity of the plants to access P from inositol phosphates in soil include, small amounts and/or poor availability of inositol phosphates, immobilization of phytase activity by sorption on soil components, and compensatory effects due to the presence of soil microorganisms (George *et al.* 2005a,b; Richardson *et al.* 2007). Irrespective of this, relatively large responses (up to 70% over the transgenic control) of plants that exude phytase have been observed under certain conditions, for example after recent P application and in manured soils (George *et al.* 2005b; George 2008; unpublished). Biochemical characteristics of different phosphatases are also important for mineralization of soil organic P (Figure 11.6). Phytase from *Aspergillus niger* (which is expressed in transgenic plants, Richardson *et al.* 2001a; George *et al.* 2004, 2005b) was less effective at mineralizing P from endogenous sources of organic P in a range of soils in comparison to a phytase from *Peniophora lycii*. A key difference between these two phytases is their mobility in soils at pH typical of the soils tested, where the *P. lycii* phytase retained greater activity within soil solution as compared to interaction with the soil solid phase (George *et al.*

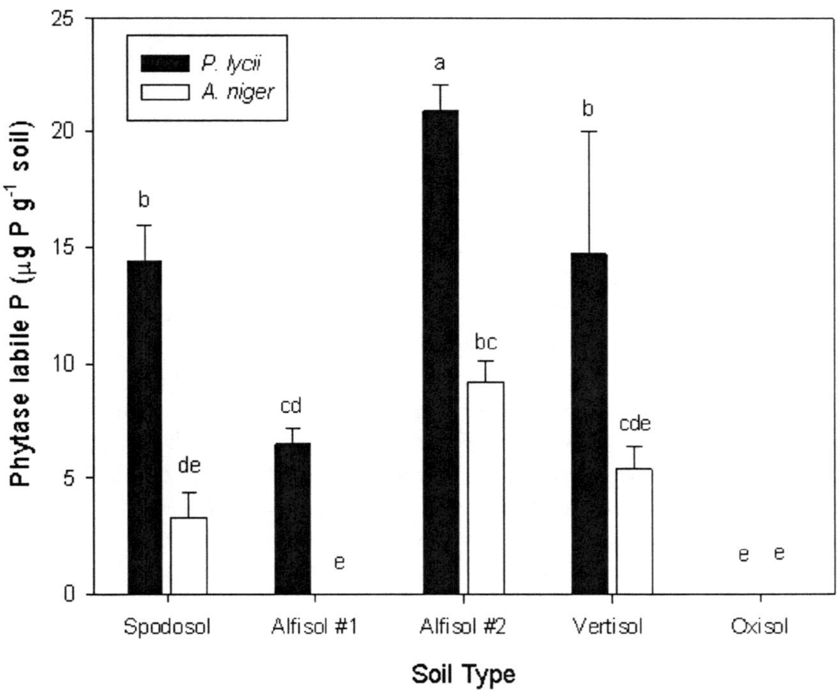

Fig. 11.6 Concentration of phytase labile P (µg P g^{-1} soil) in suspensions (1:10 w:v) of a range of Australian soils incubated with *Peniophora lycii* and *Aspergillus niger* phytase (120 nKat g^{-1} soil) for 24 hours. Data are the mean of four replicates with bars representing one standard error. Differences in P release were established using ANOVA and columns with the same letters are not significantly different (LSD, $p < 0.05$). (Taken from George *et al.* 2007.)

2007). These results further highlight the complexity inherent in attempting to improve multi-mechanistic tolerance traits by a single gene approach. Whilst potential exists for manipulating P-use efficiency at a genetic scale, success will often be limited by poor understanding of the control of the mechanisms imposed by different soil environments. It is also important that, where possible, functional redundancy and/or compatibility of alternative mechanisms be considered in order to optimize the efficacy of the target trait.

CONCLUDING REMARKS

Management of soil P remains a critical issue for the economic and environmental sustainability of agriculture globally. It is therefore essential that we have appropriate understanding of the mechanisms by which plants are able to acquire P from soil. In this chapter, various processes and physiological traits of plants that facilitate the availability and acquisition of P from soil have been outlined and some possibilities for deploying these traits into agricultural germplasm discussed. Better understanding of these processes and development of improved germplasm may ultimately improve the P-use efficiency of agriculture systems and provide valuable information for wider-scale land and resource management. However, at present it is evident that the full extent of the complexity of the gene by gene and gene by environment interactions that are associated with plant P nutrition are not well appreciated, and that our comprehension of the functional redundancy and compatibility of different mechanisms both within individual plants and between coexisting organisms is poor. Furthermore, management systems that optimize the availability of soil P to plants and maximize the benefits from P inputs, whether they are applied as soluble fertilizer, rock phosphates or as organic amendments, need to be further developed. This will require detailed knowledge of how different crops and cropping systems (e.g. strategic use of crop rotations, manure crops etc.) contribute to the soil P cycle and the relative importance of microbial processes and their interactions with plants roots. In the longer term, and as we gain better understanding of how different mechanisms interact and influence P availability, there will be opportunity to develop more P-efficient plants through the manipulation of specific traits. This might be achieved through selection and marker-assisted plant breeding for specific root traits or by direct manipulation of plants by gene technologies. In either case it is important that a systems approach to P management continues to be developed for a more sustainable agriculture.

REFERENCES

Ae N, Arihara J, Okada K (1991) Phosphorus uptake mechanisms of pigeon pea grown in alfisols and vertisols. In: Johansen C, Lee KK, Sahrawat KL (eds), Phosphorus Nutrition in Grain Legumes in the Semi-arid Tropics. ICRISAT, Patancheru, pp 91–98

Amtmann A, Hammond JP, Armengaud P, White PJ (2006) Nutrient sensing and signalling in plants: potassium and phosphorus. Adv Bot Res 43: 209–257

Anderson G (1980) Assessing organic phosphorus in soils. In: Khasawneh, FE, Sample EC, Kamprath EJ (eds), The Role of Phosphorus in Agriculture. ASA/CSSA/SSSA, Madison, WI, pp 263–310

Asmar F (1997) Variation in activity of root extracellular phytase between genotypes of barley. Plant Soil 195: 61–64

Asmar F, Gahoonia TS, Nielsen NE (1995) Barley genotypes differ in activity of soluble extracellular phosphatase and depletion of organic phosphorus in the rhizosphere soil. Plant Soil 172: 117–122

Attiwill PM, Adams MA (1993) Nutrient cycling in forests. New Phytol 124: 561–582

Aung K, Lin SI, Wu CC, Huang YT, Su CL, Chiou TJ (2006) *pho2*, a phosphate overaccumulator, is caused by a nonsense mutation in a MicroRNA399 target gene. Plant Physiol 141: 1000–1011

Barber SA (1984) Soil Nutrient Bioavailability: A Mechanistic Approach. Wiley, New York

Barley KP, Rovira AD (1970) The influence of root hairs on the uptake of phosphate. Commun Soil Sci Plant Anal 1: 287–292

Bari R, Pant BD, Stitt M, Scheible WR (2006) PHO2, MicroRNA399, and PHR1 define a phosphate-signaling pathway in plants. Plant Physiol 141: 988–999

Barrett-Lennard EG, Dracup M, Greenway H (1993) Role of extracellular phosphatases in the phosphorus-nutrition of clover. J Exp Bot 44: 1595–1600

Bar-Yosef B (1991) Root excretions and their environmental effects. Influence on availability of phosphorus. In: Waisel Y, Eshel A, Kafkafi U (eds), Plant Roots: The Hidden Half. Marcel Dekker, New York, pp 529–557

Bates TR, Lynch JP (1996) Stimulation of root hair elongation in *Arabidopsis thaliana* by low phosphorus availability. Plant Cell Environ 19: 529–538

Bieleski RL (1973) Phosphate pools, phosphate transport, and phosphate availability. Ann Rev Plant Physiol 24: 225–252

Bolan NS (1991) A critical review on the role of mycorrhizal fungi in the uptake of phosphorus by plants. Plant Soil 134: 189–207

Bradshaw AD, Chadwick MJ, Jowett D, Lodge RW, Snaydon RW (1960) Experimental investigations into the mineral nutrition of several grass species. Part III. Phosphate level. J Ecol 48: 631–637

Buso GSC, Bliss FA (1988) Variability among lettuce cultivars grown at two levels of available phosphorus. Plant Soil 111: 67–73

Caradus JR (1979) Selection for root hair length in white clover (*Trifolium repens* L.). Euphytica 28: 489–494

Caradus JR (1995) Genetic control of phosphorus uptake and phosphorus status in plants. In: Johansen C, Lee KK, Sharma KK, Subbarao GV, Kueneman EA (eds), Genetic Manipulation of Crop Plants to Enhance Integrated Nutrient Management in Cropping Systems. 1. Phosphorus. ICRISAT, Patancheru, pp 55–74

Chen CR, Condron LM, Davis MR, Sherlock RR (2002) Phosphorus dynamics in the rhizosphere of perennial ryegrass (*Lolium perenne* L.) and radiata pine (*Pinus radiata* D.Don). Soil Biol Biochem 34: 487–499

Christie EK (1975) Physiological responses of semiarid grasses. II The pattern of root growth in relation to external phosphorus concentration. Aust J Agric Res 26: 437–446

Ciarelli DM, Furlani AMC, Dechen AR, Lima M (1998) Genetic variation among maize genotypes for phosphorus uptake and phosphorus-use efficiency in nutrient solution. J Plant Nutr 21: 2219–2229

Condron LM, Frossard E, Tiessen H, Newman RH, Stewart JWB (1990) Chemical nature of organic phosphorus in cultivated and uncultivated soils under different environmental conditions. J Soil Sci 41: 41–50

Dalal RC (1977) Soil organic phosphorus. Adv Agron 29: 85–117 de la Fuente-Martínez JM, Ramirez-Rodriguez V, Cabrera-Ponce JL, Herrera-Estrella L (1997) Aluminum tolerance in transgenic plants by alteration of citrate synthesis. Science 276: 1566–1588

Delhaize E, Hebb DM, Ryan PR (2001) Expression of a *Pseudomonas aeruginosa* citrate synthase gene is not associated with either enhanced citrate accumulation or efflux. Plant Physiol 125: 2059–2067

Delhaize E, Ryan PR, Hocking PJ, Richardson AE (2003) Effects of altered citrate synthase and isocitrate dehydrogenase expression on internal citrate concentrations in tobacco (*Nicotiana tabacum* L.). Plant Soil 248: 137–144

Delhaize E, Ryan PR, Hebb DM, Yamamoto Y, Sasaki T, Matsumoto H (2004) Engineering high level aluminum tolerance in barley with the ALMT1 gene. Proc Natl Acad Sci USA 101: 15249–15254

Dinkelaker B, Hengeler C, Marschner H (1995) Distribution and function of proteoid roots and other root clusters. Bot Acta 108: 183–200

Dracup MNH, Barrett-Lennard EG, Greenway H, Robson AD (1984) Effect of phosphorus deficiency on phosphatase activity of cell walls from roots of subterranean clover. J Exp Bot 35: 466–480

Duff SMG, Sarath G, Plaxton WC (1994) The role of acid phosphatases in plant phosphorus metabolism. Physiol Plant 90: 791–800

Fageria NK, Baligar VC (1997a) Phosphorus use efficiency by corn genotypes. J Plant Nutr 20: 1267–1277

Fageria NK, Baligar VC (1997b) Upland rice genotypes evaluation for phosphorus use efficiency. J Plant Nutr 20: 499–509

Fageria NK, Wright RJ, Baligar VC (1988) Rice cultivar evaluation for phosphorus use efficiency. Plant Soil 111: 105–109

Feng G, Song YC, Li XL, Christie P (2003) Contribution of arbuscular mycorrhizal fungi to utilization of organic sources of phosphorus by red clover in calcareous soil. Appl Soil Ecol 22: 139–148

Findenegg GR, Nelemans JA (1993) The effect of phytase on the availability of phosphorus from myo-inositol hexaphosphate (phytate) for maize roots. Plant Soil 154: 189–196

Fitter AH (1985) Functional significance of root morphology and root system architecture. In: Fitter AH, Atkinson D, Read DJ, Useher MB (eds), Ecological Interactions in Soil-Plant, Microbes and Animals. Blackwell, London, pp 87–106

Föhse D, Claassen N, Jungk A (1991) Phosphorus efficiency of plants II. Significance of root radius, root hairs and cation-anion balance for phosphorus influx in seven plant species. Plant Soil 132: 261–272

Fox TR, Comerford NB (1992) Rhizosphere phosphatase activity and phosphatase hydrolyzable organic phosphorus in two forested spodosols. Soil Biol Biochem 24: 579–583

Furlani AMC, Furlani PR, Tanaka RT, Mascarenhas HAA, Delgado MDDP (2002) Variability in soybean germplasm in relation to phosphorus uptake and use efficiency. Sci Agric 59: 529–536.

Gahoonia TS, Nielsen NE (1992) The effect of root induced pH changes on the depletion of inorganic and organic phosphorus in the rhizosphere. Plant Soil 143: 185–191

Gahoonia TS, Nielsen NE (1996) Variation in acquisition of soil phosphorus among wheat and barley genotypes. Plant Soil 178: 223–230

Gahoonia TS, Nielsen NE (1997) Variation in root hairs of barley cultivars doubled soil phosphorus uptake. Euphytica 98: 177–182

Gahoonia TS, Nielsen NE, Lyshede OB (1999) Phosphorus (P) acquisition of cereal cultivars in the field at three levels of P fertilization. Plant Soil 211: 269–281

Gardner WK, Parbury DG, Barber DA (1981) Proteoid root morphology and function in *Lupinus albus*. Plant Soil 60: 143–147

Gaume A, Machler F, De León C, Narro L, Frossard E (2001) Low-P tolerance by maize (*Zea mays* L.) genotypes: significance of root growth, and organic acids and acid phosphatase root exudation. Plant Soil 228: 253–264

Gavito ME, Olsson PA (2003) Allocation of plant carbon to foraging and storage in arbuscular mycorrhizal fungi. FEMS Microbiol Ecol 45: 181–187

George TS, Gregory PJ, Robinson JS, Buresh RJ (2002) Changes in phosphorus concentrations and pH in the rhizosphere of some agroforestry and crop species. Plant Soil 246: 65–73

George TS, Richardson AE, Hadobas PA, Simpson RJ (2004) Characterisation of transgenic *Trifolium subterraneum* L. which expresses *phyA* and releases extracellular phytase: growth and P nutrition in laboratory media and soil. Plant Cell Environ 27: 1351–1361

George TS, Richardson AE, Simpson RJ (2005a) Behaviour of plant-derived extracellular phytase upon addition to soil. Soil Biol Biochem 37: 977–988

George TS, Simpson RJ, Hadobas PA, Richardson AE (2005b) Expression of a fungal phytase gene in *Nicotiana tabacum* improves phosphorus nutrition in plants grown in amended soil. Plant Biotech J 3: 129–140

George TS, Richardson AE, Smith JB, Hadobas PA, Simpson RJ (2005c) Limitations to the potential of transgenic *Trifolium subterraneum* L. plants that exude phytase when grown in soils with a range of organic P content. Plant Soil 278: 263–274.

George TS, Turner BL, Gregory PJ, Cade-Menun BJ, Richardson AE (2006) Depletion of organic phosphorus from oxisols in relation to phosphatase activities in the rhizosphere. Eur J Soil Sci 57: 47–57

George TS, Gregory PJ, Simpson RJ, Richardson AE (2007) Differential interactions of *Aspergillus niger* and *Peniophora lycii* phytases with soil particles affects the hydrolysis of inositol phosphates. Soil Biol Biochem 39: 793–803

George TS, Gregory PJ, Hocking PJ, Richardson AE (2008) Variation in root-associated phosphatase activities in wheat contributes to the utilisation of organic P substrates in vitro, but does not explain differences in the P nutrition of plants when grown in soil. Environ Exp Bot (in press)

Greenwood DJ, Stellacci AM, Meacham MC, Mead A, Broadley MR, White PJ (2006) Relative values of physiological parameters of P response of different genotypes can be measured in experiments with only two P treatments. Plant Soil 281: 159–172

Gregory PJ, George TS (2005) Soil Management for Nutrient Use Efficiency - An Overview. Proceedings 564, International Fertiliser Society, York

Hamburger D, Rezzonico E, MacDonald-Comber Petétot J, Somerville C, Poirier Y (2002) Identification and characterization of the *Arabidopsis* PHO1 gene involved in phosphate loading to the xylem. Plant Cell 14: 889–902

Hammond JP, Broadley MR, White PJ (2004) Genetic responses to phosphorus deficiency. Ann Bot 94: 323–332

Hammond JP, Bennett MJ, Bowen HC, Broadley MR, Eastwood DC, May ST, Rahn C, Swarup R, Woolaway KE, White PJ (2003) Changes in gene expression in *Arabidopsis* shoots during phosphate starvation and the potential for developing smart plants. Plant Physiol 132: 578–596

Haran S, Logendra S, Seskar M, Bratanova M, Raskin I (2000) Characterization of *Arabidopsis* acid phosphatase promoter and regulation of acid phosphatase expression. Plant Physiol 124: 615–626

Harrison AF (1987) Soil Organic Phosphorus - A Review Of World Literature. CAB International, Wallingford

Hawkes GE, Powlson DS, Randall EW, Tate KR (1984) A ^{31}P nuclear magnetic resonance study of the phosphorus species in alkali extracts of soils from long-term field experiments. J Soil Sci 35: 35–45

Hayes JE, Richardson AE, Simpson RJ (1999) Phytase and acid phosphatase activities in extracts from roots of temperate pasture grasses and legumes. Aust J Plant Physiol 26: 801–809

Hayes JE, Richardson AE, Simpson RJ (2000a) Components of organic phosphorus in soil extracts that are hydrolyzed by phytase and acid phosphatase. Biol Fertil Soils 32: 279–286

Hayes JE, Simpson RJ, Richardson AE (2000b) The growth and phosphorus utilisation of plants in sterile media when supplied with inositol hexaphosphate, glucose 1-phosphate or inorganic phosphate. Plant Soil 220: 165–174

Hayes JE, Zhu Y-G, Mimura T, Reid RJ (2004) An assessment of the usefulness of solution culture in screening for phosphorus efficiency in wheat. Plant Soil 261: 91–97

Hedley MJ, White RE, Nye PH (1982) Plant-induced changes in the rhizosphere of rape (*Brassica napus* var. Emerald) seedlings III. Changes in L value, soil phosphate fractions and phosphatase activity. New Phytol 91: 45–56
Hens M, Turner BL, Hocking PJ (2003) Chemical nature and bioavailability of soil organic phosphorus mobilized by organic anions. In: Rengel Z (ed), Proceedings of the Second International Symposium on Phosphorus Dynamics in the Soil-Plant Continuum. University of Western Australia, Uniprint, Perth, pp 16–17
Hermans C, Hammond JP, White PJ, Verbruggen N (2006) How do plants respond to nutrient shortage by biomass allocation? Trends Plant Sci 11: 610–617
Hill JO, Simpson RJ, Moore AD, Chapman DF (2006) Morphology and response of roots of pasture species to phosphorus and nitrogen nutrition. Plant Soil 286: 7–19
Hinsinger P (2001) Bioavailability of soil inorganic P in the rhizosphere as affected by root-induced chemical changes: a review. Plant Soil 237: 173–195
Hocking P (2001) Organic acids exuded from roots in phosphorus uptake and aluminium tolerance of plants in acid soils. Adv Agron 74: 63–97
Hocking PJ, Keerthisinghe G, Smith FW, Randall PJ (1997) Comparison of the ability of different crop species to access poorly-available soil phosphorus. In: Ando T, Fujita K, Mae T, Matsumoto H, Mori S, Sekiya J (eds), Plant nutrition for sustainable food production and agriculture. Kluwer Academic Publishers, Dordrecht, The Netherlands, pp 305–308
Hodge A (2004) The plastic plant: root responses to heterogeneous supplies of nutrients. New Phytol 162: 9–24
Hoffland E, Findenegg GR, Nelemans JA (1989) Solubilization of rock phosphate by rape. II. Local root exudation of organic acids as a response to P-starvation. Plant Soil 113: 161–165
Hunter DA, McManus MT (1999) Comparison of acid phosphatase in two genotypes of white clover with different responses to applied phosphate. J Plant Nutr 22: 679–692
Ishikawa S, Adu-Gyamfi JJ, Nakamura T, Yoshihara T, Watanabe T, Wagatsuma T (2002) Genotypic variability in phosphorus solubilising activity of root exudates by pigeon pea grown in low-nutrient environments. Plant Soil 245: 71–81
Itoh S, Barber SA (1983) Phosphorus uptake by six plant species as related to root hairs. Agron J 75: 457–461
Jakobsen I, Leggett ME, Richardson AE (2005) Rhizosphere microorganisms and plant phosphorus uptake. In: Sims JT, Sharpley AN (eds), Phosphorus, Agriculture and the Environment. American Society for Agronomy, Madison, WI, pp 437–494
Joner EJ, van Aarle IM, Vosatka M (2000) Phosphatase activity of extra-radical arbuscular mycorrhizal hyphae: a review. Plant Soil 226: 199–210
Jones DL (1998) Organic acids in the rhizosphere - a critical review. Plant Soil 205: 25–44
Jones DL, Darrah PR (1994) Role of root derived organic acids in the mobilization of nutrients in the rhizosphere. Plant Soil 166: 247–257
Karandashov V, Nagy R, Wegmuller S, Amrhein N, Bucher M (2004) Evolutionary conservation of a phosphate transporter in the arbuscular mycorrhizal symbiosis. Proc Natl Acad Sci USA 101: 6285–6290
Keerthisinghe G, Hocking PJ, Ryan PR, Delhaize E (1998) Effect of phosphorus supply on the formation and function of proteoid roots of white lupin (*Lupinus albus* L.). Plant Cell Environ 21: 467–478
Kirk GJD, Santos EE, Findenegg GR (1999) Phosphate solubilization by organic anion secretion from rice (*Oryza sativa* L.) growing in aerobic soil. Plant Soil 211: 11–18
Koyama H, Kawamura A, Kihara T, Hara T, Takita E, Shibata D (2000) Overexpression of mitochondrial citrate synthase in *Arabidopsis thaliana* improved growth on a phosphorus limited soil. Plant Cell Physiol 41: 1030–1037
Lan M, Comerford NB, Fox TR (1995) Organic anions effect on phosphorus release from spodic horizons. Soil Sci Soc Am J 59: 1745–1749
Leggett M, Gleddie S, Holloway G (2001) Phosphate-solubilising microorganisms and their use. In: Ae N, Arihara J, Okada K, Srinivasan A (eds), Plant Nutrient Acquisition: New Perspectives. Springer, Tokyo, pp 299–318

Li L, Tang C, Rengel Z, Zhang F (2003) Chickpea facilitates phosphorus uptake by intercropped wheat from an organic phosphorus source. Plant Soil 248: 297–303

Li M, Osaki M, Rao IM, Tadano T (1997) Secretion of phytase from the roots of several plant species under phosphorus-deficient conditions. Plant Soil 195: 161–169

Liao M, Hocking PJ, Dong B, Delhaize E, Richardson AE, Ryan PR (2008) Variation in early phosphorus uptake efficiency among wheat genotypes grown on two contrasting Australian soils. Aust J Agric Res 59: 157–166

Lipton DS, Blanchar RW, Blevins DG (1987) Citrate, malate, and succinate concentration in exudates from P-sufficient and P-stressed *Medicago sativa* L. seedlings. Plant Physiol 85: 315–317

Liu H, Trieu AT, Blaylock LA and Harrison MJ (1998) Cloning and characterization of two phosphate transporters from *Medicago truncatula* roots: Regulation in response to phosphate and to colonization by arbuscular mycorrhizal (AM) fungi. Mol Plant-Microbe Interact 11: 14–22

López-Bucio J, de la Vega OM, Guevara-Garcia A, Herrera-Estrella L (2000) Enhanced phosphorus uptake in transgenic tobacco plants that overproduce citrate. Nature Biotechnol 18: 450–453

Lopez-Hernandez D, Brossard M, Frossard E (1998) P-isotopic exchange values in relation to P mineralization in soils with very low P-sorbing capacities. Soil Biol Biochem 30: 1663–1670

Lung S-C, Chan W-L, Yip W, Wang L, Yeung EC, Lim BL (2005) Secretion of beta-propeller phytase from tobacco and *Arabidopsis* roots enhances phosphorus utilisation. Plant Sci 169: 341–349

Lynch J (1995) Root architecture and plant productivity. Plant Physiol 109: 7–13

Lynch JP (2005) Root architecture and nutrient acquisition. In: BassiriRad H (ed), Nutrient Acquisition by Plants: An Ecological Perspective. Springer, Berlin, pp 147–183

Lynch JP, Brown KM (2001) Topsoil foraging - an architectural adaptation of plants to low phosphorus availability. Plant Soil 237: 225–237

Lynch JP, van Beem JJ (1993) Growth and architecture of seedling roots of common bean genotypes. Crop Sci 33: 1253–1257

Ma Z, Walk TC, Marcus A, Lynch JP (2001) Morphological synergism in root hair length, density, initiation and geometry for phosphorus acquisition in *Arabidopsis thaliana*: a modeling approach. Plant Soil 236: 221–235

Manske GGB, Ortiz-Monasterio JI, Van Grinkel M, Rajaram S, Molina E, Vlek PLG (2000) Traits associated with improved P-uptake efficiency in CIMMYT's semidwarf spring bread wheat grown on an acid andisol in Mexico. Plant Soil 221: 189–204

Maroko JB, Buresh RJ, Smithson PC (1999) Soil phosphorus fractions in unfertilized fallow-maize systems on two tropical soils Soil Sci Soc Am J 63: 320–326

Marschner H, Römheld V, Horst WJ, Martin P (1986) Root induced changes in the rhizosphere: importance for the mineral nutrition of plants. Z Pflanz Bodenkunde 149: 441–456

Marschner P, Solaiman Z, Rengel Z (2007) Brassica genotypes differ in growth, phosphorus uptake and rhizosphere properties under P-limiting conditions. Soil Biol Biochem 39: 87–98

McGill WB, Cole CV (1981) Comparative aspects of cycling of organic C, N, S and P through soil organic matter. Geoderma 26: 267–286

McLachlan KD (1980) Acid phosphatase activity of intact roots and phosphorus nutrition of plants. I Assay conditions and phosphatase activity. Aust J Agric Res 31: 429–440

Miller SS, Liu J, Allan DL, Menzhuber CJ, Fedorova M, Vance CP (2001) Molecular control of acid phosphatase secretion into the rhizosphere of proteoid roots from phosphorus-stressed white lupin. Plant Physiol 127: 594–606

Mitsukawa N, Okumura S, Shirano Y, Sato S, Kato T, Harashima S, Shibata D (1997) Overexpression of an *Arabidopsis thaliana* high-affinity phosphate transporter gene in tobacco cultured cells enhances cell growth under phosphate-limited conditions. Proc Natl Acad Sci USA 94: 7098–7102

Mudge SR, Rae AL, Diatloff E, Smith FW (2002) Expression analysis suggests novel roles for members of the Pht1 family of phosphate transporters in *Arabidopsis*. Plant J 31: 341–353

Mudge SR, Smith FW, Richardson AE (2003) Root-specific and phosphate-regulated expression of phytase under the control of a phosphate transporter promoter enables *Arabidopsis* to grow on phytate as a sole phosphorus source. Plant Sci 165: 871–878

Neumann G, Massonneau A, Martinoia E, Römheld V (1999) Physiological adaptations to phosphorus deficiency during proteiod root development in white lupin. Planta 208: 373–382

Newman RH, Tate KR (1980) Soil phosphorus characterization by ^{31}P-nuclear magnetic resonance. Commun Soil Sci Plant Anal 11: 835–842

Nziguheba G, Palm CA, Buresh RJ, Smithson PC (1998) Soil phosphorus fractions and adsorption as affected by organic and inorganic sources. Plant Soil 198: 159–168

Oberson A, Joner EJ (2005) Microbial turnover of phosphorus in soil. In: Turner BL, Frossard E, Baldwin DS (eds), Organic Phosphorus in the Environment. CAB International, Wallingford, pp 133–164

Oberson A, Besson JM, Maire N, Sticher H (1996) Microbial processes in soil organic transformations in conventional and biological cropping systems. Biol Fertil Soils 21: 138–148

Oberson A, Friesen DK, Rao IM, Bühler S, Frossard E (2001) Phosphorus transformations in an oxisol under contrasting land-use systems: the role of the microbial biomass. Plant Soil 237: 197–210

Oehl F, Oberson A, Sinaj S, Frossard E (2001) Organic phosphorus mineralization studies using isotopic dilution techniques. Soil Sci Soc Am J 65: 780–787

Oehl F, Frossard E, Fliessbach A, Dubois D, Oberson A (2004) Basal organic phosphorus mineralization in soils under different farming systems. Soil Biol Biochem 36: 667–675

Osborne LD, Rengel Z (2002) Screening cereals for genotypic variation in the efficiency of phosphorus uptake and utilization. Aust J Agric Res 53: 295–303

Otani T, Ae N (1999) Extraction of organic phosphorus in andosols by various methods. Soil Sci Plant Nutr 45: 151–161

Otani T, Ae N, Tanaka H (1996) Uptake mechanisms of crops grown in soils with low P status. II. Significance of organic acids in root exudates of pigeon pea. Soil Sci Plant Nutr 42: 553–560

Pearse SJ, Venaklaas EJ, Cawthray G, Bolland MDA, Lambers H (2007) Carboxylate composition of root exudates does not relate consistently to a crop species' ability to use phosphorus from aluminium, iron or calcium phosphate sources. New Phytol 173: 181–190

Pearse SJ, Venaklaas EJ, Cawthray G, Bolland MDA, Lambers H (2008) Rizosphere processes do not explain variation in P acquisition from sparingly soluble forms of P among *Lupinus albus* accessions. Aust J Agri Res 59: (in press)

Polglase PJ, Attiwill PM, Adams MA (1992) Nitrogen and phosphorus cycling in relation to stand age of *Eucalyptus regnans* F. Muell. III. Labile inorganic and organic P, phosphatase activity and P availability. Plant Soil 142: 177–185

Rae AL, Jarmey JM, Mudge SR, Smith FW (2004) Over-expression of a high-affinity phosphate transporter in transgenic barley plants does not enhance phosphate uptake rates. Funct Plant Biol 31: 141–148

Raghothama KG (1999) Phosphate acquisition. Annu Rev Plant Phys 50: 665–693

Rausch C, Bucher M (2002) Molecular mechanisms of phosphate transport in plants. Planta 216: 23–37

Richardson AE (1994) Soil microorganisms and phosphorus availability. In: Pankhurst CE, Doube BM, Gupta VVSR, Grace PR (eds), Management of the Soil Biota in Sustainable Farming Systems. CSIRO, Melbourne, pp 50–62

Richardson AE (2001) Prospects for using soil microorganisms to improve the acquisition of phosphorus by plants. Aust J Plant Physiol 28: 897–906

Richardson AE, Hadobas PA, Hayes JE (2000) Phosphomonoesterase and phytase activities of wheat (*Triticum aestivum* L.) roots and utilisation of organic phosphorus substrates by seedlings grown in sterile culture. Plant Cell Environ 23: 397–405

Richardson AE, Hadobas PA, Hayes JE (2001a) Extracellular secretion of *Aspergillus* phytase from *Arabidopsis* roots enables plants to obtain phosphorus from phytate. Plant J 25: 641–649

Richardson AE, Hadobas PA, Hayes JE, O'Hara CP, Simpson RJ (2001b) Utilization of phosphorus by pasture plants supplied with myo-inositol hexaphosphate is enhanced by the presence of soil microorganisms. Plant Soil 229: 47–56

Richardson AE, George TS, Hens M, Simpson RJ (2005) Utilisation of soil organic phosphorus by higher plants. In: Turner BL, Frossard E, Baldwin DS (eds), Organic Phosphorus in the Environment. CAB International, Wallingford, pp 165–184

Richardson AE, George TS, Jakobsen I, Simpson RJ (2007) Plant utilization of inositol phosphates. In: Turner BL, Richardson AE, Mullaney EJ (eds), Inositol Phosphates: Linking Agriculture and the Environment. CAB International, Wallingford, pp 242–260

Runge-Metzger A (1995) Closing the cycle: obstacles to efficient P management for improved global security. In: Tiessen H (ed), Phosphorus in the Global Environment: Transfers, Cycles And Management. Wiley, Chichester, pp 27–42

Ryan PR, Delhaize E, Jones DL (2001) Function and mechanism of organic anion exudation from plant roots. Annu Rev Plant Physiol Plant Mol Biol 52: 527–560

Sanyal SK, De Datta SK (1991) Chemistry of phosphorus transformations in soil. Adv Soil Sci 16: 1–120

Sasaki T, Yamomoto Y, Ezaki B, Katsuhara M, Ahn SJ, Ryan PR, Delhaize E, Matsumoyo H (2004) A wheat gene encoding an aluminium-activated malate transporter. Plant J 37: 645–653

Schachtman DP, Reid RJ, Ayling SM (1998) Phosphorus uptake by plants: from soil to cell. Plant Physiol 116: 447–453

Schünmann PHD, Richardson AE, Vickers CE, Delhaize E (2004) Promoter analysis of the barley *Pht1;1* phosphate transporter gene identifies regions controlling root expression and responsiveness to phosphate deprivation. Plant Physiol 136: 4205–4214

Smith SE, Read DJ (1997) Mycorrhizal Symbiosis. Academic, San Diego, CA

Subbarao GV, Ae N, Otani T (1997) Genotypic variation in the iron- and aluminium-phosphate solubilising activity of pigeon pea root exudates under P deficient conditions. Soil Sci Plant Nutr 43: 295–305

Tadano T, Ozawa K, Sakai H, Osaki M, Matsui H (1993) Secretion of acid phosphatase by the roots of crop plants under phosphorus-deficient conditions and some properties of the enzyme secreted by lupin roots. Plant Soil 155/156: 95–98

Tarafdar JC, Claassen N (1988) Organic phosphorus compounds as a phosphorus source for higher plants through the activity of phosphatases produced by plant roots and microorganisms. Biol Fertil Soils 5: 308–312

Tarafdar JC, Marschner H (1994) Efficiency of VAM hyphae in utilisation of organic phosphorus by wheat plants. Soil Sci Plant Nutr 40: 593–600

Tarafdar JC, Yadav RS, Meena SC (2001) Comparative efficiency of acid phosphatase originated from plant and fungal sources. J Plant Nutr Soil Sci 164: 279–282

Tesfaye M, Temple SJ, Allan DL, Vance CP, Samac DA (2001) Overexpression of malate dehydrogenase in transgenic alfalfa enhances organic acid synthesis and confers tolerance to aluminum. Plant Physiol 127: 1836–1844

Ticconi CA, Abel S (2004) Short on phosphate: plant surveillance and countermeasures. Trends Plant Sci 9: 548–555

Trasar-Cepeda MC, Carballas T (1991) Liming and the phosphatase activity and mineralization of phosphorus in an acidic soil. Soil Biol Biochem 23: 209–215

Vance CP, Uhde-Stone C, Allan DL (2003) Phosphorus acquisition and use: critical adaptations by plants for securing a nonrenewable resource. New Phytol 157: 423–447

Veneklaas EJ, Stevens J, Cawthray GR, Turner S, Grigg AM, Lambers H (2003) Chickpea and white lupin rhizosphere carboxylates vary with soil properties and enhance phosphorus uptake. Plant Soil 248: 187–197

Vesterager JM, Nielsen NE, Hogh-Jensen, H (2006) Variation in phosphorus uptake and use efficiencies between pigeonpea genotypes and cowpea. J Plant Nutr 29: 1869–1888

Wakelin SA, Warren RA, Ryder MH (2004) Effect of soil properties on growth promotion of wheat by *Penicillium radicum*. Aust J Soil Res 42: 897–904

Wang L, Liao H, Yan X, Zhuang B, Dong Y (2004) Genetic variability for root hair traits as related to phosphorus status in soybean. Plant Soil 261: 77–84

Wang QR, Li JY, Li ZS, Christie P (2005) Screening Chinese wheat germplasm for phosphorus efficiency in calcareous soils. J Plant Nutr 28: 489–505

Wasaki J, Omura M, Ando M, Dateki H, Shinano T, Osaki M, Ito H, Matsui H, Tadano T (2000) Molecular cloning and root specific expression of secretory acid phosphatase from phosphate deficient lupin (*Lupinus albus* L.). Soil Sci Plant Nutr 46: 427–437

Wasaki J, Yonetani R, Kuroda S, Shinano T, Yazaki J, Fujii F, Shimbo K, Yamamoto K, Sakata K, Sasaki T, Kishimoto N, Kikuchi S, Yamagishi M, Osaki M (2003) Transcriptomic analysis of metabolic changes by phosphorus stress in rice plant roots. Plant Cell Environ 26: 1515–1523

Wassen MJ, Venterink HO, Lapshina ED, Tanneberger L (2005) Endangered plants persist under phosphorus limitation. Nature 437: 547–550

Watt M, Evans JR (1999) Linking development and determinacy with organic acid efflux from proteoid roots of white lupin grown with low phosphorus and ambient or elevated atmospheric CO_2 concentration. Plant Physiol 120: 705–716

White PJ, Hammond JP (2008) Phosphorus nutrition of terrestrial plants. In: White PJ, Hammond JP (eds), The Ecophysiology of Plant-Phosphorus Interactions. Springer, Dordrecht, The Netherlands, pp 51–81

White PJ, Broadley MR, Greenwood DJ, Hammond JP (2005) Genetic Modifications to Improve Phosphorus Acquisition by Roots. Proceedings 568, International Fertiliser Society, York

Williamson LC, Ribrioux SPCP, Fitter AH, Leyser HMO (2001) Phosphate availability regulates root system architecture in Arabidopsis. Plant Physiol 126: 875–890

Wissuwa M (2003) How do plants achieve tolerance to phosphorus deficiency? Small causes with big effects. Plant Physiol 133: 1947–1958

Wissuwa M (2005) Mapping nutritional traits in crop plants. In: Broadley MR, White PJ (eds), Plant Nutritional Genomics. Blackwell, Cambridge, pp 220–241

Wissuwa M, Ae N (2001a) Further characterization of two QTLs that increase phosphorus uptake of rice (*Oryza sativa* L.) under phosphorus deficiency. Plant Soil 237: 275–286

Wissuwa M, Ae N (2001b) Genotypic variation for tolerance to phosphorus deficiency in rice and the potential for its exploitation in rice improvement. Plant Breeding 120: 43–48

Wouterlood M, Cawthray GR, Scanlon TT, Lambers H, Veneklaas EJ (2004) Carboxylate concentrations in the rhizosphere of lateral roots of chickpea (*Cicer arietinum*) increase during plant development, but are not correlated with phosphorus status of soil or plants. New Phytol 162: 745–753

Wu P, Ma L, Hou X, Wang M, Wu Y, Liu F, Deng XW (2003) Phosphate starvation triggers distinct alterations of genome expression in *Arabidopsis* roots and leaves. Plant Physiol 132: 1260–1271

Xiao K, Harrison MJ, Wang Z-Y (2005) Transgenic expression of a novel *M. truncatula* phytase gene results in improved acquisition of organic phosphorus by *Arabidopsis*. Planta 222: 27–36

Yan X, Lynch JP, Beebe SE (1995) Genetic variation for phosphorus efficiency of common bean in contrasting soil types. I. Vegetative response. Crop Sci 35: 1086–1093

Yan X, Liao H, Beebe SE, Blair MW, Lynch JP (2004) QTL mapping of root hair and acid exudation traits and their relationship to phosphorus uptake in common bean. Plant Soil 265: 17–29

Yip W, Wang L, Cheng C, Wu W, Lung S, Lim BL (2003) The introduction of a phytase gene from *Bacillus subtilis* improved the growth performance of transgenic tobacco. Biochem Biophys Res Commun 310: 1148–1154

Zhang H, Forde BG (1998) An Arabidopsis MADS box gene that controls nutrient-induced changes in root architecture. Science 279: 407–409

Zhang WH, Ryan PR, Tyerman SD (2004) Citrate-permeable channels in the plasma membrane of cluster roots from white lupin. Plant Physiol 136: 3771–3783

Zhu J, Lynch JP (2004) The contribution of lateral rooting to phosphorus acquisition efficiency in maize (*Zea mays*) seedlings. Funct Plant Biol 31: 949–958

Zhu J, Kaeppler SM, Lynch JP (2005) Mapping of QTL controlling root hair length in maize (*Zea mays* L.) under phosphorus deficiency. Plant Soil 270: 299–310

Zimmermann P, Zardi G, Lehmann M, Zeder C, Amrhein N, Frossard E, Bucher M (2003) Engineering the root-soil interface via targeted expression of a synthetic phytase gene in trichoblasts. Plant Biotech J 1: 353–360

Chapter 12
PHOSPHORUS AND THE FUTURE

John A. Raven

INTRODUCTION

Before man took a major hand in biogeochemistry in the late Holocene, phosphorus (P) availability was a major constraint on primary productivity and diazotrophy, and this remains so over significant areas of land and ocean today. Photosynthetic organisms have a significant inheritance of strategies of prospecting, mining, transporting and storing phosphate, as well as in economizing on the use of P in some aspects of their biochemistry and in reallocating the element within the organism. The benefits in terms of P requirements of the mechanisms involved in seeking and acquiring phosphate, and in economizing on the use of P within the organism, have costs in terms of other resource requirements and of restrictions on growth rate. Restricting the use of P fertilizers is desirable for reducing leakage of the element into 'natural' environments, and will be enforced by the non-renewable nature of P reserves. Modifying crops by purely conventional and GM-assisted breeding to accommodate these requirements while maintaining or increasing productivity will be a major challenge for plant scientists.

It is perhaps unnecessary to point out that the predictable changes in plant-phosphorus interactions over the next few centuries will be dominated by man's response to increasing human population, the corresponding increased exploitation, and ultimate exhaustion, of accessible mineral phosphate reserves (Tilman *et al.* 2001, 2002), and the interactions of these with global environmental change.

The approach taken here to considering these changes, and their probable effects on what remains of 'natural' plant populations, is to first consider what is known of plant-phosphorus interactions before man's influences became significant, with current anthropogenic phosphorus mobilization now equaling natural processes (Tilman *et al.* 2001, 2002). This will give an appreciation of what plants, defined in the broad, non-phylogenetic sense as organisms that can produce oxygen by photosynthesis, did under natural conditions of phosphate supply, with an emphasis on biogeochemistry and evolution. Such considerations provide an evolutionary background of how plants from various habitats deal with changes in phosphate availability.

This background will then be used to reconsider how man's activities in response to the forcing factors mentioned above (human population increase, exhaustion of

P.J. White, J.P. Hammond (eds.) *The Ecophysiology of Plant-Phosphorus Interactions*,

mineral phosphate reserves, global environmental change) must be conditioned by the biogeochemistry of phosphate and the evolutionary history of crops. Finally, an attempt is made to assess priorities for action.

PHOSPHORUS-PLANT INTERACTIONS BEFORE MAN'S ACTIVITIES BECAME GLOBALLY SIGNIFICANT, AND THE PRESENT SITUATION

There are clear chemical reasons for why nutrient elements perform the functions that they do in living organisms (Williams 1981; Williams and Fráusta da Silva 1996). The availability of these essential elements is determined by a variety of cosmological and geological factors. The cosmological factors include the genesis, post Big Bang, of elements of higher atomic number than helium in stars larger than the sun, and the apportioning of elements during the genesis and development of our solar system (Williams and Fráusta da Silva 1996; Raven *et al.* 2004, 2005a; Pasek and Lauretta 2005; Pasek *et al.* 2007). The geological factors determining availability of elements in the lithosphere, hydrosphere and atmosphere have been increasingly modified over at least the last 3.5 billion years by the influence of organisms, i.e. geochemistry increasingly became biogeochemistry.

In the case of P we have a cosmically, and terrestrially, abundant element that is rather unavailable to organisms compared to their stoichiometric requirement for it relative to other elements. Phosphorus is a macronutrient and, while acknowledging that 'content' does not equal 'requirement' (organisms contain many non-essential elements), it is significant that it varies relatively little in organisms grown under nutrient-sufficient conditions among the major clades of marine microalgae (Bertilsson *et al.* 2003; Heldal *et al.* 2003; Ho *et al.* 2003; Quigg *et al.* 2003; Falkowski *et al.* 2004; Finkel *et al.* 2006) and of flowering plants (Broadley *et al.* 2004). Such comparisons among clades often use P content as a 'fixed point' to which the content of other elements is referred, e.g. P = 1 in considering variations in the C:N:P ratio in marine phytoplankton from the canonical Redfield Ratio of 106 C:16 N:1 P (by atoms: Redfield 1958; Falkowski and Raven 2007), as exemplified by the studies of Burkhardt *et al.* (1999), Gervais and Riebesell (2001) and Leonardos and Geider (2005).

There are interactions between the P requirements of oxygen-producing photolithotrophs and the means by which they acquire other nutrient elements. One much-investigated, but still poorly understood phenomenon, is the apparently high P requirement for diazotrophy in certain cyanobacteria, and symbioses of diazotrophic (cyano) bacteria with certain eukaryotes, compared to related organisms (Tyrrell 1999; Sañudo-Wilhelmy *et al.* 2001; Vitousek *et al.* 2002; Mills *et al.* 2004; Schulze *et al.* 2006). This higher requirement is not always seen in the elemental composition of the organisms, but is rather seen in the requirement for external P for diazotrophic growth relative to non-diazotrophic growth. This P-diazotrophy interaction is one aspect of the puzzle as to why diazotrophy is not more common in some ecosystems with substantial losses of nitrogen (N) compounds by denitrification, leaching or sedimentation (Tyrrell 1999; Vitousek *et al.* 2002; Falkowski and Raven 2007). Another area that needs further investigation is the implications for P nutrition on carbon (C) acquisition in organisms with, and without, inorganic C concentrating mechanisms (CCMs), such as the C_4 pathway, and the effects of P deficiency on expression of CCMs, where these are susceptible to varying degrees of expression, to supplement the limited information already available (Bożena *et al.* 2000; Beardall *et al.* 2005; Ghannoum and Conroy 2007).

It is worth pointing out that, although terrestrial net primary productivity expressed in C terms is greater than that in the world's oceans (Falkowski and Raven 2007), the greater C:P in terrestrial vegetation than in marine phytoplankton (Redfield 1958; McGroddy *et al.* 2004), which are by far the predominant marine primary producers, means that P acquisition and assimilation in the oceans each year exceeds that on land.

There is geological evidence consistent with variations in P availability in various environments over the last few billion years (Raven *et al.* 2005a; Elser *et al.* 2006b; Konhauser *et al.* 2007), with a major downturn in P availability in most habitats occupied by oxygen-producing photolithotrophs upon the global oxygenation permitted by the evolution and diversification of these oxygen-producers as a result of the binding of inorganic phosphate to iron in the Fe(III) form (Falkowski and Raven 2007).

There is also evidence from extant organisms of acclimatory mechanisms related to the acquisition of P from habitats of low P availability (Lambers and Poot 2003; Smith and Read 2007), and for tissue P redeployment, including the extent of resorption from structures that are no longer metabolically active (Hedin 2004; McGroddy *et al.* 2004; Gusewell 2005), as well as changes in biomass allocation (Hermans *et al.* 2006), within the organism during their ontogeny. There are also mechanisms of economizing on the use of P in metabolism, including the partial substitution of P by other elements (reviewed by Ticconi and Abel 2004). A recently discovered example of the latter is the constitutive and acclimatory (in response to P deficiency) replacement of phospholipids by glycolipids (galactolipids and sulfolipids) in plastids (where glycolipids dominate anyway) and in extraplastid membranes (Dörmann and Benning 2002; Andersson *et al.* 2003, 2005; Frentzen 2004; Gaude *et al.* 2004; Kobayashi *et al.* 2006; van Mooy *et al.* 2006). Interestingly, glycolipids occur where phospholipids would otherwise occur in peribacteroid membranes surrounding rhizobial symbionts in diazotrophic legumes (Gaude *et al.* 2004), economizing on P use in the symbiosis with its apparent high P demand as discussed above. The degree of use of sulfolipids and of galactolipids in replacing phospholipids is apparently related to the balance of neutral (galactolipid, many phospholipids) and negatively charged (sulfolipid, phosphatidyl glycerol) lipids in the membranes (Frentzen 2004) and the availability of sulfur in the environment (van Mooy *et al.* 2006; Giordano *et al.* 2007).

Straightforward reduction in the P-containing components of the organism, without elemental substitution, can be seen in the ribosomal RNA (rRNA) content. This can occur genotypically (e.g. fast and slow growing genotypes within a clade), developmentally (e.g. changes in growth rate among instars of arthropods) and environmentally (e.g. in response to P deficiency) with lower rRNA content and lower growth rate; Sterner and Elser 2002; Gillooly *et al.* 2005; Karpinets *et al.* 2006). However, as discussed elsewhere in this chapter, chloroplast rRNA can increase, and chloroplast DNA decrease, during P deficiency. Continuing with the general phenomenon of a decrease in total rRNA with P deficiency, in extreme cases the rRNA-P per cell can be as low as the DNA-P content, with corresponding impact on the growth rate, in the marine picoplanktonic cyanobacterium *Prochlorococcus marina* (Bertilsson *et al.* 2003). This is a remarkable economy in P to have as much P in the rRNA as in the DNA, considering the small gene number and high gene density in the genomes of the various strains of this organism. The genotypic correlations between growth rate and rRNA content can be related, at least in general terms, to the number of copies of the rRNA operon in the genome, since there is little variation in cell size among the bacterial genotypes examined (Klappenbach *et al.* 2000; Karpinets *et al.* 2006).

However, these correlations between growth rate and rRNA content are generally less impressive in photolithotrophs than is the case for non-photosynthetic organisms (reviewed by Ågren 2004; Raven *et al.* 2004, 2005a), although other interesting stoichiometric relations are found (Kerkhoff *et al.* 2005, 2006; Niklas *et al.* 2005; Niklas 2006; Niklas and Cobb 2006). For the copy number of rRNA operons in the genome in photosynthetic organisms, Zhu *et al.* (2005) examined 18 species of eukaryotic microalgae with a range of biomasses over almost six orders of magnitude, and found an increase in copy number with increased cell size. The paper did not cite specific growth rates, but the decrease in specific growth rate with (genotypic) increase in cell size is quite small (e.g. Banse 1982) so that, among the microalgae, cell size rather than specific growth rate is the main determinant of the rRNA operon copy number.

Overall, these data could be taken as indicating that there is a relatively greater excess of rRNA in photolithotrophs that are growing more slowly as a result of their genetic composition or limiting resource supply than is the case in many saprotrophs and phagotrophs. However, the need for protein synthesis in mature, non-growing organs could help to explain some of this apparent uncoupling of whole-organism growth from whole-organism rRNA content. An example is the non-growing but photosynthetically active leaves in flowering plants. In addition to the general turnover of proteins in the absence of net synthesis of protein, the thylakoid membranes of all oxygen-producers have a turnover of photodamaged D1 protein (*psbA* gene product) in the photosystem II reaction centre. Although this polypeptide is only a small fraction of the total leaf protein, the photodamage-related protein resynthesis can be much more rapid than the turnover of other polypeptides (Raven 1989, 1994; Schaefer *et al.* 1990; Zagdanska 1995; Scheurwater *et al.* 2000; Raven *et al.* 2002, 2004, 2005a). Furthermore, the possibility for photodamage can be temporally very variable (e.g. sunflecks incident on a forest floor plants), so that an apparent excess of rRNA may be needed relative to what is needed for the mean rate of protein (including D1) turnover (Raven 1989). The rRNA needed for D1 repair is in the chloroplasts, since *psbA* is a plastid-located gene. A possibility that has not yet been explored is the extent to which the decreased expression of certain genes (*psbO*, *psbP*) related to photosystem II and expressed on the luminal side of the thylakoid membrane under P deficiency in *Arabidopsis thaliana* (Jain *et al.* 2005) interacts with the potential for photodamage. The reduced expression of these genes may underlie the observation (Jacob 1995) of increased thermal dissipation of excitation energy by, and decreased quantum yield of, photosystem II in P-deficient plants. Yehudai-Resheff *et al.* (2007) noted that chloroplast rRNA increased in P-deficient cells of *Chlamydomonas reinhardtii* while the degree of polyploidy of chloroplasts genomes declined, i.e. the response of chloroplast rRNA was the opposite of that of chloroplast DNA. Before suggesting that the increased chloroplast rRNA in P-deficient *Chlamydomonas reinhardtii* cells is related to any increased potential for photodamage to D1 and a corresponding increase in the need for resynthesis of D1, it must be remembered that the increased content of rRNA could be a result of polyadenylation of rRNA (Yehudai-Resheff *et al.* 2007). Jain *et al.* (2005) point out that decreased effectiveness of photosystem II electron transport is a result of phosphate deficiency in *Chlamydomonas*. The requirement for rRNA in the resynthesis of proteins in maintenance processes, and its implications for the extent to which rRNA content relates to growth rate, has been recognized more recently for phagotrophs (Kyle *et al.* 2006).

The acclimatory mechanisms enabling tissue P economy can be related to a general lack of P availability, as well as to particular temporal and spatial occurrences of low P availability. Habitats with very low P availability include the oligotrophic central oceanic gyres (Bertilsson *et al.* 2003; Heldal *et al.* 2003; Paytan and McLaughlin 2007), and very old soils in Western Australia and South Africa (Lambers and Poot 2003), although greater P availabilities occur in some other old, leached soils (Birks and Birks 2004; Hedin 2004; Reich and Oleksyn 2004; McGroddy *et al.* 2004; Wardle *et al.* 2004; Smith and Read 2007; cf. Wright *et al.* 2005). The photolithotrophs in these habitats also have an elaboration of mechanisms for acquiring P.

For planktonic photolithotrophs this can involve mechanisms permitting the use of extracellular organic phosphates, usually through the use of extracellular phosphatases rather than the direct uptake of the organic phosphates, in addition to the use of extracellular inorganic phosphates (Dyhrman and Haley 2006; Martiny *et al.* 2006). There is also the possibility of the use of phosphonates in some marine phytoplankton (Dyhrman and Haley 2006; Dyhrman *et al.* 2006). These are compounds with C-P bonds, with P in a less oxidized state than in phosphates. They are natural products, for example C-P bonds occur in some membrane lipids, and have been discussed in the context of the origin of life on Earth (Pasek and Lauretta 2005; Pasek *et al.* 2007), and require further investigation. There is also a need for further investigation of the biogenesis of the completely reduced compound phosphine (PH_3, more properly termed phosphane). This compound has recently been quantified as an atmospheric trace gas and as a product of anoxic lake sediments, and may also have abiogenic sources (Glindemann *et al.* 2004; Geng *et al.* 2005; Zhu *et al.* 2006, 2007).

For terrestrial plants, the emphasis in cases of moderate P deficiency, is on increasing the volume of soil exploited for the acquisition of phosphate, which is present at low concentrations in the soil solution and which has a very low diffusion coefficient. In most cases this involves mycorrhizal symbioses, and more specifically arbuscular mycorrhizas (Smith and Read 2007). Non-mycorrhizal plants, such as members of the Brassicaceae and Chenopodiaceae, are in a significant minority. Extreme P deficiency can involve the formation of cluster roots (including dauciform roots) in the absence of mycorrhizas, as occurs in members of the Proteaceae and some Leguminosae and Cyperaceae. Here the emphasis is on chemical modification of a small volume of soil by the efflux of organic anions, with varying degrees of pH alteration, which release phosphate from calcium and ferric iron associations (Lambers and Poot 2003).

Returning to the genetic evidence for the occurrence of P deficiency in the evolutionary history of organisms, an obvious example is the up- and down-regulation of a wide range of genes upon a decrease in P supply in marine phytoplankton (Martiny *et al.* 2006) and flowering plants (Hammond *et al.* 2003, 2004; Wasaki *et al.* 2003, 2006; Amtmann *et al.* 2006; Morcuende *et al.* 2007).

Two other approaches to investigate the stoichiometric relationships between the concentrations of different elements have been taken with C, N and S. One is the analysis of the use of the element in the nucleic acid bases in the genome, and in the amino acids of the derived proteome, with a weighting for the extent of expression of different genes when considering the proteome (see review by Raven *et al.* 2005a, and later work by Elser *et al.* 2006a and Bragg and Wagner 2007). This technique can only be used for C and N in the genome and C, N and sulfur (S) in the derived proteome. It has yielded interesting results on the relative use of C and N as a function of photolithotrophy (smaller use of N relative to C) and of chemoorganotrophy (greater use of N relative to C), and the extent of use of S-containing amino-acids in the proteome (Raven *et al.* 2005a; Elser *et al.* 2006a; Bragg and Wagner 2007). The possibilities of applying this technique to P are limited. The P content of a genome of a given size is immutable, while P in the proteome is limited to post-translational modification, which requires more information than is provided by the genome sequence, the derived proteome and the degree of expression of different genes. Nevertheless, such analyses are possible from knowledge of the genes whose products are susceptible to phosphorylation and the expression of these genes. This sort of analysis would be an example of a genomic analysis of elemental use in organisms, with implications for elemental availability in the evolution of those organisms (Lahner *et al.* 2003; Zerkle *et al.* 2005). Thus, the occurrence, and where possible the extent of their expression, of genes whose products contain a given element as a catalytic cofactor in an organism or clade may be related to the availability of those elements during the evolution of that organism or clade (Lahner *et al.* 2003). Where genomic analyses are allied to elemental analyses (Lahner *et al.* 2003), they overlap with attempts to relate the quantitative occurrence of elements in organisms from different clades, to draw conclusions about the availability of various elements during the evolution of those clades (Ho *et al.* 2003; Quigg *et al.* 2003; Falkowski *et al.* 2004).

Although phosphorylated proteins only account for a very small fraction of the total P in an organism, the extent of their occurrence in different organisms could provide important evolutionary information. The variations in C and N in the genome, and of C and N, (and, to some extent, S) in the proteome, are also relatively small, but they contain important evolutionary information on the relative elemental availability to major clades during their evolution. In the case of P it is possible to argue that the regulatory significance of protein phosphorylation, and the small fraction of the total organismal P that occurs as phosphoproteins, makes them unlikely targets for P economy. As an example, the occurrence of a phosphoprotein may be essential for a regulatory process that leads to an economy in P use that far exceeds the amount of P in the particular phosphoprotein. However, such arguments should not be used to prevent attempts see if information on past P availability can be obtained from the extent of occurrence of proteins susceptible to regulatory phosphorylation in a range of genomes.

An ecological implication of changes in the P concentrations of the cells and organs of photolithotrophs is that variations in stoichiometry have consequences for herbivory (Sterner and Elser 2002; Kuang *et al.* 2004). The extent to which this will alter the impact of herbivory on photolithotroph populations depends on the time (pre- or post-ingestion) at which the grazer detects the consequences of altered P concentrations on the quality of the food item as far as that grazer's nutrition and fitness are concerned. Similar considerations apply to the effect of altered phosphorus content, and genome size, on the susceptibility of photolithotrophs to viruses and host-virus dynamics (Raven *et al.* 2005b; Raven 2006; Chasen and Elser 2007).

At present, man's activities are mobilizing as much phosphate each year from rock phosphate deposits as is mobilized 'naturally' from rocks in pedogenesis (Tilman *et al.* 2001, 2002). Despite the very low mobility of phosphate in soil, much of the anthropogenic phosphate joins the natural phosphate in rivers and, ultimately, the sea. Eventually, after a water column residence time of a few tens of thousands of years, the phosphate is sedimented (Paytan and McLaughlin 2007). Any of the applied P fertilizer that becomes unavailable to crops, either by changing chemically (even though it stays in the soil) or spatially (by being removed from the root zone) is lost as far as man is concerned, since there are no known technologies for dealing with such diffuse P losses. By contrast, there are means of recovering P, in a form that can be converted into fertilizer or P supplements to animal feed, from point sources of P loss such as human sewage and slurry from intensive animal production (Shu *et al.* 2005).

WHAT OF THE FUTURE?

The previous discussion in this chapter, and in this book as a whole, shows that recent advances have provided an excellent basis for furthering our understanding of phosphate-plant-environment interactions. Advances can be confidently expected in our understanding of the regulation of P acquisition and allocation at the molecular genetic level in the 'usual suspects', from *Prochlorococcus* to *Populus*, of oxygen-producing photolithotrophs for which there are (more or less) complete genome sequences. Equally, great advances can be expected of further observations and modeling at more integrated levels, from Plant Physiology through Macroecology to Earth System Science.

As well as the pure scientific interest of such work, it must also be used to address the pressing problems of feeding the increasing human population while retaining ecosystem services and biodiversity in the face of global environmental change, a finite supply of 'fossil' water for irrigation, and a finite supply of economically extractable rock phosphate. While what constitutes 'economically extractable' will change with increasing demand for phosphate, it would seem that phosphate reserves will be exhausted within two centuries with the projected rates of future extraction. 'Peak phosphate' (by analogy with Hubbard's concept of 'peak oil'), will be reached much sooner. However, while oil as an energy source can be replaced (albeit with significant dislocation for some energy requirements, such as road, air and sea transport) by other sources, there is no replacement for phosphate.

What can be done? Limiting human population growth is an absolute necessity, but it is beyond the scope of this chapter to address the socio-economic, political and ethical ramifications of such limitation. From the point of view of P fertilizers for agricultural production, there must be immediate implementation of recovery of phosphate from point sources of anthropogenic output (Shu *et al.* 2005). Also vital is improved agronomic technology, with better real-time estimation of P requirements of individual plants and precision placement of fertilizer. This technology is information-intensive, and may also be energy-intensive compared to current methods, but if it is applied successfully there will not only be less use of phosphate fertilizer, but also less loss of phosphate to groundwater, inland water systems and the ocean with a corresponding decrease in eutrophication. In addition to these possibilities there is, of course, the need to address the topics that have dominated this chapter. The desired outcome is minimizing the plant's P requirement, while maintaining or increasing productivity. This must be achieved without compromising the ability of the plant to respond to environmental variability or the quality of the agricultural product. It is uncertain how useful (or possible!) it would be to engineer crop plants to have cluster roots for 'phosphate mining'. While some crop plants (e.g. *Lupinus albus*, *Hippophae* spp. and *Macadamia*) have cluster roots, they were not selected, or bred, as crops for this reason. Equally, it is not known if the optimal (in terms of phosphate uptake per unit of other resource consumed) mycorrhizal symbionts occur in association with all crop plants; further research is needed in this area. Such research must take into account continuing environmental change, and the decreasing amount of water available for irrigation.

PRIORITIES FOR ACTION

- Limit human population growth
- Implement recovery of phosphate from anthropogenic point sources

- Develop and implement precision fertilizer placement technology
- Use genetic modification to minimize P requirements while maintaining or increasing yields of crops under environmental change and decreased availability of water for irrigation
- Use genetic modification to maximize phosphate uptake per unit of other resources used in producing and using the uptake apparatus
- Develop better knowledge of the global phosphorus cycle, making sure that minor players, such as the reduced components (phosphites, phosphonates and phosphane), are adequately dealt with

These items are listed in an approximate priority order, although in practice they will be progressed in parallel.

Acknowledgements The author's work on the ecophysiology of oxygen-evolving photolithotrophs is supported by the Natural Environment Research Council UK. The ideas developed in this chapter have benefited greatly from my membership of Working Groups 3 and 33 of the ARC-NZ Vegetation Function Network.

REFERENCES

Ågren GI (2004) The C:N:P stoichiometry of autotrophs – theory and observations. Ecol Lett 7: 185–191

Amtmann A, Hammond JP, Armengaud P, White PJ (2006) Nutrient sensing and signalling in plants: potassium and phosphorus. Adv Bot Res 43: 209–257

Andersson MX, Stridh MH, Larsson KE, Liljenberg C, Sandelius AS (2003) Phosphate-deficient oat replaces a major portion of the plasma membrane phospholipids with the galactolipid digalactosyldiacylglycerol. FEBS Lett 537: 128–132

Andersson MX, Larsson KE, Tjellström H, Liljenberg C, Sandelius AS (2005) Phosphate-limited oat. The plasma membrane and the tonoplast as major targets for phospholipid-to-glycolipid replacement and stimulation of phospholipases in the plasma membrane. J Biol Chem 280: 27578–27586

Banse K (1982) Cell volumes, maximal growth rates of unicellular algae and ciliates, and the role of ciliates in the marine pelagial. Limnol Oceanogr 27: 1059–1071

Beardall J, Roberts S, Raven JA (2005) Regulation of inorganic carbon acquisition by phosphorus limitation in the green alga *Chlorella emersonii*. Can J Bot 83: 859–864

Bertilsson S, Bergland O, Karl DM, Chisholm SW (2003) Elemental composition of marine *Prochlorococcus* and *Synechococcus*: implications for the ecological stoichiometry of the sea. Limnol Oceanogr 48: 1721–1731

Birks HJB, Birks HH (2004) The rise and fall of forests. Science 305: 484–485

Bożena KS, Zielinski P, Maeszewski S (2000) Involvement of glycolate metabolism in acclimation of *Chlorella vulgaris* cultures to low phosphate supply. Plant Physiol Biochem 38: 727–734

Bragg JG, Wagner A (2007) Protein carbon content evolves in response to carbon availability and may influence the fate of duplicated genes. Proc Roy Soc Lond B 274: 1063–1070

Broadley MR, Bowen HC, Cotterill HL, Hammond JP, Meacham MC, Mead A, White PJ (2004) Phylogenetic variation in the shoot mineral concentration of angiosperms. J Exp Bot 55: 321–336

Burkhardt S, Zondervan I, Riebesell U (1999) Effect of CO_2 concentration on C:N:P ratio in marine phytoplankton. Limnol Oceanogr 44: 683–690

Chasen JL, Elser JJ (2007) The effect of host *Chlorella* NC64A carbon:phosphorus ratio on the production of *Paramecium bursa Chlorella* Virus -1. Freshwater Biol 52: 112–122

Dörmann P, Benning C (2002) Galactolipids rule in seed plants. Trends Plant Sci 7: 112–118
Dyhrman ST, Haley ST (2006) Phosphorus scavenging in the unicellular marine diazotroph *Crocosphaera watsonii*. Appl Environ Microbiol 72: 1452–1458
Dyhrman ST, Chappel PD, Haley ST, Moffett JW, Orchard ED, Waterbury JB, Webb EA (2006) Phosphonate utilization by the globally important marine diazotroph *Trichodesmium*. Nature 439: 68–71
Elser JJ, Fagan WF, Subramanian S, Kumar S (2006a) Signatures of ecological resource availability in the animal and plant proteomes. Mol Biol Evol 23: 1946–1951
Elser JJ, Watts J, Schampel JH, Farmer J (2006b) Early Cambrian food webs on a trophic knife-edge? A hypothesis and preliminary data from a modern stromatolite-based ecosystem. Ecol Lett 9: 295–303
Falkowski PG, Raven JA (2007) Photosynthesis in Aquatic Plants (2nd edition). Princeton University Press, Princeton, NJ
Falkowski PG, Katz ME, Knoll AH, Quigg A, Raven JA, Schofield O, Taylor FJR (2004) The evolution of modern eukaryotic phytoplankton. Science 305: 354–360
Finkel ZV, Quigg A, Raven JA, Reinfelder JR, Schofield OE, Falkowski PG (2006) Irradiance and elemental stoichiometry of marine phytoplankton. Limnol Oceanogr 51: 2690–2701
Frentzen M (2004) Phosphatidylglycerol and sulfoquinovosyldiacylglycerol: anionic membrane lipids and phosphate regulation. Curr Opin Plant Biol 7: 270–276
Gaude N, Tippmann H, Flemetakis E, Katinakis P, Udvardi M, Dormann P (2004) The galactolipid digalactosyldiacylglycerol accumulates in the peribacteroid membrane of nitrogen-fixing nodules of soybean and *Lotus*. J Biol Chem 279: 34624–34630
Geng JJ, Niu XJ, Jin XC, Wang XR, Gu XH, Edwards M, Glindemann D (2005) Simultaneous monitoring of phosphine and of phosphorus species in Taihu lake sediments and phosphine emissions from lake sediments. Biogeochem 76: 283–298
Gervais F, Riebesell U (2001) Effect of phosphorus limitation on elemental composition and stable isotope fractionation in a marine diatom growing under different CO_2 concentrations. Limnol Oceanogr 46: 497–504
Ghannoum O, Conroy JP (2007) Phosphorus deficiency inhibits growth in parallel with photosynthesis in a C_3 (*Panicum laxum*) but not two C_4 (*P. coloratum* and *Cenchrus ciliaris*) grasses. Funct Plant Biol 34: 72–81
Gillooly JF, Allen AP, Brown JH, Elser JJ, del Rio CM, Savage VM, West GB, Woodruff WH, Woods HA (2005) The metabolic basis of whole-organism RNA and phosphorus content. Proc Natl Acad Sci USA 102: 11923–11927
Giordano M, Norici A, Ratti S, Raven JA (2007) Role of sulphur for algae: acquisition, metabolism, ecology and evolution. In: Hell R, Leustek T, Knaff D *et al.* (eds), Sulfur Metabolism in Phototrophic Organisms, Springer, Dordrecht, The Netherlands pp 405–423
Glindemann D, Edwards M, Schrems O (2004) Phosphine and methylphosphine production by simulated lightning - a study for the volatile phosphorus cycle and cloud formation in the earth atmosphere. Atmos Environ 38: 6867–6874
Gusewell S (2005) Nutrient resorption of wetland graminoids is related to the type of nutrient limitation. Funct Ecol 19: 344–354
Hammond JP, Bennett MJ, Bowen HC, Broadley MR, Eastwood DC, May ST, Rahn C, Swarup R, Woolaway KE, White PJ (2003) Changes in gene expression in Arabidopsis shoots during phosphate starvation and the potential for developing smart plants. Plant Physiol 132: 578–596
Hammond JP, Broadley MR, White PJ (2004) Genetic responses to phosphorus deficiency. Ann Bot 94: 323–332
Hedin LO (2004) Global organization of terrestrial plant-nutrient interactions. Proc Natl Acad Sci USA 101: 10849–10850
Heldal M, Scanlan DJ, Norland S, Thingstad F, Mann NH (2003) Elemental composition of single cells of various strains of marine *Prochlorococcus* and *Synechococcus* using X-ray microanalysis. Limnol Oceanogr 48: 1732–1743
Hermans C, Hammond JP, White PJ, Verbruggen N (2006) How do plants respond to nutrient shortage by biomass allocation? Trends Plant Sci 11: 610–617

Ho TY, Quigg A, Finkel ZV, Milligan AJ, Wyman K, Falkowski PG, Morel FMM (2003) The elemental composition of some marine phytoplankton. J Phycol 39: 1145–1159
Jacob J (1995) Phosphate deficiency increases the rate constant of thermal dissipation of excitation energy by photosystem II in intact leaves of sunflower and maize. Funct Plant Biol 22: 417–424
Jain A, Cao A, Karthikeyan AS, Baldwin JC, Raghothama KG (2005) Phosphate deficiency suppresses expression of light-regulated *psbO* and *psbP* genes encoding extrinsic proteins of oxygen-evolving complex of PsII. Curr Sci 89: 1592–1596
Karpinets TV, Greenwood DJ, Sams CE, Ammons JT (2006) RNA:phosphorus ratio of the unicellular organism as a characteristic of phosphorous and nitrogen stoichiometry and of the cellular requirement of ribosomes for protein synthesis. BMC Biol 4: 30
Kerkhoff AJ, Enquist BJ, Elser JJ, Fagan WF (2005) Plant allometry, stoichiometry and the temperature-dependence of primary productivity. Global Ecol Biogeogr 14: 585–598
Kerkhoff AJ, Fagan WF, Elser JJ, Enquist BJ (2006) Phylogenetic and growth form variation in the scaling of nitrogen and phosphorus in the seed plants. Am Nat 168: E103–E122
Klappenbach JA, Dunbar JM, Schmidt TM (2000) rRNA operon copy number reflects ecological strategies of bacteria. Appl Environ Microbiol 66: 1328–1333
Kobayashi K, Masuda T, Takamiya K-I, Ohta H (2006) Membrane lipid alteration during phosphate starvation is regulated by phosphate signaling and auxin/cytokinin cross-talk. Plant J 47: 238–248
Konhauser KO, Lalonde SV, Amskold L, Holland HD (2007) Was there an Archean phosphate crisis? Science 315: 1234
Kuang Y, Huisman J, Elser JJ (2004) Stoichiometric plant-herbivore models and their interpretation. Math Biosci Eng 1: 215–222.
Kyle M, Acharya K, Weider LJ, Looper K, Elser JJ (2006) Coupling of growth rate and body stoichiometry in *Daphnia*: a role for maintenance processes? Freshwater Biol 51: 2087–2095
Lahner B, Gong J, Mahmoudian M, Smith EL, Abid KB, Rogers EE, Guerinot ML, Harper JF, Ward JM, McIntyre L, Schroeder JI, Salt DE (2003) Genomic scale profiling of nutrient and trace elements in *Arabidopsis thaliana*. Nature Biotech 21: 1215–1221
Lambers H, Poot P (2003) (eds) Structure and Function of Cluster Roots and Plant Responses to Phosphate Deficiency. Kluwer, Dordrecht, The Netherlands, 376 pp.
Leonardos N, Geider RJ (2005) Elevated carbon dioxide increases organic carbon fixation by *Emiliania huxleyi* (Haptophyta) under nutrient-limited high-light conditions. J Phycol 41: 1196–1203
Martiny AC, Coleman ML, Chisholm SW (2006) Phosphate acquisition genes in *Prochlorococcus* ecotypes: evidence for genome-wide adaptation. Proc Natl Acad Sci USA 103: 12552–12557
McGroddy ME, Daufresne T, Hedin LO (2004) Scaling of C:N:P stoichiometry in forests worldwide: implications for terrestrial Redfield-like ratios. Ecology 85: 2390–2401
Mills MM, Ridame C, Davey M, La Roche J, Geider RJ (2004) Iron and phosphorus co-limitation in the eastern tropical North Atlantic. Nature 429: 292–294
Morcuende R, Bari R, Gibon Y, Zheng WM, Pant BD, Blasing O, Usadel B, Czechowski T, Udvardi MK, Stitt M, Scheible WR (2007) Genome-wide reprogramming of metabolism and regulatory networks of *Arabidopsis* in response to phosphorus. Plant Cell Environ 30: 85–112
Niklas KJ (2006) Plant allometry, leaf nitrogen and phosphorus stoichiometries, and interspecific trends in annual growth rates. Ann Bot 97: 155–163
Niklas KJ, Cobb ED (2006) Biomass partitioning and leaf N,P-stoichiometry: comparisons between tree and herbaceous current-year shoots. Plant Cell Environ 29: 2030–2042
Niklas KJ, Owens T, Reich PR, Cobb ED (2005) Nitrogen/phosphorus leaf stoichiometry and the scaling of plant growth. Ecol Lett 8: 636–642
Pasek MA, Lauretta DS (2005) Aqueous corrosion of phosphide minerals from iron meteorites: a highly reactive source of prebiotic phosphorus on the surface of the early Earth. Astrobiology 5: 515–535.

Pasek MA, Dworkin JP, Lauretta DS (2007) A radical pathway for organic phosphorylation during schreibesite corrosion with implications for the origin of life. Geochim Cosmochim Acta 71: 1721–1736
Paytan A, McLaughlin K (2007) The oceanic phosphorus cycle. Chem Rev 107: 563–576
Quigg A, Finkel ZV, Irwin AJ, Rosenthal Y, Ho TY, Reinfelder JR, Schofield O, Morel FMM, Falkowski PG (2003) The evolutionary inheritance of elemental stoichiometry in marine phytoplankton. Nature 425: 291–294
Raven JA (1989) Fight or flight: the economics of repair and avoidance of photoinhibition of photosynthesis. Funct Ecol 3: 5–19
Raven JA (1994) The cost of photoinhibition of plant communities. In: Baker NR, Bowyer JR (eds), Photoinhibition of Photosynthesis. Bioscientific, Oxford, pp. 449–464
Raven JA (2006) Aquatic viruses: the emerging story. J Mar Biol Assoc UK 86: 449–451
Raven JA, Handley LL, Andrews M (2002) Optimizing carbon-nitrogen budgets: prospects for crop improvement. In: Foyer CH, Noctor G (eds), Photosynthetic Nitrogen Assimilation and Associated Carbon and Respiratory Metabolism. Kluwer, Dordrecht, The Netherlands, pp 265–274
Raven JA, Handley LL, Andrews M (2004) Global aspects of C/N interactions determining plant-environment interactions. J Exp Bot 55: 11–25
Raven JA, Andrews M, Quigg A (2005a) The evolution of oligotrophy: implications for the breeding of crop plants for low input agricultural systems. Ann Appl Biol 146: 261–280
Raven JA, Finkel ZV, Irwin AJ (2005b) Picophytoplankton: bottom-up and top-down controls on ecology and evolution. Vie Milieu 55: 209–215
Redfield AC (1958) The biological control of chemical factors in the environment. Am Sci 46: 205–221
Reich PB, Oleksyn J (2004) Global patterns of plant leaf N and P in relation to temperature and latitude. Proc Natl Acad Sci USA 101: 11001–11006
Sañudo-Wilhelmy SA, Kustka AB, Gobler CJ, Hutchins DA, Yang M, Lwiza K, Burns J, Capone DG, Raven JA, Carpenter EJ (2001) Phosphorus limitation of nitrogen fixation by *Trichodesmium* in the central Atlantic Ocean. Nature 411: 66–69
Schaefer J, Skokut TA, Stejskal EO, McKay RA, Varner JE (1990) Estimation of protein-turnover in soybean leaves using magic angle double cross-polarization N-15 nuclear magnetic resonance. J Biol Chem 256: 11574–11579
Scheurwater I, Dünnebacke M, Eising R, Lambers H (2000) Respiratory costs and rate of protein turnover in the roots of a fast-growing (*Dactylis glomerata* L.) and a slow-growing (*Festuca ovina* L.) grass species. J Exp Bot 51: 1089–1097
Schulze J, Temple G, Temple SJ, Beschow H, Vance CP (2006) Nitrogen fixation by white lupin under phosphorus deficiency. Ann Bot 98: 731–740
Shu L, Schneider P, Jegatheesan V, Johnson J (2005) An economic evaluation of phosphorus recovery as struvite from digester supernatant. Bioresource Technol 97: 2211–2216
Smith SE, Read DJ (2007) Mycorrhizal Symbiosis (3rd edition). Elsevier, London
Sterner RW, Elser JJ (2002) Ecological Stoichiometry: The Biology of Elements from Molecules to the Biosphere. Princeton University Press, Princeton, NJ
Ticconi CA, Abel S (2004) Short on phosphate: plant surveillance and countermeasures. Trends Plant Sci 9: 548–555
Tilman D, Fargione J, Wolff B, D'Antonio C, Dobson A, Howarth R, Schindler D, Schlesinger WH, Simberloff D, Swackhamer D (2001) Forecasting agriculturally driven global environmental change. Science 292: 281–284
Tilman D, Cassman KG, Matson PA, Naylor R, Polasky S (2002) Agricultural sustainability and intensive production practices. Nature 418: 671–677
Tyrrell T (1999) The relative influence of nitrogen and phosphorus on oceanic primary productivity. Nature 400: 525–531
Van Mooy BAS, Rocap G, Fredicks HF et al (2006) Sulfolipids dramatically decrease phosphorus demand by picocyanobacteria in oligotrophic marine environments. Proc Natl Acad Sci USA 103: 8607–8612

Vitousek PM, Cassman K, Cleveland C, Crews T, Field CB, Grimm NB, Howarth RW, Marino R, Martinelli L, Rastetter EB, Sprent JI (2002) Towards an ecological understanding of nitrogen fixation. Biogeochem 57: 1–45

Wardle DA, Walker LR, Bardgett RD (2004) Ecosystem properties and forest decline in contrasting long-term chronosequences. Science 305: 509–513

Wasaki J, Yonetani R, Kuroda S, Shinano T, Yazaki J, Fujii F, Shimbo K, Yamamoto K, Sakata K, Sasaki T, Kishimoto N, Kikuchi S, Yamagishi M, Osaki M (2003) Transcriptomic analysis of metabolic changes by phosphorus stress in rice plant roots. Plant Cell Environ 36: 1515–1523

Wasaki J, Shinano T, Onishi K, Yonetani R, Yazaki J, Fujii F, Shimbo K, Ishikawa M, Shimatani Z, Nagata Y, Hashimoto A, Ohta T, Sato Y, Miyamoto C, Honda S, Kojima K, Sasaki T, Kishimoto N, Kikuchi S, Osaki M (2006) Transcriptomic analysis indicates putative metabolic changes caused by manipulation of phosphorus availability in rice leaves. J Exp Bot 57: 2049–2059

Williams RJP (1981) The natural selection of the elements. Proc Roy Soc Lond B 313: 361–397

Williams RJP, Fráusta da Silva JJR (1996) The Natural Selection of the Chemical Elements. The Environment and Life's Chemistry. Clarendon, Oxford

Wright IJ, Reich PB, Cornelissen JHC, Falster DS, Garnier E, Hikosaka K, Lamont BB, Lee W, Oleksyn J, Osada N, Poorter H, Villar R, Warton DI, Westoby M (2005) Assessing the generality of global leaf trait relationships. New Phytol 166: 485–496

Yehudai-Resheff S, Zimmer SL, Komine Y, Stern DB (2007) Integration of chloroplast nucleic acid metabolism into the phosphate deprivation response in *Chlamydomonas reinhardtii*. Plant Cell 19: 1023–1038

Zagdanska B (1995) Respiratory energy demands for protein turnover and ion transport in wheat leaves upon water deficit. Physiol Plant 95: 428–436

Zerkle AL, House CH, Brantely SL (2005) Biogeochemical signatures through time as inferred from whole microbial genomes. Am J Sci 305: 467–502

Zhu F, Massana R, Not F, Marie D, Vaulot D (2005) Mapping of picoeukaryotes in marine ecosystems with quantitative PCR of the 18S rRNA gene. FEMS Microbiol Ecol 52: 79–92

Zhu RB, Kong DM, Sun LG, Geng JJ, Wang XR, Glindemann D (2006) Tropospheric phosphine and its sources in coastal Antarctica. Environ Sci Technol 40: 7656–7661

Zhu RB, Glindemann D, Kong DM, Sun LG, Geng JJ, Wang XR (2007) Phosphine in the marine atmosphere along a hemispheric course from China to Antarctica. Atmos Environ 41: 1567–1573

INDEX

Plant Ecophysiology

1. M.J. Hawkesford and P. Buchner (eds.): *Molecular Analysis of Plant Adaptation to the Environment.* 2001 ISBN 978-1-4020-0016-2
2. D. Grill, M. Tausz and L.J. De Kok (eds.): *Significance of Glutathione to Plant Adaptation to the Environment.* 2001 ISBN 978-1-4020-0178-9
3. S. Amâncio and I. Stulen (eds.): *Nitrogen Acquisition and Assimilation in Higher Plants.* 2004 ISBN 978-1-4020-2727-3
4. H. Lambers and T.D. Colmer (eds.): *Root Physiology.* From Gene to Function. 2005 ISBN 978-1-4020-4098-9
5. G. Wieser and M. Tausz (eds.): *Trees at their Upper Limit.* Treelife Limitation at the Alpine Timberline. 2007 ISBN 978-1-4020-5073-9
6. M.J. Hawkesford and L.J. De Kok (eds.): *Sulfur in Plants.* An Ecological perspective. 2007 ISBN 978-1-4020-5886-8
7. P.J. White and J.P. Hammond (eds.): *The Ecophysiology of Plant-Phosphorus Interactions.* 2008 ISBN 978-1-4020-8434-8

Printed in the United States
124548LV00002B/121/P

9 781402 084348